Advanced Principles and Applications of Mathematics

Advanced Principles and Applications of Mathematics

Editor: Lawrence Bech

NY RESEARCH PRESS
New York

Published by NY Research Press
118-35 Queens Blvd., Suite 400,
Forest Hills, NY 11375, USA
www.nyresearchpress.com

Advanced Principles and Applications of Mathematics
Edited by Lawrence Bech

International Standard Book Number: 978-1-63238-581-9 (Hardback)

Cataloging-in-Publication Data

Advanced principles and applications of mathematics / edited by Lawrence Bech.
p. cm.
Includes bibliographical references and index.
ISBN 978-1-63238-581-9
1. Mathematics. I. Bech, Lawrence.
QA39.3 .A38 2018
510--dc23

Contents

Preface VII

Chapter 1 **Strong Masting Conjecture for Multiple Size Hexagonal Tessellation in GSM Network Design** 1
Samuel K. Amponsah, Elvis K. Donkoh, James A. Ansere, Kusi A. Bonsu

Chapter 2 **Zero-Sum Coefficient Derivations in Three Variables of Triangular Algebras** 14
Youngsoo Kim & Byunghoon Lee

Chapter 3 **Absolute Valued Algebras with Strongly One Sided Unit** 25
Alassane Diouf

Chapter 4 **Semigroup Methods for the M/G/1 Queueing Model with Working Vacation and Vacation Interruption** 29
Ehmet Kasim

Chapter 5 **Mathematical Formulation of Laminated Composite Thick Conical Shells** 51
Mohammad Zannon, Hussam Alrabaiah

Chapter 6 **The Global Formulation of the Cauchy Problem** 57
Mohammad Ali Bashir, Tarig Abdelazeem Abdelhaleem

Chapter 7 **Description of the Dependence Strength of Two Variogram Models of a Spatial Structure using Archimedean Copulas** 65
Moumouni Diallo & Diakarya Barro

Chapter 8 **Finite Element Approximation and Numerical Analysis of Three-dimensional Electrical Impedance Tomography** 75
Yirang Yuan, Jiuping Li, Changfeng Li, Tongjun Sun

Chapter 9 **γ-Max Labelings of Graphs** 94
Supaporn Saduakdee, Varanoot Khemmani

Chapter 10 **Tensor Product Of Zero-divisor Graphs with Finite Free Semilattices** 102
Kemal Toker

Chapter 11 **Computing of Z- valued Characters for the Projective Special Linear Group $L_2(2^m)$ and the Conway Group Co_3** 107
Ali Moghani

Chapter 12 **Mathematical Models of Refugee Immigration and Recommendations of Policies** 114
Qilong Cheng, Tiancheng Yu, Jingkai Yan, Ru Wang

Chapter 13 **Mixed and Hybrid Finite Element Methods for Convection-Diffusion Problems and their Relationships with Finite Volume: The Multi-Dimensional Case** 134
Michel Fortin, Abdellatif Serghini Mounim

Chapter 14 **Useful Numerical Statistics of some Response Surface Methodology Designs** 150
Iwundu M. P.

Chapter 15 **Dimension Formulae for the Polynomial Algebra as a Module over the Steenrod Algebra in Degrees Less than or Equal to 12** 178
Mbakiso Fix Mothebe, Professor Kaelo, Orebonye Ramatebele

Chapter 16 **Asymptotic Behavior of Higher Order Quasilinear Neutral Difference Equations** 187
V. Sadhasivam, Pon. Sundar, A. Santhi

Chapter 17 **Fermat's Theorem – a Geometrical View** 202
Luis Teia

Chapter 18 **Types of Derivatives: Concepts and Applications (II)** 209
Salma A. Khalil, Mohammed A. Basheer, Tarig A. Abdelhaleem

Permissions

List of Contributors

Index

Preface

The application of mathematics in various fields such as science and engineering is known as applied mathematics. It studies and works upon practical problems by studying and understanding mathematical models. Applied mathematics studies various topics such as differential equations, analysis and stochastics. This book provides significant information of this discipline to help develop a good understanding of mathematics and related fields. It is a vital tool for all those who are researching or studying applied mathematics as it gives incredible insights into emerging trends and concepts.

The researches compiled throughout the book are authentic and of high quality, combining several disciplines and from very diverse regions from around the world. Drawing on the contributions of many researchers from diverse countries, the book's objective is to provide the readers with the latest achievements in the area of research. This book will surely be a source of knowledge to all interested and researching the field.

In the end, I would like to express my deep sense of gratitude to all the authors for meeting the set deadlines in completing and submitting their research chapters. I would also like to thank the publisher for the support offered to us throughout the course of the book. Finally, I extend my sincere thanks to my family for being a constant source of inspiration and encouragement.

Editor

Strong Masting Conjecture for Multiple Size Hexagonal Tessellation in GSM Network Design

Samuel K. Amponsah[1], Elvis K. Donkoh[2], James A. Ansere[3], Kusi A. Bonsu[3]

[1]Department of Mathematics, Kwame Nkrumah University of Science & Technology, Kumasi, Ghana

[2]Department of Mathematics & Statistics, University of Energy & Natural Resources, Sunyani, Ghana

[3]Electrical/Electronics Department, Sunyani Polytechnic, Sunyani, Ghana

Correspondence: Elvis K. Donkoh, School of Sciences, Department of Mathematics & Statistics, University of Energy & Natural Resources, Sunyani, Ghana. E-mail: elvis.donkor@uenr.edu.gh

Abstract

One way to improve cellular network performance is to use efficient handover method and design pattern among other factors. The efficient design pattern has been proven geometrically to be hexagonal (Hales, 2001, pp. 1- 22) due to its maximum tessellable area coverage. But uneven geographical distribution of subscribers requires tessellable hexagons of different radii due to variation of costs of GSM masts. This will call for an overlap difference. The constraint of minimum overlap difference for multiple cell range is a new area that is untapped in cell planning. This paper addresses such multiple size hexagonal tessellation problem using a conjecture. Data from MTN River State-Nigeria, was collected. Multiple Size Hexagonal Tessellation Model (MSHTM) conjecture for masting three (3) different size MTN GSM masts in River State, accounted for least overlap difference with area of 148.3km^2 using 36 GSM masts instead of the original 21.48 km^2 for 50 GSM masts. Our conjecture generally holds for k-different ($k \geq 2$) cell range.

Keywords: disks, frequency, GSM masts, hexagon, overlap difference, tessellation.

1. Introduction

Cell planning is the most significant operations in GSM design network. It includes the choice of design pattern (triangular, square or hexagon), geographic, environmental and network parameters such as terrain and artificial structures, base station location and transmission power among others. But the hexagonal design has least overlap and hence has a strength higher than both a square and an equi-triangular polygon. The hexagon motivated circular shaped cells are produced when two or more sector signal radiated antenna are used. These circular cells overlap significantly and is crucial for subscriber movement-handover. This overlap removes signal loss due to no coverage or ensures soft handover due to better overlapping regions. How much overlap difference to be permitted in the uniform design network has been studied extensively by Donkoh E. K. et.al (2015a). Uneven geographical distribution of subscribers require varied cell ranges. This will result in variation in overlap difference. Both the design pattern and geometry of overlap is complex but the result is more economical as fewer GSM masts will be used. Realistically, how much overlap has been a thriving challenge in recent times due to variation in subscriber geographical distribution, affordability and significance of mobile telephone in this age of technology.

2. Related Literature

Antenna signal radiation in GSM network design takes many form. But the most profitable radiation pattern is known to be the sector motivated circular shape (Azad, 2012.). Due to it convenience and usefulness manufacturers of GSM antenna like Mobile Mark, Multiband Technologies, Global Source, Asian Creation among others have designed GSM antenna's with variety of sector angles including 30^0, 60^0, 90^0, 120^0 as the common ones. Nonetheless, designing GSM network due to uneven geographical distribution of subscribers requires geometry of 2-D hexagonal tessellation . Unfortunately, multiple size hexagonal design has since not been generally practiced due to the complexity in the position of the base station that offers minimum overlap difference. This paper uses the least tessellable polygonal overlap difference to conjecture a multiple size design formula for hexagons of several dimensions.

Donkoh & Opoku (2016, pp.33) emphasize with geometric proof that the hexagon has the least overlap difference of 13.4% and hence has the strongest tessellable area coverage.

Donkoh E. K et.al (2015a, pp. 5) investigated the hexagonal overlap difference for a uniform 0.6km cell range of 50 GSM masts design and obtain an overlap difference of 5.788km using 35 GSM masts instead of the original 26.884km.

This however confirms the economies of scale for design pattern minimizing overlap difference with smooth handover.

Hamad-Ameen J.J (2008, pp. 393), studied frequency and cell planning with clusters (K=7,9,13) and without clusters for uniform cell range and concluded that it is very efficient in GSM masting.

Donkoh E. K et. al (2015c), the authors considered GSM network design with two different cell range namely 1km and 3km obtaining the least overlap difference of 15.27km as compared to the original 30.95km.

In application to earth and mineral science, Raposo (2011), uses vertex-clustering of uniform hexagonal tessellation to simplify scale-specific automated map line.

Beyond applications of Christaller (1933) classic theory, hexagonal tessellation has been advocated for thematic cartography by Carr et.al (1992), and has been used to study cluster perception in animated maps (Griffin, 2006), as well as color perception (Brewer, 1996).

However, geometry of multiple size hexagonal tessellation in covering bounded areas, optimizing the overlap difference for k-different ($k \geq 2$) cell ranges have not been studied. We therefore proposed the strong multiple size masting conjecture for solving the bounded area coverage problem in GSM network design.

3. Computational Experience

We consider intersecting circular cells superimposed on non-hexagonal polygon with cell range R_1 and R_2 and [1]apothems r_1 and r_2 respectively. In Figure 1, XZ is one side of the non-hexagonal polygon and the cells overlap for covering purposes with an overlap difference(d) of $R_2 - r_2 + R_1 - r_1$.

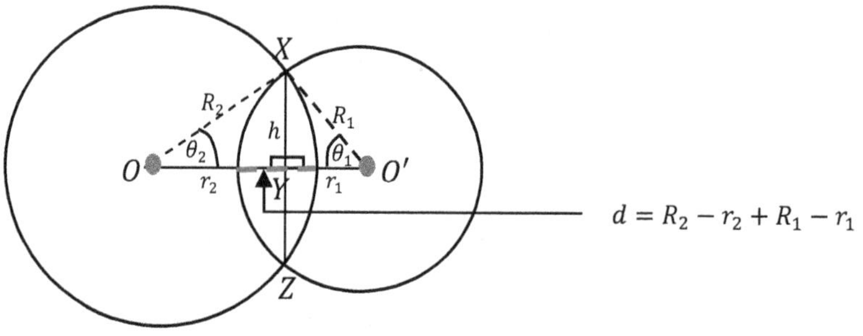

Figure 1. Overlap difference for Non-Uniform cell Range R_1, R_2.

From triangle XOY in Figure 1, $R_2^2 = h^2 + r_2^2$ where XY is a side of the right triangle.

$$h = \sqrt{R_2^2 - r_2^2} \quad (1)$$

$$\sin\theta_2 = \frac{h}{R_2}$$

$$h = R_2 \sin\theta_2 \quad (2)$$

Equating (1) and (2),

$$R_2 \sin\theta_2 = \sqrt{R_2^2 - r_2^2}$$

$$\sin\theta_2 = \sqrt{1 - \left(\frac{r_2^2}{R_2^2}\right)} \quad (3)$$

Also, $r_2 = R_2 cos\theta_2$

Similarly in triangle $XO'Y$, $R_1^2 = h^2 + r_1^2$ where h is a side of the right triangle.

$$h = \sqrt{R_1^2 - r_1^2} \quad (4)$$

$$\sin\theta_1 = \frac{h}{R_1}$$

$$h = R_1 \sin\theta_1 \quad (5)$$

Equating (4) and (5),

[1] Apothem is a line drawn from the center of a regular polygon to an edge and perpendicular to that edge. It is the perpendicular bisector of that edge and also the radius of the inscribed circle to that polygon.

$$R_1 \sin\theta_1 = \sqrt{R_1^2 - r_1^2}$$

$$\sin\theta_1 = \sqrt{1 - \left(\frac{r_1^2}{R_1^2}\right)} \tag{6}$$

Similarly in triangle $XO'Y$, $R_1^2 = h^2 + r_1^2$ where h is a side of the right triangle.
Also, $r_1 = R_1 cos\theta_1$
Equating (2) and (5) and extending it to radius R_2

$$R_1 \sin\theta_1 = R_2 \sin\theta_2$$

$$R_1 = R_2 \frac{\sin\theta_2}{\sin\theta_1}$$

$$R_1 = \frac{R_1 R_2 \sin\theta_2}{\sqrt{R_1^2 - r_1^2}}$$

$$\sin\theta_2 = \frac{\sqrt{R_1^2 - r_1^2}}{R_2} \tag{7}$$

Similarly, in triangle $XO'Y$,

$$\sin\theta_1 = \frac{\sqrt{R_2^2 - r_2^2}}{R_1} \tag{8}$$

A single overlap difference for Figure 2 is

$$d_1 = R_2 - r_2 + R_1 - r_1 \tag{9}$$

$$d_1 = \sum_{i=1}^{2} R_i - r_i$$

$$d_1 = \textstyle\sum_{i=1}^{2} R_i(1 - cos\theta_i) \tag{10}$$

Generally n non-uniform overlaps,

$$d_n = \sum_{i=1}^{2n} R_i - r_i$$

$$d_n = \textstyle\sum_{i=1}^{2n} R_i(1 - cos\theta_i) \tag{11}$$

Equation (9), (10) or (11) is used to calculate the overlap difference for non-uniform disks as shown in Figure 1. We established a formula for calculating the area of a pair of overlap for non-hexagonal polygon inscribed disks. From Figure 1 we have:

Area of single overlap difference = (Area of sector $XO'Z$ − Area of triangle $XO'Z$) + (Area of sector XOZ − Area of triangle XOZ)

$$A_d = \tfrac{1}{2} R_1^2(\theta_1 - sin\theta_1) + \tfrac{1}{2} R_2^2(\theta_2 - sin\theta_2) \tag{12}$$

Substituting (7) and (8) into (12)

$$A_d = \tfrac{1}{2} R_1^2 \left[sin^{-1}\left(\frac{\sqrt{R_2^2 - r_2^2}}{R_1}\right) - \frac{\sqrt{R_2^2 - r_2^2}}{R_1} \right] + \tfrac{1}{2} R_2^2 \left[sin^{-1}\left(\frac{\sqrt{R_1^2 - r_1^2}}{R_2}\right) - \frac{\sqrt{R_1^2 - r_1^2}}{R_2} \right] \tag{13}$$

Generally for n different overlaps,

$$A_n = \frac{1}{2} \sum_{i=1}^{2n} R_i^2(\theta_i - sin\theta_i)$$

$$A_d = \tfrac{1}{2} \textstyle\sum_{i=1}^{2n} R_i^2 \left[sin^{-1}\left(\frac{\sqrt{R_{i+1}^2 - r_{i+1}^2}}{R_i}\right) - \frac{\sqrt{R_{i+1}^2 - r_{i+1}^2}}{R_i} \right] \tag{14}$$

Equation (13) is the area of each non-uniform cell range in terms of overlap difference that is not created by hexagon. The value A_d can be calculated from Figure 1 in Autocad environment or using equation (13) as shown in column 4 and 8 in Table 2 where R_1 and R_2 are the cell ranges of the GSM masts. The equation $d = d_m - d_n$ is the overlap difference between disks with centres m and n.

3.1 Overlap for Optimal Disks Covering in Hexagonal Tessellation.

Donkoh E. K et.al (2015b, pp.26) in their study of overlap dimensions in cyclic tessellable regular polygon emphasize that hexagon has better overlap difference of 13.7% as compared to 29.3% for square and 50.0% for equi-triangular polygon.

$$\text{Overlap difference} = \begin{cases} Hexagon, & 0 < Strong \leq 13.7\% \\ Square, & 29.3\% \leq Moderate < 50\% \\ Equi-triangular, & Weak \geq 50\% \end{cases}$$

Figure 2(a) illustrates the hexagonal cell layout with inradius and circumradius of the hexagonal cell as r_1 and R_1, respectively. In Figure 2(b), cells partially overlapped because R_1 equals to the hexagon's circumradius.

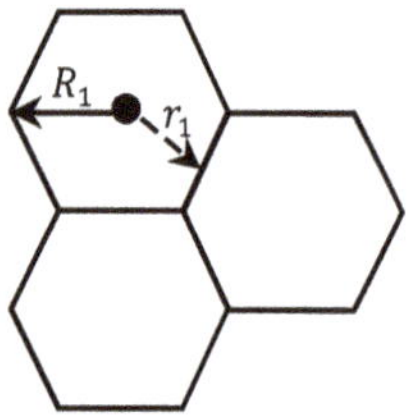

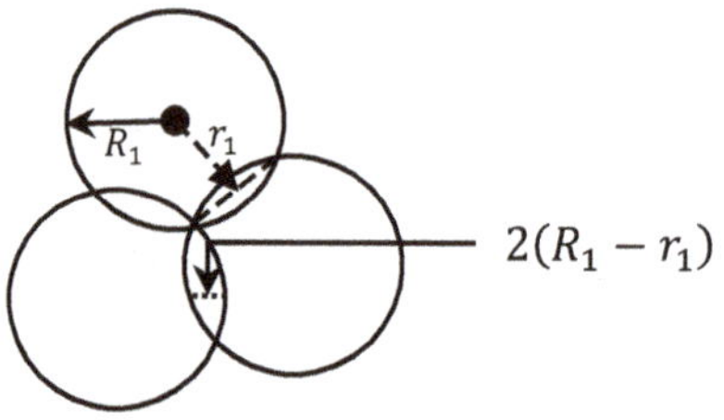

(a) Hexagonal cell layout (b) Idealized circular layout

Figure 2. Cell Layout Model's For GSM Networks.

3.2 Generalized Overlap Difference in Hexagon-Inscribed Disks.

Overlap in cell planning ensures smooth handover in GSM network with a differential effect such as increasing the number of GSM masts, environmental destruction etc and therefore must be kept as minimal as possible. A typical overlap may arise as a result of uniform or non-uniform cell range. Ample research work has been done on the uniform cell range (Donkoh E. K, et al 2015a). We minimize the overlap difference for non-uniform cell range of two different GSM antenna masts with radii R_1 and R_2 $(R_2 > R_1)$ and corresponding apothem r_1 and r_2 to be $R_1 - r_1 + 2R_2 - r_2$. A mixture of non-uniform cell radius with wrong frequency assignment results in large overlap difference and in effect increases interference (such as cross talk, background noise, error in digital signaling-missed calls, blocked calls, dropped calls, etc). It also has the differential advantage of reducing cost of coverage area as multiple cell range are factored in the design. Figure 4 shows 1: k side length, for k∈ ℕ ≥ 2 overlap difference for two different radii R_1 and R_2.

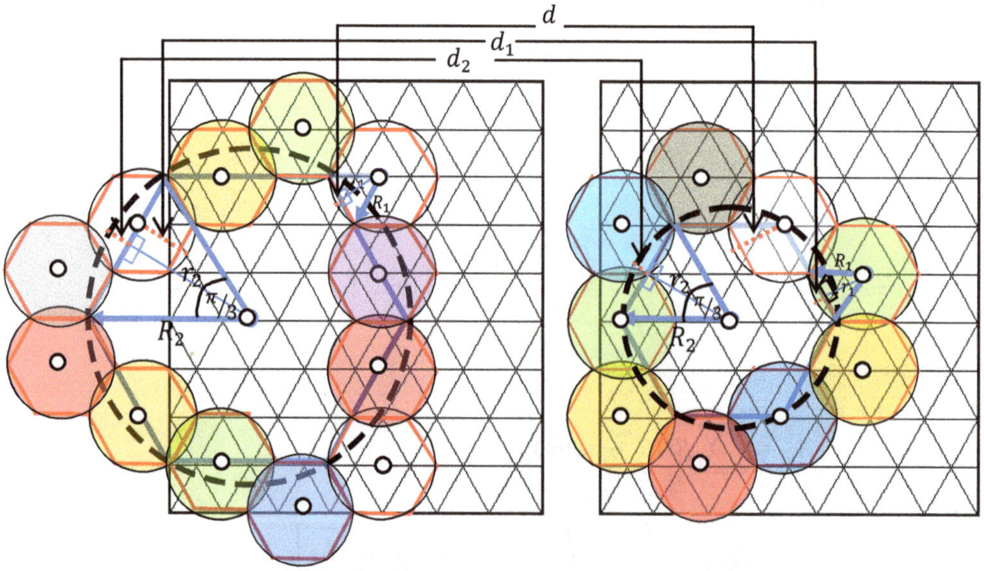

Figure 4. Overlap difference for non-uniform disks, represented by ▪▪▪▪, and GSM cell by ○.

Let $d_{R_2,R_1} = d + d_1 + d_2$ represents the global least overlap difference for two different size disks superimposed on tessellable hexagon in the ratio 1:k, where

$$d_{R_2,R_1} = R_1 - r_1 + R_1 + R_2 - r_2$$

$$d_{R_2,R_1} = R_2 - r_2 + 2R_1 - r_1 \qquad (18)$$

Theorem: The apothem r_n created by n sided tessellable regular polygon inscribed in a disk of radius R_1 is

$$r_n = R_1 Cos\left(\frac{\pi}{n}\right).$$

Proof.

Donkoh et al (2016, pp.33-34) gave a formal proof of this theorem. Equation (18) then becomes

$$d_{R_2,R_1} = R_2 - R_2 cos\left(\frac{\pi}{n}\right) + 2R_1 - R_1 cos\left(\frac{\pi}{n}\right)$$

$$d_{R_2,R_1} = R_2\left[1 - cos\left(\frac{\pi}{n}\right)\right] + R_1\left[2 - cos\left(\frac{\pi}{n}\right)\right] \qquad (19)$$

Since the polygon is a hexagon $n = 6$ sided but with different radii. Thus

$$d_{R_2,R_1} = R_2\left[1 - cos\left(\frac{\pi}{6}\right)\right] + R_1\left[2 - cos\left(\frac{\pi}{6}\right)\right]$$

$$d_{R_2,R_1} = \frac{1}{2}\left[(2-\sqrt{3})R_2 + (4-\sqrt{3})R_1\right] \qquad (20)$$

Equation (20) is the least overlap difference for GSM network design using two different radii since the $1{:}n$ size hexagon tile completely.

3.3 Masting Conjecture

Generally for k different tessellable regular polygons $n_1, n_2, \ldots n_k$ inscribed in disks with respective radii $R_k, R_{k-1}, \ldots, R_1$ $(where\ R_k > R_{k-1})$ the least overlap difference is

$$d_{n_k,n_{k-1},\ldots n_1} = R_k\left[1 - cos\left(\frac{\pi}{n_1}\right)\right] + R_{k-1}\left[2 - cos\left(\frac{\pi}{n_2}\right)\right] + \cdots + R_1\left[k - cos\left(\frac{\pi}{n_k}\right)\right].$$

$$d_{n_k,n_{k-1},\ldots n_1} = \sum_{l=1}^{k} R_{k-l+1}\left[l - cos\left(\frac{\pi}{n_k}\right)\right] \qquad (21)$$

Since we are tilling with hexagon $n_k = 6, \forall\, k \in \mathbb{R}$

$$d_{n_k,n_{k-1},\ldots n_1} = \sum_{l=1}^{k} R_{k-l+1}\left[l - cos\left(\frac{\pi}{6}\right)\right]$$

$$d_{n_k,n_{k-1},\ldots n_1} = \sum_{l=1}^{k} R_{k-l+1}\left[l - \frac{\sqrt{3}}{2}\right] \qquad (22)$$

For three different tessellable regular hexagon the least overlap difference can be obtain from equation (22) to be

$$d_{R_3,R_2,R_1} = \sum_{l=1}^{k=3} R_{k-l+1}\left[l - \frac{\sqrt{3}}{2}\right]$$

$$= R_3\left(1 - \frac{\sqrt{3}}{2}\right) + R_2\left(2 - \frac{\sqrt{3}}{2}\right) + R_1\left(3 - \frac{\sqrt{3}}{2}\right)$$

$$= R_3 - R_3\frac{\sqrt{3}}{2} + 2R_2 - R_2\frac{\sqrt{3}}{2} + 3R_1 - R_1\frac{\sqrt{3}}{2}$$

$$= R_3 - R_3 cos\left(\frac{\pi}{6}\right) + 2R_2 - R_2 cos\left(\frac{\pi}{6}\right) + 3R_1 - R_1 cos\left(\frac{\pi}{6}\right)$$

$$d_{n_3,n_2,n_1} = R_3 - r_3 + 2R_2 - r_2 + 3R_1 - r_1 \qquad (23)$$

This is the one sided least overlap difference of triple non-uniform hexagonal tessellation for masting in GSM network. Equation (23) significantly informs cell planners how to design the GSM network for least overlap difference. Figure 6 illustrate this concept.

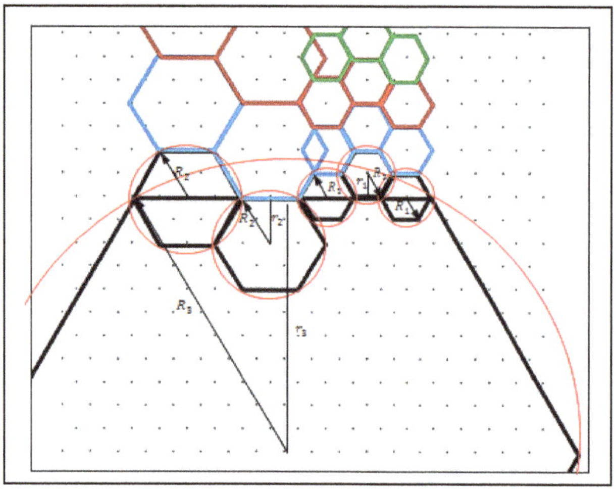

6(a): Triple different size hexagonal tessellation

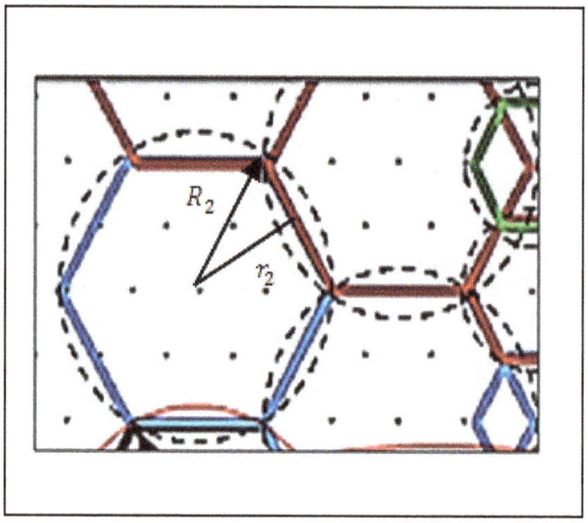

6(b): Section of hexagonal tiling with radius R_2

Figure 6. Triple Size Hexagonal Tessellation

Continuous upward tiling will lead us to the following overlap differences .

Case I: Hexagon with Radius R_2 as in 6(b)

Overlap difference will be

$$d = 2(R_2 - r_2)$$

$$d = (2 - \sqrt{3})R_2$$

For k overlaps, the difference will be

$$d_k = (2 - \sqrt{3})kR_2 \tag{24}$$

Case II: Hexagons with radii R_2 and R_1

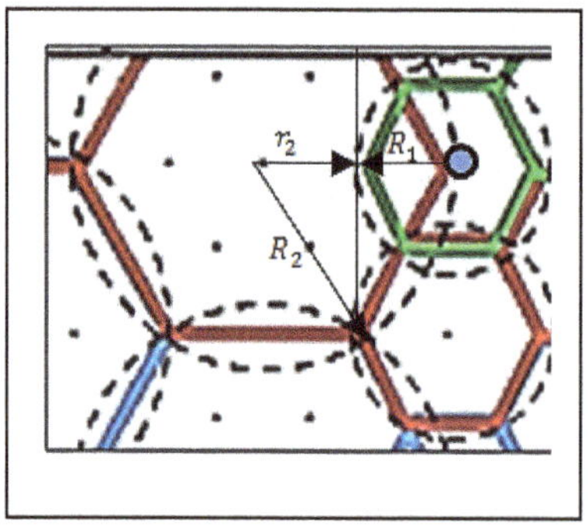

6(c): Section of hexagon tilling with radii R_2,R_1

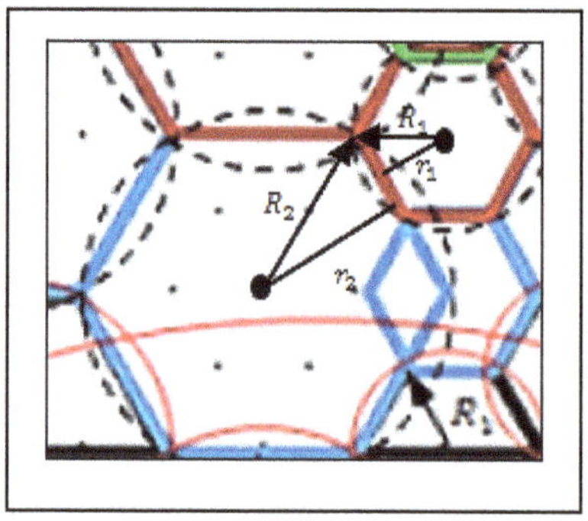

6(d): Section of hexagon tiling with radii R_2, R_1

Overlap difference will be

$$d = R_2 - r_2 + R_1 - r_1$$

$$d = R_2\left(1 - cos\frac{\pi}{6}\right) + R_1\left(1 - cos\frac{\pi}{6}\right)$$

$$d = \frac{1}{2}(2 - \sqrt{3})(R_1 + R_2)$$

For m such overlaps, the difference will be

$$d_m = \frac{1}{2}(2-\sqrt{3})(R_1+R_2)m \quad (25)$$

Case III: Hexagons with Radii R_2 and R_1

Overlap difference is

$$d = R_2 - r_2 + R_1 - r_1$$

$$d = R_2\left[1-\cos\left(\frac{\pi}{6}\right)\right] + R_1\left[1-\cos\left(\frac{\pi}{6}\right)\right]$$

$$d = \frac{1}{2}(2-\sqrt{3})(R_1+R_2)$$

For p such overlaps, the difference will be

$$d_p = \frac{1}{2}(2-\sqrt{3})(R_1+R_2)p \quad (26)$$

Case IV: Hexagons with Radius R_1

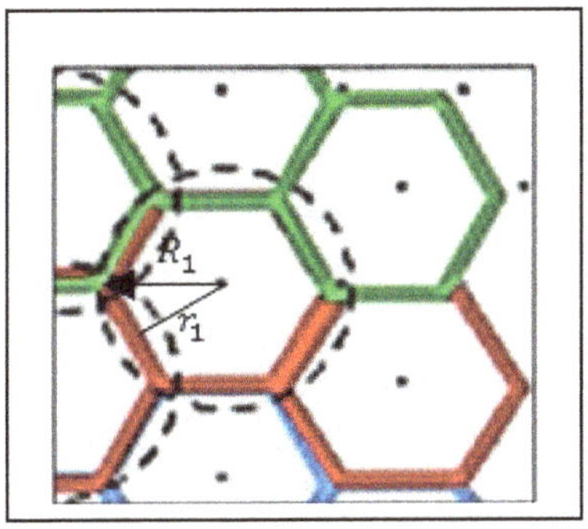

Figure 6(e). Section of hexagonal tilling with radius R_1

Overlap difference will be

$$d = 2(R_1 - r_1)$$

$$d = 2\left(R_1 - R_1\cos\frac{\pi}{6}\right)$$

For n such overlaps, the difference will be

$$d_n = (2-\sqrt{3})nR_1 \quad (27)$$

Generally, the various cases put together gives us total overlap difference

$$d = 1(R_3 - r_3 + 2R_2 - r_2 + 3R_1 - r_1) + (2-\sqrt{3})kR_2 + \frac{1}{2}(2-\sqrt{3})(R_1+R_2)p + \frac{1}{2}(2-\sqrt{3})(R_1+R_2)m + (2-\sqrt{3})nR_1$$

$$d = \left[R_3\left(1-\frac{\sqrt{3}}{2}\right) + R_2\left(2-\frac{\sqrt{3}}{2}\right) + R_1\left(3-\frac{\sqrt{3}}{2}\right)\right] + (2-\sqrt{3})(kR_2+nR_1) + \frac{1}{2}(2-\sqrt{3})(R_1+R_2)(m+p) \quad (28)$$

Table 1. WGS-84 coordinates of 50 MTN GSM masts in River State, Nigeria.

Location	Geographical Coordinates		Grid Coordinates	
	Latitude	Longitudes	Easterns (Xm)	Northings(Ym
1. AGUDAMA STR., BY GARRISON	4^0 48′ 14.28446″ N	7^0 00′ 24.89204″ E	278949.137	531317.354
2. 11B ELECHI BEACH, DIOBU	4^0 46′ 59.28462″ N	6^0 59′ 41.57230″E	277607.260	529016.990
3. 16 NHEDIOHANMA, DIOBU, PHC	4^0 47′ 41.24449″ N	6^0 59′ 18.15230 E	276889.199	530308.265
4. APEX MILL LTD. TRANS AMADI INDUS. AREA	4^0 4′ 33.60453″ N	7^0 01′ 36.11181″ E	281145.856	531904.560
5. BY 3 KING'S AVENUE, ABULOMA, OZUBOKO	4^0 47′ 01.14485″ N	7^0 02′ 18.95184″ E	282458.036	529060.136
6. EJUAN COMMUNITY, ABULOMA	4^0 47′ 33.90477″ N	7^0 02′ 30.59174″ E	282819.658	530065.587

7. NZIMIRO STREET SHELL RA, OLD GRA	4^0 47′ 20.16467″ N	7^0 00′ 53.33206″ E	279820.847	529652.068
8. OCO MILLER IND.. SERV. LTD., TRANS AMADI	4^0 48′ 30.12458″ N	7^0 02′ 12.23173″ E	282258.753	531794.438
9. OFF JOHN OGBODA STREET, NGWOR STR.	4^0 48′ 18.00450″ N	7^0 00′ 48.47199″ E	279676.210	531429.531
10. OKIS AWO CLOSE, RAINBOW	4^0 47′ 50.34468″ N	7^0 01′ 50.39185″ E	281582.136	530574.219
11. OLD GRA FORCE AVENU	4^0 46′ 45.00473″ N	7^0 00′ 32.87214″ E	279187.121	528573.665
12. OPP GIGGLES CYBER CAFE, BOROKIRI	4^0 44′ 54.54525″ N	7^0 02′ 30.05200″ E	282789.100	525169.673
13. OPP NANA'S HOTELS, MOORE HOUSE STR.	4^0 45′ 26.40512″ N	7^0 02′ 14.45200″ E	282311.053	526149.861
14. ORUTA COMPOUND, OZUBOKO-AM	4^0 46′ 26.64496″ N	7^0 02′ 35.21184″ E	282956.163	527998.775
15. PLOT 305 BOROKIRI, FILLAREA	4^0 45′ 10.62516″ N	7^0 02′ 02.39203″ E	281937.962	525666.112
16. NKPOGU BYE-PASS, TOKI HOTEL ROAD	4^0 48′ 12.94455″ N	7^0 01′ 12.45193″ E	280414.825	531271.930
17. ST MARY'S CATHOLIC CH. LAGOS BUS STOP	4^0 45′ 43.56496″ N	7^0 01′ 08.57216″ E	280282.014	526682.863
18. CHINDAH ESTATE, UST, PORT HARCOURT	4^0 48′ 11.54438″ N	6^0 59′ 11.65228″ E	276691.599	531239.780
19. 23 DICK TIGER, STREET, DIOBU	4^0 47′ 21.42455″ N	6^0 59′ 22.43231″ E	277019.328	529698.935
20. MILE ONE POLICE STATION	4^0 47′ 27.60457″ N	6^0 59′ 51.17222″ E	277905.683	529886.217
21. OPP 100 ABEL JUMBO ST, MILE 2 DIOBU PHC	4^0 47′ 35.44447″ N	6^0 59′ 05.95232″ E	276512.658	530131.171
22. OMEGA BEACH EASTERN BY PASS	4^0 46′ 47.94479″ N	7^0 01′ 19.55203″ E	280626.117	528659.839
23. NNOKAM., RUMUOKWOKUNU VILL.	4^0 48′ 25.64431″ N	6^0 59′ 00.35228″ E	276344.602	531674.010
24. OPP 3 DICK NWOKE STR, OGBUNABALI	4^0 47′ 37.26461″ N	7^0 00′ 44.75204″ E	279557.926	530178.201
25. 4 NZIMIRO STREET, PORT HARCOURT	4^0 47′ 45.30456″ N	7^0 00′ 20.21210″ E	278802.305	530427.412
26. IMMACULATE CATH. HT PARISH,MILE 3, DIOBU	4^0 47′ 59.04442″ N	6^0 59′ 13.75227″ E	276755.193	530855.544
27. 9 EZEBUNWO CLOSE, OROWURUKWO	4^0 48′ 41.76436″ N	7^0 00′ 12.53205″ E	278570.664	532162.743
28. BY ADARI-OBU LANE, ABULOMA	4^0 46′ 48.72494″ N	7^0 02′ 56.45173″ E	283612.721	528675.271
29. ROAD E, UST CAMPUS	4^0 47′ 21.14448″ N	6^0 58′ 44.25243″ E	275842.544	529693.789
30. CHIEF ODUM CLOSE, EASTERN BYE-PASS	4^0 47′ 52.56457″ N	7^0 00′ 48.89200″ E	279686.886	530647.896
31. ROAD 3, AGIP ESTATE	4^0 48′ 34.14426″ N	6^0 58′ 36.15233″ E	275599.515	531937.367
32. 11 WONODI STREET, GRA PHASE III	4^0 48′ 34.14439″ N	6^0 59′ 56.35210″ E	278071.310	531930.090
33. 31B FORCES AVENUE, OLD GRA	4^0 47′ 03.72470″ N	7^0 00′ 32.99212″ E	279192.486	529148.794
34. CHRISTIAN COUNCIL COLL., ELEKAHIA	4^0 48′ 39.18446″ N	7^0 00′ 57.89194″ E	279968.423	532079.405
35. OUR LADY FATIMAH'S COL., NEW MKT LAYOUT	4^0 45′ 27.06508″ N	7^0 01′ 53.93203″ E	281678.645	526171.937
36. BY RESURRECTION MINISTRIES, BOROKIRI	4^0 44′ 39.12525″ N	7^0 02′ 18.53204″ E	282432.687	524696.934
37. 5 ABUJA BYPASS, MILE 3, DIOBU	4^0 48′ 10.04441″ N	6^0 59′ 39.35220″ E	277545.195	531191.187
38. 6 ABOBIRI STREET, OFF INDUSTRY ROAD,	4^0 45′ 48.66494″ N	7^0 00′ 49.97219″ E	279709.176	526841.199
39. ABONNEMA WHARF RD, RCCG KIDNEY PARISH	4^0 46′ 32.40475″ N	7^0 00′ 11.99225″ E	278542.448	528188.416
40. BY MARINE BASE COMM. BANK PREMISES	4^0 46′ 08.22491″ N	7^0 01′ 18.65207″ E	280594.871	527439.601
41. 8RECLAMATION LAYOUT OFF HARBOUR ROAD	4^0 45′ 30.48496″ N	7^0 00′ 42.65224″ E	279481.951	526283.300
42. 3 ENWENABURU AVENUE, ELIOGBOLU	4^0 51′ 46.38391″ N	7^0 00′ 55.25165″ E	279903.855	537831.021
43. AGGREY RD HOUSING ESTATE, PORT HARC.	4^0 45′ 45.24502″ N	7^0 01′ 57.05202″ E	281776.402	526730.203
44. BEHIND CHINDAH BAR,IHUNWO OROGBUM RD	4^0 48′ 29.44433″ N	6^0 59′ 24.75220″ E	277096.965	531788.546
45. FACULTY OF LAW BUILD., UST CAMPUS	4^0 47′ 53.14438″ N	6^0 58′ 41.25237″ E	275752.978	530677.224
46. UST CAMPUS	4^0 47′ 52.54438″ N	6^0 58′ 37.35240″ E	275632.722	534326.027
47. INSIDE EL-SHADDAI INT'L INC. PREMISES, AMADI	4^0 49′ 52.30000″ N	7^0 01′01.68000″ E	280091.597	530775.220
48. OPP 6A WOKE LANE OGBUNABALI	4^0 47′ 56.88454″ N	7^0 00′ 32.21205″ E	279173.186	530782.113

49.	CHIEF AKAROLO ESTATE, ELEKAHIA	$4^0\,49'\;04.98441''$ N	$7^0\,01'\;29.27181''$ E	280937.839	532869.255
50.	DANGOTE PREM., TRANS AMADI, OGINIGBA	$4^0\,49'\;16.44446''$ N	$7^0\,02'\;15.95166''$ E	282377.492	533217.187

A plot of the 50 MTN GSM masts coordinates is shown in Figure 7(a). We write a matlab code for the covering as shown in Figure 7(b).

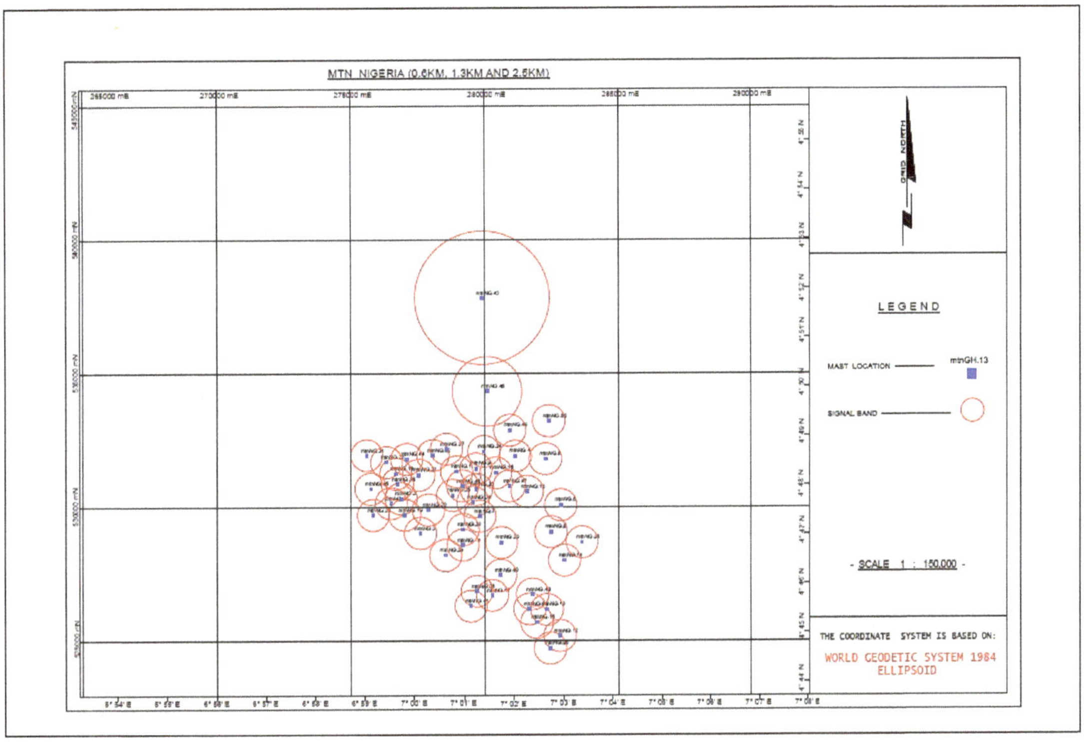

Figure 7(a). Position of 50 MTN GSM cell coordinates with cell band in River State, Nigeria

A Matlab code for 50 MTN GSM Masts position, has been covered with 36 hexagons, in River State-Nigeria. This is written below.

```
>>load set3
>>scatter(set3(:,1),set3(:,2),'fill')
>>n=5.1;
>>m=0.96;
>>figure(1),hold on
>>for i=5.0008:.015:n
    >>for j = 0.81325:0.015:m
        >>hexagon(0.005,i,j)
  >> end
>>end
 >>for i=5.0008:.015:n
    >>for j = 0.81325:0.015:m
        >>hexagon(0.005,i+0.0075,j+0.0075)
    >>end
 >>end
>> axis([5 5.1 0.8 0.96])
 >>xlabel('Easterns (xkm)')
 >>ylabel('Northerns (ykm)')
```

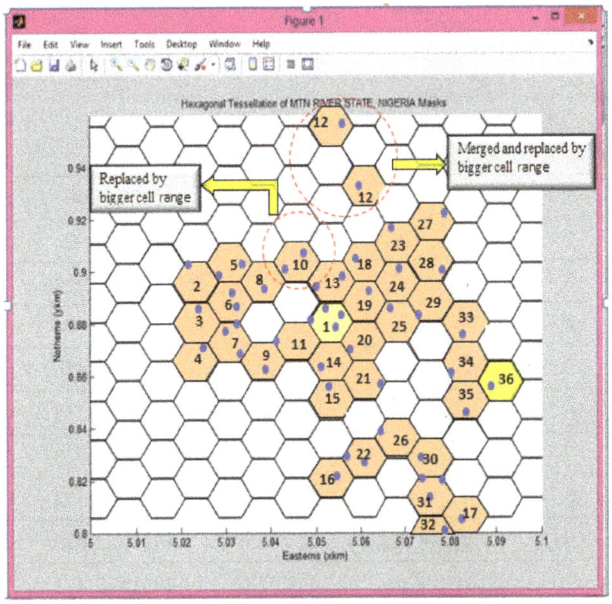

Figure 7(b). Graph of GSM masts nodes

>>title('Hexagonal Tessellation of MTN RIVER STATE, NIGERIA Masts').

This is shown in Figure 7(b). We plot the Figure 7(b) on Autocad and compute the overlap difference and overlap area shown in Table 2. Figure 8 shows the plot.

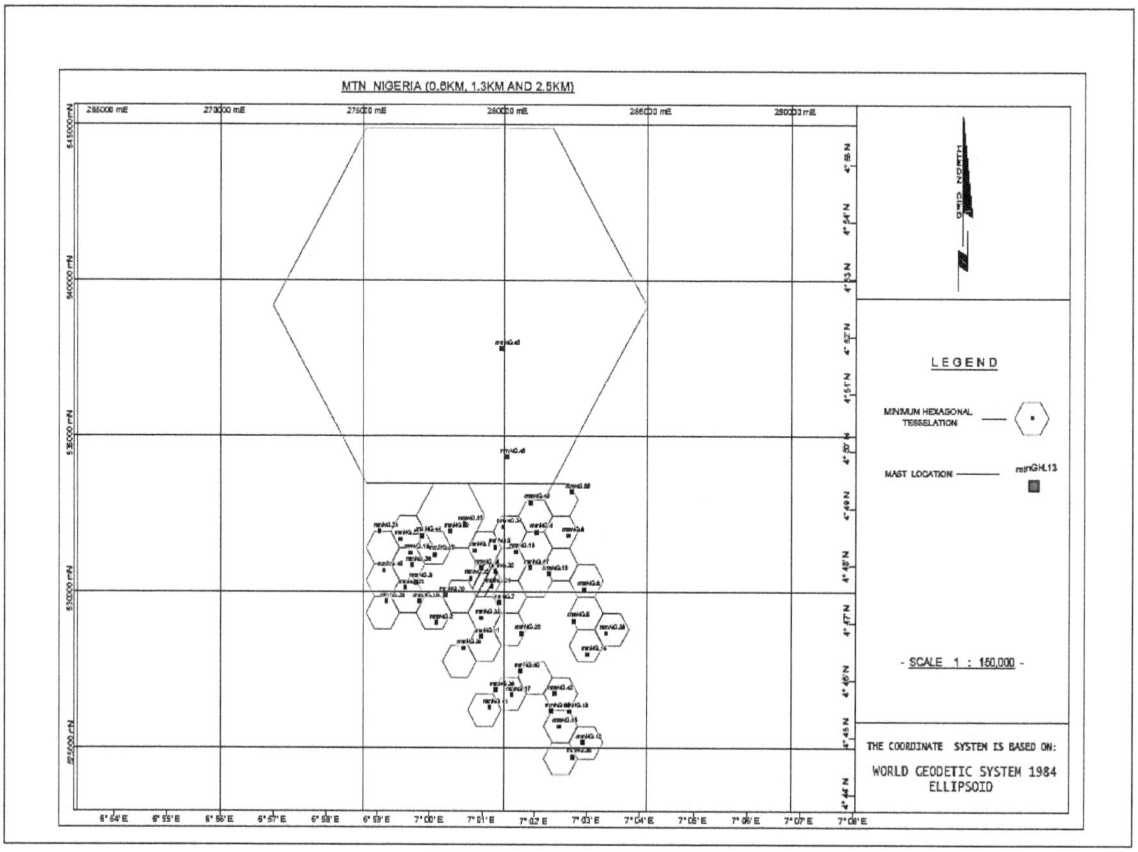

Figure 8. Optimal disks covering for triple size hexagons using MTN masts, River State – Nigeria.

Table 2 shows the overlap difference $(d)$and overlap area (A_d) obtain as a result of Figure 7(b).

Table 2. Overlap difference for 0.6km,1.3km, 2.5km MTN cell range – River State, Nigeria

Serial	Overlap Difference $d = d_m - d_n$	Value (m)	Area of overlap(A_d)	Serial	Overlap Differen $d = d_m - d_n$	Value (m)	Area of overlap(A_d)
1.	$d_{42} - d_{46}$	289.9813	222044.00	22.	$d_{43} - d_{15}$	123.7141	38621.1256
2.	$d_{46} - d_{49}$	215.2717	124885.00	23.	$d_{13} - d_{35}$	567.2068	811838.0168
3.	$d_{49} - d_4$	212.9525	114433.1922	24.	$d_{13} - d_{12}$	109.4508	30229.0436
4.	$d_4 - d_8$	81.6679	16830.20104	25.	$d_{13} - d_{15}$	589.0909	875691.5421
5.	$d_4 - d_{16}$	233.2389	137274.0656	26.	$d_{15} - d_{35}$	567.2068	811838.0168
6.	$d_4 - d_{47}$	47.7039	5742.411344	27.	$d_{15} - d_{12}$	214.6637	116279.6587
7.	$d_4 - d_{34}$	9.6558	235.2681028	28.	$d_{12} - d_{36}$	607.9591	932685.5619
8.	$d_{16} - d_{47}$	493.6746	614990.0552	29.	$d_{31} - d_{23}$	409.7396	423645.3124
9.	$d_{16} - d_{34}$	277.3459	194102.0076	30.	$d_{23} - d_{44}$	438.9687	486243.3069
10.	$d_{16} - d_9$	444.7582	499153.8841	31.	$d_{23} - d_{26}$	810.5369	1657799.9
11.	$d_{16} - d_{30}$	241.1916	146794.8718	32.	$d_{23} - d_{45}$	40.8618	4213.2916
12.	$d_{47} - d_{10}$	505.1518	643917.6609	33.	$d_{44} - d_{32}$	215.4276	117108.7144

13.	$d_6 - d_5$	131.4954	43632.25393	34.	$d_{44} - d_{37}$	453.1748	518224.6067
14.	$d_5 - d_{14}$	27.5591	1916.534266	35.	$d_{44} - d_{26}$	206.3699	107468.0168
15.	$d_{40} - d_{38}$	131.1031	43372.29959	36.	$d_{32} - d_1$	129.4741	42301.16504
16.	$d_{40} - d_{17}$	381.1399	366568.6824	37.	$d_{32} - d_{37}$	292.9307	216529.1175
17.	$d_{43} - d_{35}$	633.2396	1011865.197	38.	$d_{27} - d_1$	273.7579	189112.3313
18.	$d_{43} - d_{13}$	633.2396	1011865.197	39.	$d_{34} - d_9$	487.4520	599584.2833
19.	$d_9 - d_{30}$	418.2921	441515.3955	40.	$d_{24} - d_{25}$	404.3435	412560.3237
20.	$d_9 - d_{48}$	669.0557	1129564.61	41.	$d_{24} - d_{48}$	483.9451	590988.0793
21.	$d_9 - d_1$	464.3242	544037.9123	42.	$d_{29} - d_{21}$	399.7777	403295.761
43.	$d_{45} - d_{26}$	182.0447	83626.24686	63.	$d_{29} - d_{19}$	23.2047	1358.7465
44.	$d_{45} - d_{21}$	264.4319	176446.9753	64.	$d_{19} - d_{20}$	294.0752	218224.4106
45.	$d_{45} - d_{29}$	212.4948	113941.817	65.	$d_{19} - d_{21}$	534.0102	719590.8493
46.	$d_{26} - d_{37}$	341.6531	294548.8004	66.	$d_{19} - d_{26}$	13.6140	467.6899
47.	$d_{26} - d_3$	636.5535	1022483.593	67.	$d_{20} - d_2$	280.9723	199211.103
48.	$d_{26} - d_{19}$	13.6140	467.6899306	68.	$d_7 - d_{33}$	394.9396	393593.4759
49.	$d_3 - d_{21}$	783.8926	1550599.56	69.	$d_{33} - d_{11}$	624.8460	985218.4021
50.	$d_3 - d_{20}$	99.3801	24922.14313	70.	$d_{33} - d_{39}$	40.3124	4100.7551
51.	$d_3 - d_{19}$	576.9297	839909.1656	71.	$d_{39} - d_{11}$	448.9873	508691.6872
52.	$d_3 - d_{37}$	100.0536	25261.08301	72.	$d_{41} - d_{38}$	597.6027	901180.1747
53.	$d_3 - d_{45}$	5.3750	72.90267503	73.	$d_{41} - d_{17}$	305.7118	235836.4513
54.	$d_{25} - d_1$	298.0264	224127.9406	74.	$d_{38} - d_{40}$	131.1031	43372.2996
55.	$d_{25} - d_{48}$	686.8085	1190303.882	75.	$d_{18} - d_{44}$	517.7495	676434.7394
56.	$d_{25} - d_{30}$	288.3549	209817.2539	76.	$d_{18} - d_{23}$	644.1559	1047052.613
57.	$d_{25} - d_{20}$	152.7068	58844.14814	77.	$d_{18} - d_{37}$	345.0220	300386.2877
58.	$d_{24} - d_{30}$	712.9229	1282542.18	78.	$d_{18} - d_{26}$	810.5369	1657799.9
59.	$d_{24} - d_7$	611.8305	944601.809	79.	$d_{18} - d_3$	247.7574	154895.8552
60.	$d_{18} - d_{45}$	105.7063	28196.04989	80.	$d_{18} - d_{21}$	77.0424	14977.7347
	Total $(\sum d) = 26{,}412.518$m				Total $(\sum A_d) = 32{,}834{,}104.29\text{m}^2$		

Let R_i be disks with radius , N_{R_i} be number of cells with radius i then

Area of cells = sum of area of all disks− sum of all overlap areas of cells

$$\text{Area} = \sum_{i=1}^{n} N_{R_i} \times \pi R_i^2 - \sum A_d \tag{29}$$

Where the overlap area is calculated using the inclusive exclusive formula

$$\left|\bigcup_{i=1}^{n} A_i\right| = Area \sum_i |A_i| - Area \sum_{i<j} |A_i \cap A_j| + Area \sum_{i<j<k} |A_i \cap A_j \cap A_k| - \cdots + (-1)^{n+1} Area \left|\bigcap_{i=1}^{n} A_i\right|$$

where the first sum is over all i the second sum is over all pairs $i.j$ with $i < j$, the third sum is over all triples i, j, k with $i < j < k$ and so fourth.

$$= (48 \times \pi \times 600^2 + 1 \times \pi \times 1300^2 + 1 \times \pi \times 6600^2 - 32{,}834{,}104.29)m^2$$

$$\text{Area} = 21.48 km^2$$

We compute the total coverage area of the proposed multiple size hexagonal tessellation employed in the design and compare with the existing coverage area.

Total area of hexagon = sum of Number of hexagons × unit area

$$= \sum_{i=1}^{n} N_{R_i} \times \frac{3\sqrt{3}}{2} R_i^2 \quad (30)$$

$$= \left(34 \times \frac{3\sqrt{3}}{2} \times 600^2 + 1 \times \frac{3\sqrt{3}}{2} \times 1200^2 + 1 \times \frac{3\sqrt{3}}{2} \times 6600^2\right) m^2$$

$$\text{Area} = 148.71 km^2$$

Table 3. Comparative Analysis of the MSHT model to the Original Layout method for MTN River State, Nigeria.

Case Study	Original Layout		Multiple Size Hexagonal Tessellation	
	Number of masts	Area covered (km^2)	Number of masts	Area covered (km^2)
MTN Nigeria, River State R_1 =0.6km, R_2 =1.3km, R_3 =2.5km	50	21.48	36	148.71
Number of overlaps	80		k = 1, m = 0, p = 1, n = 52	
$d = \left[R_3\left(1-\frac{\sqrt{3}}{2}\right) + R_2\left(2-\frac{\sqrt{3}}{2}\right) + R_1\left(3-\frac{\sqrt{3}}{2}\right)\right] + (2-\sqrt{3})kR_2 + \frac{1}{2}(2-\sqrt{3})(R_1+R_2)p + \frac{1}{2}(2-\sqrt{3})(R_1+R_2)m + (2-\sqrt{3})nR_1$	26.413km		12.368km	
Ratio	1mast	0.43km^2	1mast	4.13km^2
Number of GSM masts based on cell range	$N_{R_1=0.6km}$-48, $N_{R_2=1.3km}$-1, $N_{R_3=2.5km}$-1		$N_{R_1=0.6km}$- 34, $N_{R_2=1.2km}$ − 1, $N_{R_3=6.6km}$ - 1	

5. Discussion

Designing of multiple size hexagonal tessellation with minimum overlap difference is a new area in cell planning in telecommunication network design. Our study conjectures an algorithm for efficient masting with least overlap difference for multiple cell range. Application of this formula to MTN River State GSM network solution resulted in an overlap difference of 12.368km which is a 53.2% reduction over the original overlap difference of 26.413km. The formula also uses 36GSM masts, covering an area of 148.715km^2 compared to the cell engineers original design of 50 masts covering an area of 21.48km^2. This is equivalent to using 1GSM masts to cover 4.13km^2 in the multiple size hexagonal tessellation model instead of 1GSM masts for 0.43km^2 using the original design. Table 3 shows the results of the computation.

6. Conclusion

Our study provide an optimal multiple size hexagonal tessellation design with least overlap difference of 12.368km and total coverage area of 148.71km^2. The number of GSM masts obtained from the MSHTM is 36 as compared to the original design of 50 GSM masts. This gives a 28% reduction over the original number of GSM masts. We used geometry of hexagonal tessellation approach to geometric disks covering for multiple cell range to reach optimality and it is the first study that uses multiple size hexagonal tessellation for covering of point sets to arrive at minimum overlap difference and overlap area.

Acknowledgment

The authors would like to acknowledge the Operations Manager of ATC Ghana Limited, Mr. Alexander Owuahene, and Mr. Isaac Yankson of the Geodetic department of KNUST, who enlightened us on the use of GIS and AutoCAD software.

References

Azad, I. R.M. (2012). Multiple Antenna Technique (MIMO). Bachelor's Thesis Thesis, Helsinki Metropolia University of Applied Sciences. Access: https://www.theseus.fi/bitstream/handle/10024/46175/Thesis.pdf?sequence=1

Brewer, C. A. (1996). Prediction of Simultaneous Contrast Between Map Colors with Hunt's Model of Color

Appearance. *Color Research and Application,* *21*(3), 221-235. http://dx.doi.org/10.1002/(SICI)1520-6378(199606)21:33.0.CO;2-U.

Carr, D. B., Olsen, A. R., & White, D. (1992). Hexagon Mosaic Maps for Display of Univariate and Bivariate Geographical Data. *Cartography and Geographic Information Science,* *19*(4), 228-236. http://dx.doi.org/10.1559/152304092783721231

Christaller, W. (1933). Die zentralen Orte in Suddeutschland. Jena: Gustav Fischer.

Donkoh, E. K, Amponsah, S. K, Ansere, J. A, & Bonsu, K. A. (2015c). Application of Least Optimal Overlap Difference for Multiple Size Hexagonal Tessellation in GSM Network Design. *International Journal of Mathematical Sciences,* Recent Science Journal, Article ID: 27704275, UK.

Donkoh, E. K., & Opoku, A. A. (2016). Optimal Geometric Disks Covering using Tesselable Regular Polygon. *Journal of Mathematics Research,* *8*(2), Published by Canadian Center of Science and Education. http://dx.doi.org/10.5539/jmr.v8n2p25

Donkoh, E. K., Amponsah, S. K, Opoku, A. A. (2015b). Overlap Dimensions in Cyclic Tessellable Regular Polygons. 7(2), 26, ISSN: 2042-2024, e-ISSN: 2040-7505. Access: http://maxwellsci.com/print/rjms/v7-11-16.pdf

Donkoh, E. K., Amponsah, S. K., Opoku, A. A., & Buabeng, I. (2015a). Hexagonal Tessellation Model For Masting GSM Antenna: Case Study of MTN Kumasi East-Ghana. *International Journal of Applied Mathematics,* *30*(1), 5. https://www.researchgate.net/publication/282909632_Hexagonal_Tessellation_Model_for_Masting_GSM_Antenna_Case_Study_of_MTN_Kumasi_East-Ghana.

Griffin, A. L., MacEachren, A. M., Hardisty, F., Steiner, E., & Li, B. (2006). A Comparison of Animated Maps with Static Small-Multiple Maps for Visually identifying Space-Time Clusters. *Annals of the Association of American Geographers,* *96*(4), 740-753. http://dx.doi.org/10.1111/j.1467-8306.2006.00514.x

Hales, T. C. (2001). The Honeycomb Conjecture. *Discrete and Computational Geometry,* *25*(1), 1-22. http://dx.doi.org/10.1007/s004540010071.

Hamad-Ameen, J. J. (2008). Cell Planning in GSM Mobile. WSEAS Transactions on Communications. ISSN: 1109-2742, Issue 5, Volume 7, pp. 393-398.

Roposo, P. (2011). Scale-Specific Automated Map Line Simplification by Vertex-Clustering on Hexagonal Tessellation. Master's Thesis in Geography, Department of Earth and Mineral Science, Pennsylvania State University, USA.

2

Zero-Sum Coefficient Derivations in Three Variables of Triangular Algebras

Youngsoo Kim[1] & Byunghoon Lee[2]

[1] Department of Mathematics, Tuskegee University, United States

[2] Department of Mathematics, Tuskegee University, United States

Correspondence: Youngsoo Kim, Department of Mathematics, Tuskegee University, Tuskegee, AL 36088, United States. E-mail: kimy@mytu.tuskegee.edu

Abstract

Under mild assumptions Benkovič showed that an f-derivation of a triangular algebra is a derivation when the sum of the coefficients of the multilinear polynomial f is nonzero. We investigate the structure of f-derivations of triangular algebras when f is of degree 3 and the coefficient sum is zero. The zero-sum coeffient derivations include Lie derivations (degree 2) and Lie triple derivations (degree 3), which have been previously shown to be not necessarily derivations but in standard form, i.e., the sum of a derivation and a central map. In this paper, we present sufficient conditions on the coefficients of f to ensure that any f-derivations are derivations or are in standard form.

Keywords: derivation, f-derivation, Lie derivation, Lie triple derivation, triangular algebra

1. Introduction

Let $\mathcal{R}$ be a commutative ring with identity, $\mathcal{A}$ and $\mathcal{B}$ two algebras over $\mathcal{R}$ with units $1_{\mathcal{A}}$ and $1_{\mathcal{B}}$, respectively, and let $\mathcal{M}$ be an $(\mathcal{A}, \mathcal{B})$-bimodule. We assume throughout the article that $\mathcal{M}$ is faithful as a left $\mathcal{A}$-module and as a right $\mathcal{B}$-module. Let $\mathcal{T}$ be the matrix algebra

$$\mathcal{T} = \left\{ \begin{bmatrix} a & m \\ 0 & b \end{bmatrix} \middle| \, a \in \mathcal{A}, b \in \mathcal{B}, m \in \mathcal{M} \right\}.$$

An algebra isomorphic to $\mathcal{T}$ is called a *triangular algebra*. We assume $\mathcal{T}$ is 2-torsion free for the purpose of this article. Upper triangular matrix rings and nest algebras are typical examples of triangular algebras. See the references for recent results on maps of triangular algebras.

The structure of various types of derivations of $\mathcal{T}$ has been studied in a series of papers: (Benkovič, 2015), (Benkovič, 2016), (Benkovič & Eremita, 2012), (Cheung, 2003), (Ji, Liu, & Zhao, 2012), (Wang, Wang, & Du, 2013), (Xiao & Wei, 2012), (Yu & Zhang, 2010), and (Zhang & Yu, 2006). A derivation of $\mathcal{T}$ is an $\mathcal{R}$-linear map d such that $d(xy) = d(x)y + xd(y)$ for any $x, y \in \mathcal{T}$. There are variations of this definition. For example, a Jordan derivation J is defined by the property

$$J(xy + yx) = J(x)y + xJ(y) + J(y)x + yJ(x),$$

a Lie derivation L by

$$L([x, y]) = [L(x), y] + [x, L(y)],$$

and a Lie triple derivation L by

$$L([[x, y], z]) = [[L(x), y], z] + [[x, L(y)], z] + [[x, y], L(z)].$$

The most general notion of this type is that of f-derivations. Let f be a multilinear polynomial of degree $n \geq 2$ over $\mathcal{R}$ with noncommutative indeterminate variables.

$$f(x_1, x_2, \ldots, x_n) = \sum_{\pi \in S_n} \alpha_\pi x_{\pi(1)} x_{\pi(2)} \cdots x_{\pi(n)}, \quad \alpha_\pi \in \mathcal{R}$$

where the sum is over S_n, the symmetric group. An $\mathcal{R}$-linear map $L : \mathcal{T} \to \mathcal{T}$ satisfying

$$d(f(x_1, \ldots, x_n)) = \sum_{i=1}^{n} f(x_1, \ldots, x_{i-1}, d(x_i), x_{i+1}, \ldots, x_n)$$

is called an *f-derivation* or a *derivation with respect to f*. We get the usual notion of derivation when $f = x_1x_2$, and the notion of Lie derivation when $f = x_1x_2 - x_2x_1$, and so on. Obviously, a derivation is an f-derivation for any f. The converse is true for certain classes of f.

Let α be the sum of all coefficients of f. Benkovič(Benkovič, 2015) proved that in case $\alpha \neq 0$, every f-derivation is a derivation if α is $\mathcal{T}$*-regular* and $\mathcal{T}$ is $(n-1)$-torsion free. He left the case $\alpha = 0$ as an open problem. In the latter case, an f-derivation L need not be a derivation, but it could be in standard form, that is, $L = d + h$, where d is a derivation and h is a linear map into the center $\mathcal{Z}(\mathcal{T})$ satisfying $h(f(\mathcal{T}, \ldots, \mathcal{T})) = 0$. Special cases of this problem have been previously studied in (Benkovič & Eremita, 2012; Cheung, 2003; Ji et al., 2012; Xiao & Wei, 2012; Zhang & Yu, 2006). Under mild assumptions, Cheung (Cheung, 2003) proved that a Lie derivation is of standard form and Xiao and Wei (Xiao & Wei, 2012) proved that Lie triple derivations are of standard form. In (Xiao & Wei, 2012), the following conditions are assumed. Refer to (Xiao & Wei, 2012) and (Cheung, 2003) for the discussion on these conditions.

(♣) $\pi_{\mathcal{A}}(\mathcal{Z}(\mathcal{T})) = \mathcal{Z}(\mathcal{A})$ and $\pi_{\mathcal{B}}(\mathcal{Z}(\mathcal{T})) = \mathcal{Z}(\mathcal{B})$.

(♠) $[a, a'] \in \mathcal{Z}(\mathcal{A})$ for all $a' \in \mathcal{A}$ implies $a \in \mathcal{Z}(\mathcal{A})$ and $[b, b'] \in \mathcal{Z}(\mathcal{B})$ for all $b' \in \mathcal{B}$ implies $b \in \mathcal{Z}(\mathcal{B})$.

In this article, under similar assumptions as in the aforementioned papers, we will discuss the structure of f-derivations when f is of degree 3 and $\alpha = 0$, that is, when

$$f(x, y, z) = rxyz + sxzy + tyxz + uyzx + vzxy + wzyx$$

with $r, s, t, u, v, w \in \mathcal{R}$, $r + s + t + u + v + w = 0$. We will examine sufficient coefficient conditions on which an f-derivation is a derivation or is in standard form. In the process, we leave certain special cases unsolved.

2. Preliminaries

We will identify $\mathcal{A}$ with the subalgebra of $\mathcal{T}$ with elements of the form $\begin{bmatrix} a & 0 \\ 0 & 0 \end{bmatrix}$. Similarly, we will identify $m \in \mathcal{M}$ with $\begin{bmatrix} 0 & m \\ 0 & 0 \end{bmatrix}$ and $b \in \mathcal{B}$ with $\begin{bmatrix} 0 & 0 \\ 0 & b \end{bmatrix}$. Under this identification, $\mathcal{T} = \mathcal{A} + \mathcal{M} + \mathcal{B}$ and every element of $\mathcal{T}$ is written uniquely as $a + m + b$ for some $a \in \mathcal{A}$, $m \in \mathcal{M}$, and $b \in \mathcal{B}$. We denote the projections from the triangular algebra $\mathcal{T}$ to $\mathcal{A}$, $\mathcal{M}$, and $\mathcal{B}$ by $\pi_{\mathcal{A}}$, $\pi_{\mathcal{M}}$, and $\pi_{\mathcal{B}}$, respectively. We have the following evident rules, which will be used extensively throughout the article.

Lemma 1. *For any $a \in \mathcal{A}$, $m, m' \in \mathcal{M}$, $b \in \mathcal{B}$, and $t, t' \in \mathcal{T}$,*

1. $ab = ba = [a, b] = [b, a] = 0$,
2. $ma = 0$ *and* $[a, m] = am$,
3. $bm = 0$ *and* $[m, b] = mb$,
4. $mm' = 0$ *and* $[m, m'] = 0$,
5. $1_{\mathcal{T}} = 1_{\mathcal{A}} + 1_{\mathcal{B}}$,
6. $[1_{\mathcal{A}}, a + m + b] = m$ *and* $[a + m + b, 1_{\mathcal{B}}] = m$,
7. $[a + m + b, m'] = am' - m'b \in \mathcal{M}$,
8. $[a, t] \in \mathcal{A} + \mathcal{M}$, *or equivalently* $\pi_{\mathcal{B}}[a, t] = 0$,
9. $[t, b] \in \mathcal{M} + \mathcal{B}$, *or equivalently* $\pi_{\mathcal{A}}[t, b] = 0$,
10. $\pi_{\mathcal{A}}(tt') = \pi_{\mathcal{A}}(t)\pi_{\mathcal{A}}(t')$, $\pi_{\mathcal{B}}(tt') = \pi_{\mathcal{B}}(t)\pi_{\mathcal{B}}(t')$.

We will use the convention that, unless stated otherwise, elements in small letters belong to the sets in the corresponding capital letters. For example, a, a', a_1, a_2, etc. should be understood as elements of $\mathcal{A}$.

Proposition 2 (Proposition 3 (Cheung, 2003)). *The center $\mathcal{Z}(\mathcal{T})$ of the triangular algebra $\mathcal{T}$ is $\{a+b \mid am = mb$ for all $m \in \mathcal{M}\}$.*

Proof. Suppose $a+m'+b \in \mathcal{Z}(\mathcal{T})$, then we have $0 = [1_{\mathcal{A}}, a+m'+b] = m'$. We also have $0 = [a+m'+b, m] = am - mb$ for any $m \in \mathcal{M}$.

Conversely, suppose $am = mb$ for all $m \in \mathcal{M}$. Then $a \in \mathcal{Z}(\mathcal{A})$ because for any $a' \in \mathcal{A}$ and $m \in \mathcal{M}$,

$$(aa')m = a(a'm) = (a'm)b = a'(mb) = a'(am) = (a'a)m$$

and by the faithfulness of $\mathcal{M}$, $aa' = a'a$. Similarly, $b \in \mathcal{Z}(\mathcal{B})$. Now for any $a' + m' + b' \in \mathcal{T}$,

$$\begin{aligned}[a+b, a'+m'+b'] &= [a,a'] + [a,m'] + [a,b'] + [b,a'] + [b,m'] + [b,b'] \\ &= [a,m'] + [b,m'] = am' - m'b = 0.\end{aligned}$$

□

For $a+b \in \mathcal{Z}(\mathcal{T})$, the elements a and b are a pair. If $a+b_1 \in \mathcal{Z}(\mathcal{T})$ and $a+b_2 \in \mathcal{Z}(\mathcal{T})$, then $am = mb_1 = mb_2$ for any $m \in \mathcal{M}$, and by faithfulness of $\mathcal{M}$, we have $b_1 = b_2$. Using this property, we can construct an isomorphism

$$\phi : \pi_{\mathcal{A}}(\mathcal{Z}(\mathcal{T})) \to \pi_{\mathcal{B}}(\mathcal{Z}(\mathcal{T}))$$

by sending an element to the other element in the pair. It is straightforward to verify that ϕ respects algebra operations. See Proposition 3 in (Cheung, 2003) for the proof. The following formulas follow from the definition.

Lemma 3. *For any $a \in \pi_{\mathcal{A}}(\mathcal{Z}(\mathcal{T}))$, $b \in \pi_{\mathcal{B}}(\mathcal{Z}(\mathcal{T}))$, and $m \in \mathcal{M}$,*

1. $am = m\phi(a)$,
2. $\phi^{-1}(b)m = mb$.

Lemma 4. *Let $t \in \mathcal{Z}(\mathcal{T})$. If $tm = mt = 0$ for all $m \in \mathcal{M}$, then $t = 0$.*

Proof. Since $t \in \mathcal{Z}(\mathcal{T})$, $t = a + b$ for some $a \in \mathcal{A}$ and $b \in \mathcal{B}$. If $tm = mt = 0$, then $am = 0$ and $mb = 0$. Since m is arbitrary and $\mathcal{M}$ is faithful, $a = 0$ and $b = 0$. □

Definition 5. An element $r \in \mathcal{R}$ is called *$\mathcal{T}$-regular* or simply *regular* if $rt = 0$, $t \in \mathcal{T}$ implies $t = 0$. Equivalently, r is regular if $rt_1 = rt_2$, $t_1, t_2 \in \mathcal{T}$ implies $t_1 = t_2$.

3. Discussion of the Problem

Now we discuss the main problem. Suppose L is an f-derivation where

$$f(x,y,z) = rxyz + sxzy + tyxz + uyzx + vzxy + wzyx$$

with $r,s,t,u,v,w \in \mathcal{R}$, and $r+s+t+u+v+w = 0$. We assume the sum or difference of any combination of these coefficients is either 0 or regular. The regularity condition is necessary to cancel coefficients. For example, if r is regular, then L is a derivation if and only if L is a derivation with respect to $f(x,y) = rxy$.

Lemma 6. *Let $x,y,z \in \mathcal{T}$.*

1. *If $x \in \mathcal{Z}(\mathcal{T})$, then $f(x,y,z) = (r+t+u)x[y,z]$.*
2. *If $y \in \mathcal{Z}(\mathcal{T})$, then $f(x,y,z) = (u+v+w)y[z,x]$.*
3. *If $z \in \mathcal{Z}(\mathcal{T})$, then $f(x,y,z) = (r+s+v)z[x,y]$.*

In particular, if $x = 1_{\mathcal{T}}$, $y = 1_{\mathcal{T}}$, or $z = 1_{\mathcal{T}}$, then $f(x,y,z)$ is the constant multiple of a commutator.

Proof. We prove the first one. Others are similar. If x is in the center, it can be factored out. Therefore,

$$\begin{aligned}f(x,y,z) =& x(ryz + szy + tyz + uyz + vzy + wzy) \\ =& x((r+t+u)yz + (s+v+w)zy) \\ =& x((r+t+u)yz - (r+t+u)zy) \\ =& (r+t+u)x[y,z].\end{aligned}$$

□

Lemma 7. *For any* $x, y, z \in \mathcal{T}$,

1. $(r + s - u - w)[x, L(1_{\mathcal{T}})] = 0,$
2. $(t + u - s - v)[y, L(1_{\mathcal{T}})] = 0,$
3. $(v + w - r - t)[z, L(1_{\mathcal{T}})] = 0.$

Proof. For any $x \in \mathcal{T}$, $f(x, 1_{\mathcal{T}}, 1_{\mathcal{T}}) = (r + s + t + u + v + w)x = 0$, thus,

$$\begin{aligned} 0 = L(f(x, 1_{\mathcal{T}}, 1_{\mathcal{T}})) &= f(L(x), 1_{\mathcal{T}}, 1_{\mathcal{T}}) + f(x, L(1_{\mathcal{T}}), 1_{\mathcal{T}}) + f(x, 1_{\mathcal{T}}, L(1_{\mathcal{T}})) \\ &= 0 + (r + s + v)[x, L(1_{\mathcal{T}})] + (u + v + w)[L(1_{\mathcal{T}}), x] \\ &= (r + s - u - w)[x, L(1_{\mathcal{T}})]. \end{aligned}$$

Others are derived similarly. □

If one of the coefficients in this lemma is nonzero and regular, then we can cancel it to say $L(1_{\mathcal{T}})$ is in the center.

Proposition 8. *The following are equivalent.*

1. $r + s = u + w, t + u = s + v,$ *and* $v + w = r + t.$
2. $r + t + u = r + s + v = u + v + w.$

Proof. The equivalence is evident. □

By the proposition, we only need to study the following mutually exclusive cases.

- Case 1: $r + s \neq u + w$, $t + u \neq s + v$, or $v + w \neq r + t$.
- Case 2: $r + t + u = u + v + w = r + s + v = 0$.
- Case 3: $r + t + u = u + v + w = r + s + v \neq 0$.

The first case is examined in the following section. The second case results in a generalization of the theorem by (Xiao & Wei, 2012) on triple Lie derivations. It is discussed in Section 5. The third case will be left unsolved. The difficulty in the third case lies in the fact that $L(1_{\mathcal{T}})$ may not belong to the center. It complicates the effort to derive meaningful properties of f-derivations.

4. Case 1

In this case, Lemma 7 and the regularity of coefficients condition imply that $L(1_{\mathcal{T}}) \in \mathcal{Z}(\mathcal{T})$. Applying Lemma 6 to

$$L(f(1_{\mathcal{T}}, y, z)) = f(L(1_{\mathcal{T}}), y, z) + f(1_{\mathcal{T}}, L(y), z) + f(1_{\mathcal{T}}, y, L(z))$$

yields

$$(r + t + u)L([y, z]) = (r + t + u)(L(1_{\mathcal{T}})[y, z] + [L(y), z] + [y, L(z)]).$$

Similarly with $y = 1_{\mathcal{T}}$ or $z = 1_{\mathcal{T}}$, we get

$$\begin{aligned} (u + v + w)L([z, x]) &= (u + v + w)(L(1_{\mathcal{T}})[z, x] + [L(z), x] + [z, L(x)]), \\ (r + s + v)L([x, y]) &= (r + s + v)(L(1_{\mathcal{T}})[x, y] + [L(x), y] + [x, L(y)]). \end{aligned}$$

Since not all of $r + t + u$, $u + v + w$, and $r + s + v$ are zeros in Case 1, by the regularity of coefficients,

$$L([x, y]) = L(1_{\mathcal{T}})[x, y] + [L(x), y] + [x, L(y)].$$

The next proposition describes the structure of L.

Proposition 9. *Let $\mathcal{T}$ be a triangular algebra satisfying (♣). Suppose $c \in \mathcal{Z}(\mathcal{T})$ and $L : \mathcal{T} \to \mathcal{T}$ is a linear map such that*

$$L([x,y]) = c[x,y] + [L(x),y] + [x,L(y)]$$

for all $x,y \in \mathcal{T}$. Then $L = d + h - ci$ where d is a derivation, h is a central map vanishing on commutators, and i is the identity map.

Proof. Let $L' = L + ci$. Then L' is a Lie derivation since

$$\begin{aligned} L'([x,y]) =& L([x,y]) + c[x,y] \\ =& c[x,y] + [L(x),y] + [x,L(y)] + c[x,y] \\ =& [L(x) + cx, y] + [x, L(y) + cy] \\ =& [L'(x),y] + [x,L'(y)]. \end{aligned}$$

We apply Theorem 11 in (Cheung, 2003) assuming (♣). Then $L' = d + h$ where d is a derivation, h is a central map vanishing on commutators. Therefore, $L = d + h - ci$. □

Theorem 10. *Let L be an f-derivation on $\mathcal{T}$ where*

$$f(x,y,z) = rxyz + sxzy + tyxz + uyzx + vzxy + wzyx$$

with $r,s,t,u,v,w \in \mathcal{R}$, $r+s+t+u+v+w = 0$. Assume condition (♣). If $r+s \neq u+w$, $t+u \neq s+v$, or $v+w \neq r+t$, then $L(1_{\mathcal{T}}) \in \mathcal{Z}(\mathcal{T})$ and

$$L = d + h - L(1_{\mathcal{T}})i$$

where $d : \mathcal{T} \to \mathcal{T}$ is a derivation, $h : \mathcal{T} \to \mathcal{Z}(\mathcal{T})$ satisfies $h([x,y]) = 0$ for all $x,y \in \mathcal{T}$, and $i : \mathcal{T} \to \mathcal{T}$ is the identity map. Furthermore, in one of the following three cases, $L(1_{\mathcal{T}}) = 0$ and $h = 0$, thus L is a derivation.

1. *$r+s = u+w \neq 0$, $t+u \neq s+v$, and $v+w \neq r+t$.*
2. *$r+s \neq u+w$, $t+u = s+v \neq 0$, and $v+w \neq r+t$.*
3. *$r+s \neq u+w$, $t+u \neq s+v$, and $v+w = r+t \neq 0$.*

Proof. We have seen that $c = L(1_{\mathcal{T}}) \in \mathcal{Z}(\mathcal{T})$ and $L = d + h - ci$ by Proposition 9. It remains to show $L = d$ in one of three special cases. We may assume the second case where $t+u = s+v \neq 0$ without loss of generality. Note that for any $m \in \mathcal{M}$, $h(m) = h([1_{\mathcal{A}}, m]) = 0$ and $h(1_{\mathcal{T}}) = (L - d + ci)(1_{\mathcal{T}}) = c - 0 + c = 2c$.

Since both L and d are f-derivations, so is $L - d = h - ci$. For any $x,y,z \in \mathcal{T}$,

$$(h-ci)(f(x,y,z)) = f((h-ci)(x),y,z) + f(x,(h-ci)(y),z) + f(x,y,(h-ci)(z)).$$

Simplifying the equation, we get

$$h(f(x,y,z)) = f(h(x),y,z) + f(x,h(y),z) + f(x,y,h(z)) - 2cf(x,y,z).$$

Substitute $x = 1_{\mathcal{A}}, y = m, z = 1_{\mathcal{B}}$ where m is an arbitrary element of $\mathcal{M}$. Then $f(1_{\mathcal{A}}, m, 1_{\mathcal{B}}) = rm \in \mathcal{M}$. Hence, the left-hand side is zero. We can simplify the right-hand side using Lemma 6. Then we have

$$\begin{aligned} 0 &= (r+t+u)h(1_{\mathcal{A}})[1_{\mathcal{A}}, m] + (r+s+v)h(1_{\mathcal{B}})[1_{\mathcal{A}}, m] - 2c(rm) \\ &= (r+t+u)h(1_{\mathcal{A}})m + (r+s+v)h(1_{\mathcal{B}})m - (h(1_{\mathcal{A}}) + h(1_{\mathcal{B}}))(rm) \\ &= ((t+u)h(1_{\mathcal{A}}) + (s+v)h(1_{\mathcal{B}}))m. \end{aligned}$$

By Lemma 4, we get $(t+u)h(1_{\mathcal{A}}) + (s+v)h(1_{\mathcal{B}}) = 0$. Since $t+u = s+v \neq 0$ and they are regular, we have $h(1_{\mathcal{A}}) + h(1_{\mathcal{B}}) = 0$, or $2c = h(1_{\mathcal{T}}) = 0$. Since $\mathcal{T}$ is 2-torsion free, $c = 0$.

Next, we substitute $x = 1_{\mathcal{A}}$, $y = 1_{\mathcal{A}}$, and $z = m$. Then $f(1_{\mathcal{A}}, 1_{\mathcal{A}}, m) = rm + tm \in \mathcal{M}$. So the left hand side is zero again. Then

$$\begin{aligned} 0 &= (r+t+u)h(1_{\mathcal{A}})[1_{\mathcal{A}}, m] + (u+v+w)h(1_{\mathcal{A}})[m, 1_{\mathcal{A}}] \\ &= (r+t-v-w)h(1_{\mathcal{A}})m, \end{aligned}$$

which implies $h(1_{\mathcal{A}}) = 0$ since $r + t - v - w$ is nonzero, $h(1_{\mathcal{A}}) \in \mathcal{Z}(\mathcal{T})$, and m is arbitrary. We also have $h(1_{\mathcal{B}}) = h(1_{\mathcal{T}}) - h(1_{\mathcal{A}}) = 0$.

Finally, we show that $h(a) = h(b) = 0$ for any $a \in \mathcal{A}$ and $b \in \mathcal{B}$. If we substitute $x = a$, $y = 1_{\mathcal{A}}$, and $z = m$, then $(r+t+u)h(a)[1_{\mathcal{A}}, m] = (r+t+u)h(a)m = 0$, which implies $(r+t+u)h(a) = 0$ by Lemma 4. Rotating the values of x, y, and z, we also get $(u+v+w)h(a) = 0$ and $(r+s+v)h(a) = 0$. Not all three of the coefficients are zeros, therefore $h(a) = 0$. Similarly, $h(b) = 0$. □

Example 11. Let $f(x,y,z) = x[y,z] + [x,y]z$. Any f-derivation on $\mathcal{T}$ is a derivation because $r + s = -2$, $u + w = 0$, $t + u = s + v = 1$, $v + w = 0$, and $r + t = -1$.

5. Case 2

In this case $r+t+u = u+v+w = r+s+v = 0$ and $r+s+t+u+v+w = 0$. Solving the system of linear equations we get $v = t, w = r, u = s$, and $t = -r - s$. Therefore,

$$\begin{aligned} f(x,y,z) &= r(xyz - yxz - zxy + zyx) + s(xzy - yxz + yzx - zxy) \\ &= r[[x,y],z] + s[[x,z],y]. \end{aligned}$$

If $r + s = 0$, then $f(x,y,z) = r([[x,y],z] - [[x,z],y]) = r[[z,y],x]$ by the Jacobi identity. Then L is a Lie triple derivation. So this case is resolved by Theorem 2.1 in (Xiao & Wei, 2012). The case when $r + s \neq 0$ is resolved by the next theorem, which is a generalization of Theorem 2.1 in (Xiao & Wei, 2012).

Theorem 12. *Let $\mathcal{T}$ be a 2-torsion free triangular algebra $\begin{bmatrix} \mathcal{A} & \mathcal{M} \\ 0 & \mathcal{B} \end{bmatrix}$, and assume $\mathcal{M}$ is faithful as a left $\mathcal{A}$-module and as a right $\mathcal{B}$-module. Suppose that $\mathcal{T}$ satisfies conditions (♣) and (♠). Let $r, s \in \mathcal{R}$ and assume r, s, and $r+s$ are $\mathcal{T}$-regular. If L is an f-derivation where $f(x,y,z) = r[[x,y],z] + s[[x,z],y]$, then L is of standard form, that is, there exist a derivation $d : \mathcal{T} \to \mathcal{T}$ and a linear map $h : \mathcal{T} \to \mathcal{Z}(\mathcal{T})$ such that $L = d + h$ and $h(r[[x,y],z] + s[[x,z],y]) = 0$ for all $x, y, z \in \mathcal{T}$.*

The theorem is proved after a series of lemmas, which are direct extensions of results by Xiao and Wei. We assume all conditions in the theorem for all subsequent lemmas, and we let $\alpha = \pi_{\mathcal{A}}L$, $\mu = \pi_{\mathcal{M}}L$, and $\beta = \pi_{\mathcal{B}}L$. Then $L = \alpha + \mu + \beta$.

Lemma 13. $L(1_{\mathcal{T}}) \in \mathcal{Z}(\mathcal{T})$

Proof. Since $1_{\mathcal{T}}$ is in the center, $[1_{\mathcal{T}}, x] = 0$ for all $x \in \mathcal{T}$. Therefore,

$$\begin{aligned} 0 =& L(r[[1_{\mathcal{T}},y],z] + s[[1_{\mathcal{T}},z],y]) \\ =& r[[L(1_{\mathcal{T}}),y],z] + r[[1_{\mathcal{T}},L(y)],z] + r[[1_{\mathcal{T}},y],L(z)] \\ & + s[[L(1_{\mathcal{T}}),z],y] + s[[1_{\mathcal{T}},L(z)],y] + s[[1_{\mathcal{T}},z],L(y)] \\ =& r[[L(1_{\mathcal{T}}),y],z] + s[[L(1_{\mathcal{T}}),z],y]. \end{aligned}$$

Let $L(1_{\mathcal{T}}) = a + m' + b$. We want to show that $m' = 0$ and $am = mb$ for all $m \in \mathcal{M}$. Substitute $y = z = 1_{\mathcal{B}}$. Then

$$0 = r[[a + m' + b, 1_{\mathcal{B}}], 1_{\mathcal{B}}] + s[[a + m' + b, 1_{\mathcal{B}}], 1_{\mathcal{B}}] = (r+s)m'.$$

Therefore, $m' = 0$ since $r + s$ is regular. Next, let $y = m$ and $z = 1_{\mathcal{B}}$. Then

$$\begin{aligned} 0 =& r[[a+b,m],1_{\mathcal{B}}] + s[[a+b,1_{\mathcal{B}}],m] \\ =& r[am - mb, 1_{\mathcal{B}}] + s[0,m] \\ =& r(am - mb). \end{aligned}$$

Since r is regular, $am = mb$ for all $m \in \mathcal{M}$. □

Lemma 14. *For any $m \in \mathcal{M}$, $L(m) \in \mathcal{M}$.*

Proof. For $x = m$, $y = 1_{\mathcal{B}}$, and $z = 1_{\mathcal{B}}$,

$$L(f(m, 1_{\mathcal{B}}, 1_{\mathcal{B}})) = f(L(m), 1_{\mathcal{B}}, 1_{\mathcal{B}}) + f(m, L(1_{\mathcal{B}}), 1_{\mathcal{B}}) + f(m, 1_{\mathcal{B}}, L(1_{\mathcal{B}})).$$

The left-hand side is $(r+s)L(m)$ since $[[m, 1_{\mathcal{B}}], 1_{\mathcal{B}}] = m$. On the other hand, $f(L(m), 1_{\mathcal{B}}, 1_{\mathcal{B}})$, $f(m, L(1_{\mathcal{B}}), 1_{\mathcal{B}})$, and $f(m, 1_{\mathcal{B}}, L(1_{\mathcal{B}}))$ belong to $\mathcal{M}$ because $[t, 1_{\mathcal{B}}] \in \mathcal{M}$ and $[m, t] \in \mathcal{M}$ for any $t \in \mathcal{T}$ by Lemma 1 (6) and (7). Since $r + s$ is regular, $L(m) \in \mathcal{M}$. □

Lemma 15. *$\alpha(1_{\mathcal{A}}) + \beta(1_{\mathcal{A}}) \in \mathcal{Z}(\mathcal{T})$ and $\alpha(1_{\mathcal{B}}) + \beta(1_{\mathcal{B}}) \in \mathcal{Z}(\mathcal{T})$. Thus, $\phi\alpha(1_{\mathcal{A}}) = \beta(1_{\mathcal{A}})$ and $\phi\alpha(1_{\mathcal{B}}) = \beta(1_{\mathcal{B}})$.*

Proof. Let $L(1_{\mathcal{B}}) = a + m' + b$ where $a = \alpha(1_{\mathcal{B}})$, $m' = \mu(1_{\mathcal{B}})$, and $b = \beta(1_{\mathcal{B}})$. As in the previous lemma, let $x = m$, $y = 1_{\mathcal{B}}$, and $z = 1_{\mathcal{B}}$. Then

$$L(f(m, 1_{\mathcal{B}}, 1_{\mathcal{B}})) = f(L(m), 1_{\mathcal{B}}, 1_{\mathcal{B}}) + f(m, L(1_{\mathcal{B}}), 1_{\mathcal{B}}) + f(m, 1_{\mathcal{B}}, L(1_{\mathcal{B}})).$$

The left-hand side is $(r + s)L(m)$. On the other side, by Lemma 1 and 14,

$$\begin{aligned} f(L(m), 1_{\mathcal{B}}, 1_{\mathcal{B}}) &= (r + s)[[L(m), 1_{\mathcal{B}},]1_{\mathcal{B}}] = (r + s)L(m), \\ f(m, L(1_{\mathcal{B}}), 1_{\mathcal{B}}) &= r[[m, a + m' + b], 1_{\mathcal{B}}] + s[[m, 1_{\mathcal{B}}], a + m' + b] \\ &= (r + s)(-am + mb), \\ f(m, 1_{\mathcal{B}}, L(1_{\mathcal{B}})) &= r[[m, 1_{\mathcal{B}}], a + m' + b] + s[[m, a + m' + b], 1_{\mathcal{B}}] \\ &= (r + s)(-am + mb). \end{aligned}$$

Thus, $(r + s)L(m) = (r + s)(L(m) - 2(am - mb))$. Since $r + s$ is regular and $\mathcal{T}$ is 2-torsion free, $am = mb$ for all $m \in \mathcal{M}$, i.e., $a + b = \alpha(1_{\mathcal{B}}) + \beta(1_{\mathcal{B}}) \in \mathcal{Z}(\mathcal{T})$. Similarly, we can prove $\alpha(1_{\mathcal{A}}) + \beta(1_{\mathcal{A}}) \in \mathcal{Z}(\mathcal{T})$ by substituting $x = m$, $y = 1_{\mathcal{A}}$, and $z = 1_{\mathcal{A}}$. □

Lemma 16. *$\beta(a) \in \mathcal{Z}(\mathcal{B})$ and $\alpha(b) \in \mathcal{Z}(\mathcal{A})$ for any $a \in \mathcal{A}$ and $b \in \mathcal{B}$.*

Proof. Note that $f(a, m, 1_{\mathcal{B}}) = r[[a, m], 1_{\mathcal{B}}] + s[[a, 1_{\mathcal{B}}], m] = ram$. Then

$$L(ram) = r\left([[L(a), m], 1_{\mathcal{B}}] + [[a, L(m)], 1_{\mathcal{B}}] + [[a, m], L(1_{\mathcal{B}})]\right) + s\left([[L(a), 1_{\mathcal{B}}], m] + [[a, L(1_{\mathcal{B}})], m] + [[a, 1_{\mathcal{B}}], L(m)]\right).$$

Each component on the right side is computed as follows.

$$\begin{aligned} &[[L(a), m], 1_{\mathcal{B}}] = [[\alpha(a) + \mu(a) + \beta(a), m], 1_{\mathcal{B}}] = \alpha(a)m - m\beta(a), \\ &[[a, L(m)], 1_{\mathcal{B}}] = aL(m) \text{ by Lemma 14}, \\ &[[a, m], L(1_{\mathcal{B}})] = [am, \alpha(1_{\mathcal{B}}) + \mu(1_{\mathcal{B}}) + \beta(1_{\mathcal{B}})] = [am, \mu(1_{\mathcal{B}})] = 0 \text{ by Lemma 15}, \end{aligned}$$

$$\begin{aligned} &[[L(a), 1_{\mathcal{B}}], m] = [[\alpha(a) + \mu(a) + \beta(a), 1_{\mathcal{B}}], m] = [\mu(a), m] = 0, \\ &[[a, L(1_{\mathcal{B}})], m] = 0 \text{ since } [a, L(1_{\mathcal{B}})] = [a, \mu(1_{\mathcal{B}})] \in \mathcal{M} \text{ by Lemma 15}, \\ &[[a, 1_{\mathcal{B}}], L(m)] = [0, L(m)] = 0. \end{aligned}$$

Then $rL(am) = L(ram) = r(\alpha(a)m - m\beta(a) + aL(m))$. Since r is regular,

$$L(am) = \alpha(a)m - m\beta(a) + aL(m).$$

Similarly, with $x = b$, $y = m$, and $z = 1_{\mathcal{A}}$, we get

$$L(mb) = m\beta(b) - \alpha(b)m + L(m)b.$$

Now we compute $L(amb)$ in two ways. First,

$$\begin{aligned} L(a(mb)) &= \alpha(a)mb - mb\beta(a) + aL(mb) \\ &= \alpha(a)mb - mb\beta(a) + am\beta(b) - a\alpha(b)m + aL(m)b. \end{aligned}$$

Similarly,

$$\begin{aligned} L((am)b) &= am\beta(b) - \alpha(b)am + L(am)b \\ &= am\beta(b) - \alpha(b)am + \alpha(a)mb - m\beta(a)b + aL(m)b. \end{aligned}$$

Comparing two results, we obtain

$$\alpha(b)am - a\alpha(b)m = mb\beta(a) - m\beta(a)b$$

or

$$[\alpha(b), a]m = m[b, \beta(a)],$$

which leads to $[\alpha(b), a] + [b, \beta(a)] \in \mathcal{Z}(\mathcal{T})$. Then, $[\alpha(b), a] \in \mathcal{Z}(\mathcal{A})$ and $[b, \beta(a)] \in \mathcal{Z}(\mathcal{B})$. Since a and b are arbitrary, $\alpha(b) \in \mathcal{Z}(\mathcal{A})$, $\beta(a) \in \mathcal{Z}(\mathcal{B})$ by condition (♠). □

The consequence of the lemma is that $\alpha(b) \in \pi_{\mathcal{A}}(\mathcal{Z}(\mathcal{T}))$ and $\beta(a) \in \pi_{\mathcal{B}}(\mathcal{Z}(\mathcal{T}))$ by condition (♣). Therefore, $\alpha\pi_{\mathcal{B}} + \phi\alpha\pi_{\mathcal{B}}$ and $\phi^{-1}\beta\pi_{\mathcal{A}} + \beta\pi_{\mathcal{A}}$ are maps into the center $\mathcal{Z}(\mathcal{T})$. Now we define $h : T \to \mathcal{Z}(\mathcal{T})$ to be the sum of those central maps

$$h = \alpha\pi_B + \phi\alpha\pi_B + \phi^{-1}\beta\pi_A + \beta\pi_A,$$

and define $d = L - h$. It remains to prove that h vanishes on $f(x, y, z)$ and d is a derivation.

Lemma 17. *$h(f(x, y, z)) = 0$ for any $x, y, z \in \mathcal{T}$*

Proof. By Lemma 1(10), $\pi_{\mathcal{B}}(f(x, y, z)) = f(\pi_{\mathcal{B}}x, \pi_{\mathcal{B}}y, \pi_{\mathcal{B}}z)$. Then

$$\begin{aligned}
\alpha\pi_{\mathcal{B}}(f(x, y, z)) &= \pi_{\mathcal{A}}L\pi_{\mathcal{B}}(f(x, y, z)) \\
&= \pi_{\mathcal{A}}L(f(\pi_{\mathcal{B}}x, \pi_{\mathcal{B}}y, \pi_{\mathcal{B}}z)) \\
&= \pi_{\mathcal{A}}(f(L\pi_{\mathcal{B}}x, \pi_{\mathcal{B}}y, \pi_{\mathcal{B}}z) + f(\pi_{\mathcal{B}}x, L\pi_{\mathcal{B}}y, \pi_{\mathcal{B}}z) + f(\pi_{\mathcal{B}}x, \pi_{\mathcal{B}}y, L\pi_{\mathcal{B}}z)) \\
&= f(\pi_{\mathcal{A}}L\pi_{\mathcal{B}}x, 0, 0) + f(0, \pi_{\mathcal{A}}L\pi_{\mathcal{B}}y, 0) + f(0, 0, \pi_{\mathcal{A}}L\pi_{\mathcal{B}}z) \\
&= 0.
\end{aligned}$$

Similarly, other components of $h(f(x, y, z))$ are zeros as well. □

Lemma 18. *d has the following properties for any $a \in \mathcal{A}$, $m \in \mathcal{M}$, and $b \in \mathcal{B}$.*

1. $d(1_{\mathcal{T}}) = 0$,
2. $d(m) = L(m) \in \mathcal{M}$,
3. $d(a) \in \mathcal{A} + \mathcal{M}$,
4. $d(b) \in \mathcal{B} + \mathcal{M}$.

Proof. First, by Lemma 13,

$$L(1_{\mathcal{T}}) = \alpha(1_{\mathcal{T}}) + \beta(1_{\mathcal{T}}) = \alpha(1_{\mathcal{A}}) + \alpha(1_{\mathcal{B}}) + \beta(1_{\mathcal{A}}) + \beta(1_{\mathcal{B}}).$$

By Lemma 15, $\phi\alpha(1_{\mathcal{B}}) = \beta(1_{\mathcal{B}})$ and $\phi^{-1}\beta(1_{\mathcal{A}}) = \alpha(1_{\mathcal{A}})$, thus,

$$\begin{aligned}
h(1_{\mathcal{T}}) =& \alpha(1_{\mathcal{B}}) + \phi\alpha(1_{\mathcal{B}}) + \phi^{-1}\beta(1_{\mathcal{A}}) + \beta(1_{\mathcal{A}}) \\
=& \alpha(1_{\mathcal{B}}) + \beta(1_{\mathcal{B}}) + \alpha(1_{\mathcal{A}}) + \beta(1_{\mathcal{A}}).
\end{aligned}$$

Therefore, $d(1_{\mathcal{T}}) = 0$. Second, since $h(m) = 0$ by definition, $d(m) = L(m) \in \mathcal{M}$ by Lemma 14. Lastly,

$$\begin{aligned}
\pi_B d(a) =& \pi_B L(a) - \pi_B h(a) \\
=& \beta(a) - \pi_B(\phi^{-1}\beta(a) + \beta(a)) \\
=& \beta(a) - \beta(a) = 0.
\end{aligned}$$

Thus $d(a) \in \mathcal{A} + \mathcal{M}$. Similarly, $d(b) \in \mathcal{B} + \mathcal{M}$. □

To prove that d is a derivation, we prove that it is a derivation component-wise, then we combine the results together.

Lemma 19. *$d(am) = d(a)m + ad(m)$ and $d(mb) = d(m)b + md(b)$ for any $a \in \mathcal{A}$, $m \in \mathcal{M}$, and $b \in \mathcal{B}$.*

Proof. As in the proof of Lemma 16,

$$L(am) = \alpha(a)m - m\beta(a) + aL(m)$$

and by Lemma 18 (2),

$$d(am) = \alpha(a)m - m\beta(a) + ad(m).$$

On the other hand,

$$\begin{aligned} d(a)m &= \pi_{\mathcal{A}} d(a)m && \text{(by Lemma 18 (3))} \\ &= \pi_{\mathcal{A}} L(a)m - \pi_{\mathcal{A}} h(a)m \\ &= \alpha(a)m - \pi_{\mathcal{A}}(\phi^{-1}\beta(a) + \beta(a))m \\ &= \alpha(a)m - \phi^{-1}\beta(a)m \\ &= \alpha(a)m - m\beta(a). && \text{(by Lemma 3)} \end{aligned}$$

Hence, $d(am) = d(a)m + ad(m)$. Similarly, $d(mb) = d(m)b + md(b)$.

□

Lemma 20. *$d(a_1a_2) = d(a_1)a_2 + a_1d(a_2)$ and $d(b_1b_2) = d(b_1)b_2 + b_1d(b_2)$ for any $a_1, a_2 \in \mathcal{A}$ and $b_1, b_2 \in \mathcal{B}$.*

Proof. By the previous lemma,

$$d((a_1a_2)m) = d(a_1a_2)m + a_1a_2d(m).$$

We also have

$$\begin{aligned} d(a_1(a_2m)) &= d(a_1)a_2m + a_1d(a_2m) \\ &= d(a_1)a_2m + a_1d(a_2)m + a_1a_2d(m). \end{aligned}$$

Comparing two results, we have

$$d(a_1a_2)m = (d(a_1)a_2 + a_1d(a_2))m.$$

By the faithfulness of $\mathcal{M}$,

$$d(a_1a_2) = d(a_1)a_2 + a_1d(a_2).$$

Similarly,

$$d(b_1b_2) = d(b_1)b_2 + b_1d(b_2).$$

□

Lemma 21. *$d(a)b + ad(b) = 0$ for any $a \in \mathcal{A}$ and $b \in \mathcal{B}$.*

Proof. Obviously, $f(a, b, 1_{\mathcal{B}}) = r[[a, b], 1_{\mathcal{B}}] + s[[a, 1_{\mathcal{B}}], b] = 0$. Then

$$\begin{aligned} 0 = L(f(a, b, 1_{\mathcal{B}})) &= f(L(a), b, 1_{\mathcal{B}}) + f(a, L(b), 1_{\mathcal{B}}) + f(a, b, L(1_{\mathcal{B}})) \\ &= f(d(a), b, 1_{\mathcal{B}}) + f(a, d(b), 1_{\mathcal{B}}) + f(a, b, d(1_{\mathcal{B}})) \\ &\quad + f(h(a), b, 1_{\mathcal{B}}) + f(a, h(b), 1_{\mathcal{B}}) + f(a, b, h(1_{\mathcal{B}})) \\ &= f(d(a), b, 1_{\mathcal{B}}) + f(a, d(b), 1_{\mathcal{B}}) + f(a, b, d(1_{\mathcal{B}})) \end{aligned}$$

since h is a central map. We compute each component of the last line one by one. Note that for any $a \in \mathcal{A}$ and $b \in \mathcal{B}$, $bd(a) = 0$, $d(b)a = 0$, and $d(b)1_{\mathcal{B}} = d(b)$ by Lemma 18 (3) and (4). With this,

$$\begin{aligned} f(d(a), b, 1_{\mathcal{B}}) &= r[[d(a), b], 1_{\mathcal{B}}] + s[[d(a), 1_{\mathcal{B}}], b] \\ &= r[d(a)b, 1_{\mathcal{B}}] + s[d(a)1_{\mathcal{B}}, b] \\ &= (r + s)d(a)b, \\ f(a, d(b), 1_{\mathcal{B}}) &= r[[a, d(b)], 1_{\mathcal{B}}] + s[[a, 1_{\mathcal{B}}], d(b)] \\ &= r[ad(b), 1_{\mathcal{B}}] \\ &= rad(b), \\ f(a, b, d(1_{\mathcal{B}})) &= r[[a, b], d(1_{\mathcal{B}})] + s[[a, d(1_{\mathcal{B}})], b] \\ &= sad(1_{\mathcal{B}})b. \end{aligned}$$

As $d(b) = d(1_{\mathcal{B}}b) = d(1_{\mathcal{B}})b + 1_{\mathcal{B}}d(b)$ by the previous lemma, the last one is equal to $sa(d(b) - 1_{\mathcal{B}}d(b)) = sad(b)$. Therefore, taking the sum of three equations, we have $(r + s)(d(a)b + ad(b)) = 0$. Since $r + s$ is regular, $d(a)b + ad(b) = 0$. □

Lemma 22. $d(t_1t_2) = d(t_1)t_2 + t_1d(t_2)$ *for any* $t_1, t_2 \in \mathcal{T}$.

Proof. Let $t_1 = a_1 + m_1 + b_1$ and $t_2 = a_2 + m_2 + b_2$. Then, by Lemma 19 and Lemma 20,

$$\begin{aligned} d(t_1t_2) =& d((a_1 + m_1 + b_1)(a_2 + m_2 + b_2)) \\ =& d(a_1a_2 + a_1m_2 + m_1b_2 + b_1b_2) \\ =& d(a_1)a_2 + a_1d(a_2) + d(a_1)m_2 + a_1d(m_2) \\ &+ d(m_1)b_2 + m_1d(b_2) + d(b_1)b_2 + b_1d(b_2). \end{aligned}$$

On the other hand, by Lemma 18 and Lemma 21,

$$\begin{aligned} d(t_1)t_2 + t_1d(t_2) =& (d(a_1) + d(m_1) + d(b_1))(a_2 + m_2 + b_2) \\ &+ (a_1 + m_1 + b_1)(d(a_2) + d(m_2) + d(b_2)) \\ =& d(a_1)a_2 + d(a_1)m_2 + d(a_1)b_2 + d(m_1)b_2 + d(b_1)b_2 \\ &+ a_1d(a_2) + a_1d(m_2) + a_1d(b_2) + m_1d(b_2) + b_1d(b_2) \\ =& d(a_1)a_2 + d(a_1)m_2 + d(m_1)b_2 + d(b_1)b_2 \\ &+ a_1d(a_2) + a_1d(m_2) + m_1d(b_2) + b_1d(b_2). \end{aligned}$$

Therefore, $d(t_1t_2) = d(t_1)t_2 + t_1d(t_2)$. □

This completes the proof of Theorem 12.

Acknowledgements

The authors would like to thank the referee for reviews and comments.

References

Benkovič, D. (2009). Biderivations of triangular algebras. *Linear Algebra and Its Applications, 431*(9), 1587-1602. http://dx.doi.org/10.1016/j.laa.2009.05.029

Benkovič, D. (2015). A note on f-derivations of triangular algebras. *Aequationes Mathematicae, 89*(4), 1207-1211. http://dx.doi.org/10.1007/s00010-014-0298-y

Benkovič, D. (2016). Jordan ?-derivations of triangular algebras. *Linear and Multilinear Algebra, 64*(2), 143-155. http://dx.doi.org/10.1080/03081087.2015.1027646

Benkovič, D., & Eremita, D. (2004). Commuting traces and commutativity preserving maps on triangular algebras. *Journal of Algebra, 280*(2), 797-824. http://dx.doi.org/10.1016/j.jalgebra.2004.06.019

Benkovič, D., & Eremita, D. (2012). Multiplicative Lie n-derivations of triangular rings. *Linear Algebra and its Applications, 436*(11), 4223-4240. http://dx.doi.org/10.1016/j.laa.2012.01.022

Cheung, W. S. (2001). Commuting maps of triangular algebras. *Journal of the London Mathematical Society. Second Series, 63*(1), 117-127. http://dx.doi.org/10.1112/S0024610700001642

Cheung, W. S. (2003). Lie derivations of triangular algebras. *Linear and Multilinear Algebra, 51*(3), 299-310. http://dx.doi.org/10.1080/0308108031000096993

Du, Y., & Wang, Y. (2012). k-commuting maps on triangular algebras. *Linear Algebra and its Applications, 436*(5), 1367-1375. http://dx.doi.org/10.1016/j.laa.2011.08.024

Eremita, D. (2013). Functional identities of degree 2 in triangular rings. *Linear Algebra and its Applications, 438*(1), 584-597. http://dx.doi.org/10.1016/j.laa.2012.07.028

Ji, P., Liu, R., & Zhao, Y. (2012). Nonlinear Lie triple derivations of triangular algebras. *Linear and Multilinear Algebra, 60*(10), 1155-1164. http://dx.doi.org/10.1080/03081087.2011.652109

Wang, Y., Wang, Y., & Du, Y. (2013). n-derivations of triangular algebras. *Linear Algebra and its Applications, 439*(2), 463-471. http://dx.doi.org/10.1016/j.laa.2013.03.032

Wong, T. L. (2005). Jordan isomorphisms of triangular rings. *Proceedings of the American Mathematical Society, 133*(11), 3381-3388 (electronic). http://dx.doi.org/10.1090/S0002-9939-05-07989-X

Xiao, Z., & Wei, F. (2010). Jordan higher derivations on triangular algebras. *Linear Algebra and its Applications, 432*(10), 2615-2622. http://dx.doi.org/10.1016/j.laa.2009.12.006

Xiao, Z., & Wei, F. (2012). Lie triple derivations of triangular algebras. *Linear Algebra and its Applications, 437*(5), 1234-1249. http://dx.doi.org/10.1016/j.laa.2012.04.015

Yu, W., & Zhang, J. (2010). Nonlinear Lie derivations of triangular algebras. *Linear Algebra and its Applications, 432*(11), 2953-2960. http://dx.doi.org/10.1016/j.laa.2009.12.042

Zhang, J. H., Feng, S., Li, H. X., & Wu, R. H. (2006). Generalized biderivations of nest algebras. *Linear Algebra and its Applications, 418*(1), 225-233. http://dx.doi.org/10.1016/j.laa.2006.02.001

Zhang, J. H., & Yu, W. Y. (2006). Jordan derivations of triangular algebras. *Linear Algebra and its Applications, 419*(1), 251-255. http://dx.doi.org/10.1016/j.laa.2006.04.015

Zhao, Y., Wang, D., & Yao, R. (2009). Biderivations of upper triangular matrix algebras over commutative rings. *International Journal of Mathematics, Game Theory, and Algebra, 18*(6), 473-478.

3

Absolute Valued Algebras with Strongly One Sided Unit

Alassane Diouf[1]

[1] Département de Mathématiques et Informatiques, Faculté des Sciences et Techniques, Université Cheikh Anta Diop, Dakar, Sénégal

Correspondence: Alassane Diouf, Département de Mathématiques et Informatiques, Faculté des Sciences et Techniques, Université Cheikh Anta Diop, Dakar, Sénégal. E-mail: dioufalassane@hotmail.fr

Abstract

We classify the absolute valued algebras with strongly left unit of dimension ≤ 4. Also we prove that every 8-dimensional absolute valued algebra with strongly left unit contain a 4-dimensional subalgebra, next we determine the form of theirs algebras by the duplication process.

Keywords: absolute valued algebra, strongly left unit, duplication process

Mathematics Subject Classification: 17A35, 17A36

1. Introduction

The absolute valued algebras are introduced by Ostrowski 1918. It's the normed algebra A such that $||xy|| = ||x||||y||$ for all x, y in A. For an element a in an algebra A, we denote by $L_a : A \to A\ x \mapsto ax$ and $R_a : A \to A\ x \mapsto xa$. The algebra is called division if and only if R_a and L_a are bijective for all a in A. We denote by O the orthogonal group of linear isometries of Euclidean space $\mathbb{H}$. We recall O^+ the subgroup of proper linear isometries and O^- the subset of improper linear isometries. Let A be an absolute valued algebra with unit, then A is isomorphic to $\mathbb{R}$, $\mathbb{C}$, $\mathbb{H}$ or $\mathbb{O}$ (Urbanik & Wright, 1960). The absolute valued algebras with left unit satisfying to $(x^2, x^2, x^2) = 0$, for all $x \in A$ is classified in (Diankha & all, 2013_2). These algebras are finite dimensional and isomorphic to $\mathbb{R}$, $\mathbb{C}$, ${}^\star\mathbb{C}$, $\mathbb{H}$, ${}^\star\mathbb{H}$, ${}^\star\mathbb{H}(i, 1)$, $\mathbb{O}$, ${}^\star\mathbb{O}$, ${}^\star\mathbb{O}(i, 1)$, $\widetilde{\mathbb{O}}$ or $\widetilde{\mathbb{O}}(i)$ and the element e satisfy to $L_e = R_e^2 = I_A$. The algebras $\mathbb{H}_i$, $\mathbb{O}_i$ (Diankha & all, 2013_1), satisfy to $L_e = R_e^2 = I_A$ and not satisfy to $(x^2, x^2, x^2) = 0$. In this paper we give a classification of the absolute valued algebras with strongly left unit of dimension ≤ 4. We proves that if A is 8-dimensional absolute valued algebra with strongly left unit, then A contain a 4-dimensional subalgebra and A is obtained by the duplication process. Otherwise A is of the form $\mathbb{H} \times \mathbb{H}_{(\varphi,\psi)}$ with $\varphi, \psi : \mathbb{H} \to \mathbb{H}$ are linear isometries such that $\varphi(1) = 1$ and $(\varphi, \psi)^2 = (\varphi, \psi)$. The algebras $\mathbb{R}$, $\mathbb{C}$, ${}^\star\mathbb{C}$, $\mathbb{H}$, ${}^\star\mathbb{H}$, $\mathbb{H}_i$, ${}^\star\mathbb{H}(i, 1)$, $\mathbb{O}$, ${}^\star\mathbb{O}$, $\mathbb{O}_i$, ${}^\star\mathbb{O}(i, 1)$, $\widetilde{\mathbb{O}}$ are absolute valued algebras with strongly left unit. This list is completed by new algebras.

2. Preliminary

In this section we recall the some interest results:

Theorem 1 *The finite-dimensional absolute valued real algebras with a left unit are precisely those of the form $\mathbb{A}_\varphi$, where $\mathbb{A} \in \{\mathbb{R}, \mathbb{C}, \mathbb{H}, \mathbb{O}\}$ and φ is an isometric of the euclidien espace $\mathbb{A}$ fixes 1, and $\mathbb{A}_\varphi$ denotes the absolute-valued real algebra obtained by endowing the normed space of $\mathbb{A}$ with the product $x \odot y := \varphi(x)y$. Moreover, given linear isometries $\varphi, \phi : \mathbb{A} \to \mathbb{A}$ fixing 1, the algebras $\mathbb{A}_\varphi$ and $\mathbb{A}_\phi$ are isomorphic if and only if there exists an algebra automorphism ψ of $\mathbb{A}$ satisfying $\phi = \psi \circ \varphi \circ \psi^{-1}$* (Rochdi, 2003).

Lemma 1 *Let A be an absolute valued algebra with strongly left unit. The following equalities hold for all $x \in A$.*

1. $[(xe)x]e = x(xe)$
2. $[x(xe)]e = (xe)x$
3. $[xe, x] =< e, x > [e, x - xe]$
   If, moreover, x is orthogonal to e, then
4. $[xe, x] = 0$
5. $(xe)x^2 = 2 < e, x^2 > x - ||x||^2 xe$
6. $(xe)^2 = 2 < e, x^2 > e - x^2$
7. $x^2 x = -||x||^2 xe$ (Chandid & Rochdi, 2008).

The group G_2 acts transitively on the sphere $S(Im(\mathbb{O})) := S^6$, that is the mapping $G_2 \rightarrow S^6 \ \Phi \mapsto \Phi(i)$ is surjective (Postnikov, 1985).

Let $\mathbb{A}$ be one of the unital absolute valued algebras $\mathbb{R}$, $\mathbb{C}$, $\mathbb{H}$ of dimension m. Consider the caley dickson product $\odot$ in $\mathbb{A} \times \mathbb{A}$, we define on the space $\mathbb{A} \times \mathbb{A}$ the product

$$(x, y) \star (x', y') = (f_1(x), f(x)) \odot (g_1(x'), g(y')).$$

With f_1, g_1, f, g be linear isometries of $\mathbb{A}$ and $f_1(1) = g_1(1) = 1$. We obtain a $2m$-dimensional absolute valued real algebra $\mathbb{A} \times \mathbb{A}_{(f_1,f),(g_1,g)}$. The process is called duplication process. Note that the algebra is left unit if $g_1 = g = I_{\mathbb{A}}$ and this case we not the algebra by $\mathbb{A} \times \mathbb{A}_{(f_1,f)}$. We have the following result (Calderon & all, 2011):

Theorem 2 *Let A be an 8-dimensional absolute valued algebra, then the following are equivalent:*

1. *A contains a 4-dimensional subalgebra.*
2. *A is obtained by the duplication process.*
3. *Aut(A) contains a reflexion.*

Lemma 2 *Let* $\mathcal{I}^+ = \{f \in O^+ : f \text{ involutive}\}$, $\mathcal{I}^- = \{f \in O^- : f \text{ involutive}\}$, $\mathcal{I}_1^+ = \{f \in \mathcal{I}^+ : f(1) = 1\}$ *and* $\mathcal{I}_1^- = \{f \in \mathcal{I}^- : f(1) = 1\}$. *We have:*

1. $O^+ = \{T_{a,b} : a, b \in S(\mathbb{H})\}$
2. $O^- = \{T_{a,b} \circ \sigma_{\mathbb{H}} : a, b \in S(\mathbb{H})\} := O^+ \circ \sigma_{\mathbb{H}}$
3. $O_1^+ = \{T_{a,\bar{a}} : a \in S(\mathbb{H})\}$
4. $O_1^- = \{T_{a,\bar{a}} \circ \sigma_{\mathbb{H}} : a \in S(\mathbb{H})\} := O_1^+ \circ \sigma_{\mathbb{H}}$
5. $\mathcal{I}^+ = \{\pm I_{\mathbb{H}}\} \cup \{T_{a,b} : a, b \in S(Im(\mathbb{H}))\}$
6. $\mathcal{I}^- = \{\pm T_{a,\bar{a}} : a \in S(\mathbb{H})\}$
7. $\mathcal{I}_1^+ = \{I_{\mathbb{H}}\} \cup \{T_{a,\bar{a}} : a \in S(Im(\mathbb{H}))\}$
8. $\mathcal{I}_1^- = \{\sigma_{\mathbb{H}}\} \cup \{T_{a,\bar{a}} \circ \sigma_{\mathbb{H}} : a \in S(Im(\mathbb{H}))\} := \mathcal{I}_1^+ \circ \sigma_{\mathbb{H}}$ *(Diankha & all, 2013₂).*

Corollary 1 *Let A be an absolute valued algebra with left unit satisfying to* $(x^p, x^q, x^r) = 0$ *with* $\{p, q, r\} \in \{1, 2\}$. *Then A contains a strongly left unit.*

Proof. Lemma 1 (Diankha & all, 2013_2) and proof of Proposition 4.8 (Chandid & Rochdi, 2008)..

The converse of Corollary 1 is false, an effect the algebra $A := \mathbb{O}_i$ is an absolute valued algebra with strongly left unit and A not satisfy to $(x^2, x^2, x^2) = 0$.

3. Absolute Valued Algebras with Strongly Left Unit

Definition 1 *An element* $e \in A$ *is called strongly left unit, if it's left unit and square root of right unit* ($L_e = R_e^2 = I_A$).

Theorem 3 *Let A be an absolute valued algebra with strongly left unit. Then A is finite dimensional. Moreover if* $dim(A) \leq 4$, *then A is isomorphic to* $\mathbb{R}$, $\mathbb{C}$, ${}^\star\mathbb{C}$, $\mathbb{H}$, ${}^\star\mathbb{H}$, $\mathbb{H}(i, 1)$ *or* ${}^\star\mathbb{H}(i, 1)$.

Proof. The algebra A is left unit, hence A is left division (Rodriguez, 2004). Morover the assertion $R_e^2 = I_A$ imply that A is right division, then A is finite dimensional. Also A is of the form $\mathbb{A}_\varphi$, with φ a linear isometric fixed 1 and $\mathbb{A} \in \{\mathbb{R}, \mathbb{C}, \mathbb{H}, \mathbb{O}\}$ (Theorem 1). If $dim(A) \leq 2$, it's clear that A is isomorphic to $\mathbb{R}$, $\mathbb{C}$ or ${}^\star\mathbb{C}$.
Assume now $dim(A) \geq 4$, then the assertion $R_e^2 = I_A$ imply:

$$\begin{aligned} x &= (x \odot 1) \odot 1 \\ &= \varphi^2(x). \end{aligned}$$

Then φ is an involutive linear isometric $\varphi^2 = I_{\mathbb{A}}$.
If $dim(A) = 4$, we have:
$\varphi \in \mathcal{I}_1^+ \cup \mathcal{I}_1^- = \{I_{\mathbb{H}}\} \cup \{T_{a,\bar{a}} : a \in S(Im(\mathbb{H}))\} \cup \{\sigma_{\mathbb{H}}\} \cup \{T_{a,\bar{a}} \circ \sigma_{\mathbb{H}} : a \in S(Im(\mathbb{H}))\}$ (Lemma 2).

- If $\varphi = I_{\mathbb{H}}$, then A is isomorphic to $\mathbb{H}$.
- If $\varphi = \sigma_{\mathbb{H}}$, then A is isomorphic to $^{\star}\mathbb{H}$.
- If $\varphi = T_{a,\bar{a}}$: $a \in S(Im(\mathbb{H}))$, there exist $v \in S(\mathbb{H})$ such that $va\bar{v} = i$ and let the automorphism $\Phi = T_{v,\bar{v}}$ of $\mathbb{H}$ with $\Phi^{-1} = T_{\bar{v},v}$, we have $\Phi \circ T_{a,\bar{a}} \circ \Phi^{-1} = T_{i,\bar{i}}$. Then A is isomorphic to $\mathbb{H}_{T_{i,\bar{i}}}$ (Theorem 1) and the map $\Phi : \mathbb{H}(i,1) \to \mathbb{H}_{T_{i,\bar{i}}}$ $x \mapsto xi$ is an isomorphism algebras.
- If $\varphi = T_{a,\bar{a}} \circ \sigma_{\mathbb{H}}$: $a \in S(Im(\mathbb{H}))$, there exist $u \in S(\mathbb{H})$ such that $ua\bar{u} = i$ and let the automorphism $\Phi = T_{u,\bar{u}}$ of $\mathbb{H}$, we have:

$$\begin{aligned}\Phi \circ T_{a,\bar{a}} \circ \sigma_{\mathbb{H}} \circ \Phi^{-1} &= T_{u,\bar{u}} \circ T_{a,\bar{a}} \circ \sigma_{\mathbb{H}} \circ T_{\bar{u},u} \\ &= T_{u,\bar{u}} \circ T_{a,\bar{a}} \circ T_{\bar{u},u} \circ \sigma_{\mathbb{H}} \\ &= T_{ua\bar{u},\overline{ua\bar{u}}} \circ \sigma_{\mathbb{H}} \\ &= T_{i,\bar{i}} \circ \sigma_{\mathbb{H}}.\end{aligned}$$

Then A is isomorphic to $\mathbb{H}_{T_{i,\bar{i}} \circ \sigma_{\mathbb{H}}}$ (Theorem 1) and the map $\Phi :^{\star} \mathbb{H}(i,1) \to \mathbb{H}_{T_{i,\bar{i}} \circ \sigma_{\mathbb{H}}}$ $x \mapsto \bar{i}x$ is an isomorphism of algebras.∎

If $dim(A) = 4$, the last result can be obtained so by using the identity $R_e^2 = I_{\mathbb{H}}$ and the principal isotopes of $\mathbb{H}$: $\mathbb{H}(a,1)$, $^{\star}\mathbb{H}(a,1)$, where $a \in S(\mathbb{H})$. For the first isotope $e = \bar{a}$, and for the second isotope $e = a$.

For all alternative algebra A, Artin's theorem (Schafer, 1996) shows that for any $x, y \in A$, the set $\{x, y, \bar{x}, \bar{y}\}$ is contained in an associative subalgebra of A. We note by $T(x) = x + \bar{x}$ the tace of $x \in A$ and we have $x^2 - T(x)x + \|x\|^2 e = 0$ for all $x \in A$. As A is real alternative quadratic algebra, we have $A = \mathbb{R}e \oplus Im(A)$ (Frobenius decomposition) and their exist a unique linear form $\lambda : A \to \mathbb{R}$ such that $\lambda(1) = 1$, $ker(\lambda) = Im(A)$ and $< x, y >= \lambda(x\bar{y}) = \lambda(\bar{x}y)$ for all $x, y \in A$ (∗) (Koecher & Remmert, 1991). Otherwise for all $x, y \in Im(A)$ we have $xy + yx = -2 < x, y > e$ (★) and the identity $xyx = 2\lambda(xy)x - \|x\|^2\bar{y}$ for all $x, y \in Im(A)$ is called the triple product identity (**TPI**).

In 8-dimensional, by the duplication process we recover theirs algebras.

Theorem 4 *Let A be an 8-dimensional absolute valued algebra with strongly left unit. Then A contains a four-dimensional subalgeba.*

Proof. We have $\mathbb{O} = \mathbb{R} \oplus Im(\mathbb{O})$ and their exist a unique linear form $\lambda : \mathbb{O} \to \mathbb{R}$ such that $\lambda(1) = 1$ and $ker(\lambda) = Im(\mathbb{O})$. Let $u \in 1^{\perp}$, we have $0 =< 1, u >=< \Phi^n(1), \Phi^n(u) >=< 1, \Phi^n(u) >$. Then we have $\varphi^n(1^{\perp}) \subseteq 1^{\perp}$, for all $n \in \mathbb{N}$. The algebra A is of the form $\mathbb{O}_{\Phi}$ with $\Phi(1) = 1$ and $\Phi^2 = I_{\mathbb{O}}$ (Theorem 1 and Theorem 3). Otherwise we have $i \odot i = \Phi(i)i$ and $i \odot 1 = \Phi(i)$. Using the equality (★) we have $i\Phi(i) + \Phi(i)i = -2 < i, \Phi(i) > 1$. Also using Lemma 1 (7), we have $\Phi[\Phi(i)i] = \Phi(i)i$. Using the **TPI** we have

$$\Phi(i)i\Phi(i) = 2\lambda[\Phi(i)i]\Phi(i) + i = -2 < i, \Phi(i) > \Phi(i) + i. \quad (*)$$

and

$$i\Phi(i)i = 2\lambda[i\Phi(i)]i + \Phi(i) = -2 < i, \Phi(i) > i + \Phi(i). \quad (*)$$

Hence we have the products,

$$\Phi(i) \odot 1 = \Phi^2(i) = i.$$

$$\Phi(i) \odot i = \Phi^2(i)i = -1.$$

$$\Phi(i) \odot \Phi(i) = \Phi^2(i)\Phi(i) = i\Phi(i) = -2 < i, \Phi(i) > 1 - \Phi(i)i. \quad (\star)$$

$$\Phi(i) \odot \Phi(i)i = \Phi^2(i)\Phi(i)i = i\Phi(i)i = -2 < i, \Phi(i) > i + \Phi(i). \quad \textbf{(TPI)}$$

$$\Phi(i)i \odot 1 = \Phi[\Phi(i)i] = \Phi(i)i.$$

$$\Phi(i)i \odot i = \Phi[\Phi(i)i]i = \Phi(i)i^2 = -\Phi(i).$$

$$\Phi(i)i \odot \Phi(i) = \Phi[\Phi(i)i]\Phi(i) = \Phi(i)i\Phi(i) = -2 < i, \Phi(i) > \Phi(i) + i. \quad \textbf{(TPI)}$$

$$\Phi(i)i \odot \Phi(i)i = \Phi[\Phi(i)i]\Phi(i)i = (\Phi(i)i)^2 = T[\Phi(i)i]\Phi(i)i - 1.$$

$\odot$	1	i	$\Phi(i)$	$\Phi(i)i$
1	1	i	$\Phi(i)$	$\Phi(i)i$
i	$\Phi(i)$	$\Phi(i)i$	-1	$-i$
$\Phi(i)$	i	-1	$-2 < i, \Phi(i) > 1 - \Phi(i)i$	$-2 < i, \Phi(i) > i + \Phi(i)$
$\Phi(i)i$	$\Phi(i)i$	$-\Phi(i)$	$-2 < i, \Phi(i) > \Phi(i) + i$	$T[\Phi(i)i]\Phi(i)i - 1$

Then the algebra A contains a four-dimensional sub-algebra..

Theorem 5 *Let A be an 8-dimensional absolute valued algebra with strongly left unit. Then A is of the form $\mathbb{H} \times \mathbb{H}_{(\varphi,\psi)}$ where (φ, ψ) are linear isometries of $\mathbb{H}$ belong to $\mathbb{S}_1 \cup \mathbb{S}_2 \cup \mathbb{S}_3 \cup \mathbb{S}_4$ with:*

$$
\begin{aligned}
\mathbb{S}_1 &= \{I_{\mathbb{H}}\} \times \{\pm I_{\mathbb{H}},\ T_{a,b} : a, b \in S^2,\ \pm T_{c,\bar{c}} \circ \sigma_{\mathbb{H}} : c \in S^3\} \\
\mathbb{S}_2 &= \{\sigma_{\mathbb{H}}\} \times \{\pm I_{\mathbb{H}},\ T_{a,b} : a, b \in S^2,\ \pm T_{c,c} \circ \sigma_{\mathbb{H}} : c \in S^3\} \\
\mathbb{S}_3 &= \{T_{a,\bar{a}} : a \in S^2\} \times \{\pm I_{\mathbb{H}},\ T_{b,c} : b, c \in S^2,\ \pm T_{d,\bar{d}} \circ \sigma_{\mathbb{H}} : d \in S^3\} \\
\mathbb{S}_4 &= \{T_{a,\bar{a}} \circ \sigma_{\mathbb{H}} : a \in S^2\} \times \{\pm I_{\mathbb{H}},\ T_{b,c} : b, c \in S^2,\ \pm T_{d,\bar{d}} \circ \sigma_{\mathbb{H}} : d \in S^3\}.
\end{aligned}
$$

Proof. Using the Theorem 2 and Theorem 4, the algebra A is obtained by the duplication process. It's clear that the algebra A is of the form $\mathbb{H} \times \mathbb{H}_{(\varphi,\psi)}$, with $\varphi(1) = 1$. The linear isometric (φ, ψ) is involitive, then $\varphi^2 = \psi^2 = I_A$. We have $\varphi \in \mathcal{I}_1^+ \cup \mathcal{I}_1^-$ and $\psi \in \mathcal{I}^+ \cup \mathcal{I}^-$. Then the lemma 2 gives the result..

Problem 1 *In dimension 8, it will be interesting to specify these algebras by reducing the isomorphism classes.*

Acknowledgements

I thank the reviewers for the relevant remarks and suggestions.

References

Albert, A. A. (1947). Absolute valued real algebras. *Ann. Math., 48,* 495-501.

Calderon, A., Kaidi, A., Martin, C., Morales, A., Ramirez, M., & Rochdi, A. (2011). Finite - dimensional absolute - valued algebras. *israel journal of mathematics*, 184, 193 - 220.

Chandid, A. & Rochdi, A. (2008). A survey on absolute valued algebras satisfying $(x^i, x^j, x^k) = 0$. *Int. J. Algebra, 2,* 837-852.

Diankha, O., Diouf, A., & Rochdi, A. (2013_1). A brief statement on the absolute-valued algebras with one-sided Unit. *Int. J. Algebra, 7*(17), 833-838.

Diankha, O., Diouf, A., Ramirez, M. I., & Rochdi, A. (2013_2). Absolute-valued algebras with one-sided unit satisfying $(x^2, x^2, x^2) = 0$. *Int. J. Algebra, 7*(19), 935-958.

Koecher, M., & Remmert. (1991). Numbers. *Springer Verlag.*

Postnikov, M. (1985), Leçons de Géométrie. Groupes et algèbres de Lie. *Editions Mir.*

Rochdi, A. (2003). Eight-dimensional real absolute valued algebras with left unit whose automorphism group is trivial. *IJMMS, 70,* 4447-4454.

Rodriguez, A. (1992). One-sided division absolute valued algebras. *Publ. Math., 36,*925-954.

Rodriguez, A. (2004). Absolute valued algebras, and absolute valuable Banach spaces. *Advanced courses of mathematical analysis I*, 99-155, World Sci. Publ., Hackensack, NJ.

Ramirez, M. I. (1999). On four absolute valued algebras. *Proceedings of the International Conference on Jordan Structures* (Malaga, 1997), 169-173.

Schafer, R. (1996). An Introduction to Nonassociative Algebras. *Academic Press.*

Urbanik, K., & Wright£F. B. (1960). Absolute valued algebras. *Proc. Amer. Math. Soc. 11,* 861-866.

4

Semigroup Methods for the M/G/1 Queueing Model with Working Vacation and Vacation Interruption

Ehmet Kasim[1]

[1] College of Mathematics and Systems Science, Xinjiang University, Urumqi 830046, P.R.China

Correspondence: Ehmet Kasim, College of Mathematics and Systems Science, Xinjiang University, Urumqi 830046, P.R.China. E-mail: ehmetkasim@163.com

Abstract

By using the strong continuous semigroup theory of linear operators we prove that the M/G/1 queueing model with working vacation and vacation interruption has a unique positive time-dependent solution which satisfies probability conditions. When the both service completion rate in a working vacation period and in a regular busy period are constant, by investigating the spectral properties of an operator corresponding to the model we obtain that the time-dependent solution of the model strongly converges to its steady-state solution.

Keywords: M/G/1 queueing model with working vacation and vacation interruption, C_0− semigroup, dispersive operator, resolvent set, eigenvalue

1. Introduction

According to (Zhang & Hou, 2010), the M/G/1 queueing system with working vacation and vacation interruption can be described by the following system of partial differential equations:

$$\begin{aligned}
\frac{dp_{0,0}(t)}{dt} &= -\lambda p_{0,0}(t) + \int_0^\infty \mu_0(x)p_{1,0}(x,t)dx + \int_0^\infty \mu_1(x)p_{1,1}(x,t)dx,\\
\frac{\partial p_{1,0}(x,t)}{\partial t} + \frac{\partial p_{1,0}(x,t)}{\partial x} &= -[\lambda + \theta + \mu_0(x)]p_{1,0}(x,t),\\
\frac{\partial p_{n,0}(x,t)}{\partial t} + \frac{\partial p_{n,0}(x,t)}{\partial x} &= -[\lambda + \theta + \mu_0(x)]p_{n,0}(x,t) + \lambda p_{n-1,0}(x,t),\ \forall n \geq 2,\\
\frac{\partial p_{1,1}(x,t)}{\partial t} + \frac{\partial p_{1,1}(x,t)}{\partial x} &= -[\lambda + \mu_1(x)]p_{1,1}(x,t),\\
\frac{\partial p_{n,1}(x,t)}{\partial t} + \frac{\partial p_{n,1}(x,t)}{\partial x} &= -[\lambda + \mu_1(x)]p_{n,1}(x,t) + \lambda p_{n-1,1}(x,t),\ \forall n \geq 2,
\end{aligned} \tag{1.1}$$

with boundary conditions:

$$\begin{aligned}
p_{1,0}(0,t) &= \lambda p_{0,0}(t),\\
p_{n,0}(0,t) &= 0, \quad \forall n \geq 2,\\
p_{n,1}(0,t) &= \theta \int_0^\infty p_{n,0}(x,t)dx + \int_0^\infty \mu_0(x)p_{n+1,0}(x,t)dx\\
&\quad + \int_0^\infty \mu_1(x)p_{n+1,1}(x,t)dx,\ \forall n \geq 1,
\end{aligned} \tag{1.2}$$

and initial condition:

$$p_{0,0}(0) = 1, p_{0,1}(0) = 0, p_{m,0}(x,0) = p_{m,1}(x,0) = 0, \quad \forall m \geq 1. \tag{1.3}$$

Where, $(x,t) \in [0,\infty) \times [0,\infty)$; $p_{0,0}(t)$ represents the probability that there is no customer in the system and the server is in a working vacation period at time t; $p_{n,0}(x,t)dx$ $(n \geq 1)$ is the probability that at time t the server is in a working vacation period and there are n customers in the system with elapsed service time of the customer undergoing service lying in $(x, x+dx]$; $p_{n,1}(x,t)dx$ $(n \geq 1)$ is the probability that at time t the server is in a regular busy period and there are n customers in the system with elapsed service time of the customer undergoing service lying in $(x, x+dx]$; λ is the mean arrival rate of customers; θ is the vacation duration rate of the server; $\mu_0(x)$ is the service rate of the server while the server is in a working vacation period and satisfies

$$\mu_0(x) \geq 0, \qquad \int_0^\infty \mu_0(x)dx = \infty.$$

$\mu_1(x)$ is the service rate of the server while the server is in a regular busy period and satisfying

$$\mu_1(x) \geq 0, \quad \int_0^\infty \mu_1(x)dx = \infty.$$

Queuing situations in which the idle server may take vacations encounter in computer, communication and manufacturing systems, etc. In a classical vacation queue, a server may completely stop the service or do some additional work during a vacation. Proposing various vacation policies provides more flexibility for optimal design and operation control of the system. Therefore, many researchers studied such queueing system, see (Doshi, 1986; Takagi, 1990; Madan, 1992; Gupur, 2002; Gupur, 2010; Gupur & Guo, 2002; Lu & Gupur, 2010), for instance.

In many real life congestion situations, the server can be utilized for ancillary work and a different rate during the vacation period. Such a queueing situation is called queue with working vacation. (Servi & Finn, 2002) first studied the M/M/1 queueing system with multiple working vacation and obtained the transform formulae for the distribution of the number of customers in the system and the sojourn time in a steady state. Since then, Queueing models with working vacation have been studied by several researchers , see (Jain & Agrawal, 2007; Kim, Choi & Chae, 2003; Li, Tian, & Ma, 2007). Moreover, (Wu & Takagi, 2006) extended Servi and Finn's (Servi & Finn, 2002) M/M/1 queueing system to an M/G/1 queueing system with multiple working vacation, where the service times during regular service period and working vacation period. (Zhang & Hou, 2010) considered the M/G/1 queueing system with working vacation and vacation interruption where the server enters into vacations when there are no customers and it can take service at a lower rate during the vacation period. If there are customers in the system at the instant of a service completion during the vacation period, the server will come back to the normal working level no matter whether or not the vacation has ended. otherwise, it continues the vacation. Firstly, by using supplementary variable technique they established the above model and gave the Laplace-Stieltjes transform of the stationary waiting time. Then, they obtain the queue length distribution and service status at an arbitrary epoch in steady state condition under following hypothesis:

$$\lim_{t\to\infty} p_{0,0}(t) = p_{0,0}, \quad \lim_{t\to\infty} p_{n,0}(x,t) = p_{n,0}(x), \quad \lim_{t\to\infty} p_{n,1}(x,t) = p_{n,1}(x), \quad n \geq 1$$

In addition, they also perform some numerical examples to study the effect of various parameters on the system's characteristics. By reading the paper we have found that the above hypothesis implies the following two hypotheses:

Hypothesis 1 The model has a unique time-dependent solution.

Hypothesis 2 The time-dependent solution converges to its steady-state solution.

So far, any results about this model have not been found in the literature. In this paper, we do dynamic analysis for the queueing model by using the idea of (Gupur, Li, & Zhu, 2001), that is, we investigate above two hypotheses. First of all, we convert the model into an abstract Cauchy problem by choosing a suitable Banach space as a state space and introducing an operator corresponding to the model and its domain. By using the Hille-Yosida theorem and Phillips theorem as well as Fattorini theorem we prove that the model has a unique positive time-dependent solutions and therefore we obtain that the Hypothesis 1 is hold. Next, when the both service rate in a working vacation period and service rate in a regular busy period are constant, we study the asymptotic behavior of its time-dependent solution, i.e., we study the Hypothesis 2. Firstly, we determine the expression of the adjoint operator of the operator corresponding to the model and deduce the resolvent set of the operator. Then, we verify that 0 is eigenvalue of the underlying operator and its adjoint operator with geometric multiplicity one. Thus, by using Theorem 14 in (Gupur, Li, & Zhu, 2001) we obtain that the time-dependent solution of the model strongly converges to its steady-state solution.

For simplicity, we introduce some notations as follows:

$$\Gamma_1 = \begin{pmatrix} \lambda & 0 & 0 & 0 & \cdots \\ 0 & 0 & 0 & 0 & \cdots \\ 0 & 0 & 0 & 0 & \cdots \\ \vdots & \vdots & \vdots & \vdots & \ddots \end{pmatrix} \quad \Gamma_2 = \begin{pmatrix} 0 & \theta & 0 & 0 & \cdots \\ 0 & 0 & \theta & 0 & \cdots \\ 0 & 0 & 0 & \theta & \cdots \\ \vdots & \vdots & \vdots & \vdots & \ddots \end{pmatrix}.$$

$$\Gamma_3 = \begin{pmatrix} 0 & 0 & \mu_0(x) & 0 & 0 & \cdots \\ 0 & 0 & 0 & \mu_0(x) & 0 & \cdots \\ 0 & 0 & 0 & 0 & \mu_0(x) & \cdots \\ \vdots & \vdots & \vdots & \vdots & \vdots & \ddots \end{pmatrix} \quad \Gamma_4 = \begin{pmatrix} 0 & \mu_1(x) & 0 & 0 & \cdots \\ 0 & 0 & \mu_1(x) & 0 & \cdots \\ 0 & 0 & 0 & \mu_1(x) & \cdots \\ \vdots & \vdots & \vdots & \vdots & \ddots \end{pmatrix}$$

Take a state space as follows:

$$X = \left\{(p_0, p_1) \middle| p_0 \in Y_1, p_1 \in Y_2, \|(p_0, p_1)\| = \|p_0\|_{Y_1} + \|p_1\|_{Y_2} < \infty\right\},$$

$$Y_1 = \left\{ p_0 \in \mathbb{R} \times L^1[0,\infty) \times \cdots \,\middle|\, \|p_0\| = |p_{0,0}| + \sum_{n=1}^{\infty} \|p_{n,0}\|_{L^1[0,\infty)} < \infty \right\},$$

$$Y_2 = \left\{ p_1 \in L^1[0,\infty) \times L^1[0,\infty) \cdots \,\middle|\, \|p_1\| = \sum_{n=1}^{\infty} \|p_{n,1}\|_{L^1[0,\infty)} < \infty \right\}.$$

It is obvious that X is a Banach space. Now we define operators and their domain as follows.

$$A\left(\begin{pmatrix} p_{0,0} \\ p_{1,0}(x) \\ p_{2,0}(x) \\ p_{3,0}(x) \\ \vdots \end{pmatrix}, \begin{pmatrix} p_{1,1}(x) \\ p_{2,1}(x) \\ p_{3,1}(x) \\ p_{4,1}(x) \\ \vdots \end{pmatrix}\right) = \left(\begin{pmatrix} -\lambda & 0 & 0 & 0 & \cdots \\ 0 & -\frac{d}{dx} & 0 & 0 & \cdots \\ 0 & 0 & -\frac{d}{dx} & 0 & \cdots \\ 0 & 0 & 0 & -\frac{d}{dx} & \cdots \\ \vdots & \vdots & \vdots & \vdots & \ddots \end{pmatrix}\begin{pmatrix} p_{0,0} \\ p_{1,0}(x) \\ p_{2,0}(x) \\ p_{3,0}(x) \\ \vdots \end{pmatrix}, \begin{pmatrix} -\frac{d}{dx} & 0 & 0 & 0 & \cdots \\ 0 & -\frac{d}{dx} & 0 & 0 & \cdots \\ 0 & 0 & -\frac{d}{dx} & 0 & \cdots \\ 0 & 0 & 0 & -\frac{d}{dx} & \cdots \\ \vdots & \vdots & \vdots & \vdots & \ddots \end{pmatrix}\begin{pmatrix} p_{1,1}(x) \\ p_{2,1}(x) \\ p_{3,1}(x) \\ p_{4,1}(x) \\ \vdots \end{pmatrix}\right),$$

$$D(A) = \left\{ (p_0, p_1) \in X \,\middle|\, \begin{array}{l} \frac{dp_{n,0}}{dx} \in L^1[0,\infty),\ \frac{dp_{n,1}}{dx} \in L^1[0,\infty),\ p_{n,0}(x) \text{ and } p_{n,1}(x) \\ (n \geq 1) \text{ are absolutely continuous and } p_0(0) = \Gamma_1 p_0; \\ p_1(0) = \int_0^\infty \Gamma_2 p_0\, dx + \int_0^\infty \Gamma_3 p_0\, dx + \int_0^\infty \Gamma_4 p_1\, dx \end{array} \right\}.$$

$$U\left(\begin{pmatrix} p_{0,0} \\ p_{1,0}(x) \\ p_{2,0}(x) \\ p_{3,0}(x) \\ \vdots \end{pmatrix}, \begin{pmatrix} p_{1,1}(x) \\ p_{2,1}(x) \\ p_{3,1}(x) \\ p_{4,1}(x) \\ \vdots \end{pmatrix}\right) = \left(\begin{pmatrix} 0 & 0 & 0 & 0 & \cdots \\ 0 & \mathcal{D}_0 & 0 & 0 & \cdots \\ 0 & \lambda & \mathcal{D}_0 & 0 & \cdots \\ 0 & 0 & \lambda & \mathcal{D}_0 & \cdots \\ \vdots & \vdots & \vdots & \vdots & \ddots \end{pmatrix}\begin{pmatrix} p_{0,0} \\ p_{1,0}(x) \\ p_{2,0}(x) \\ p_{3,0}(x) \\ \vdots \end{pmatrix}, \begin{pmatrix} \mathcal{D}_1 & 0 & 0 & 0 & \cdots \\ \lambda & \mathcal{D}_1 & 0 & 0 & \cdots \\ 0 & \lambda & \mathcal{D}_1 & 0 & \cdots \\ 0 & 0 & \lambda & \mathcal{D}_1 & \cdots \\ \vdots & \vdots & \vdots & \vdots & \ddots \end{pmatrix}\begin{pmatrix} p_{1,1}(x) \\ p_{2,1}(x) \\ p_{3,1}(x) \\ \vdots \end{pmatrix}\right), \quad D(U) = X.$$

here

$$\mathcal{D}_0 = -(\lambda + \theta + \mu_0(x)), \qquad \mathcal{D}_1 = -(\lambda + \mu_1(x)).$$

$$E\left(\begin{pmatrix} p_{0,0} \\ p_{1,0}(x) \\ p_{2,0}(x) \\ \vdots \end{pmatrix}, \begin{pmatrix} p_{1,1}(x) \\ p_{2,1}(x) \\ p_{3,1}(x) \\ \vdots \end{pmatrix}\right) = \left(\begin{pmatrix} \int_0^\infty \mu_0(x) p_{1,0}(x)\, dx + \int_0^\infty \mu_1(x) p_{1,1}(x)\, dx \\ 0 \\ 0 \\ \vdots \end{pmatrix}, \begin{pmatrix} 0 \\ 0 \\ 0 \\ \vdots \end{pmatrix}\right), \quad D(E) = X.$$

Then the above system of equations (1.1)-(1.3) can be written as an abstract Cauchy problem in Banach space X.

$$\begin{cases} \frac{d(p_0, p_1)(t)}{dt} = (A + U + E)(p_0, p_1)(t), \quad t \in (0,\infty), \\ (p_0, p_1)(0) = \left(\begin{pmatrix} 1 \\ 0 \\ \vdots \end{pmatrix}, \begin{pmatrix} 0 \\ 0 \\ \vdots \end{pmatrix}\right). \end{cases} \tag{1.4}$$

2. Well-posedness of the System (1.4)

In this section, in order to obtain well-posedness of the system (1.4), we first need to prove that $A + U + E$ generates a positive contraction C_0- semigroup $T(t)$ on X.

Theorem 2.1 *If $\mu_0(x)$ and $\mu_1(x)$ are measurable functions and satisfy $\overline{\mu_0} = \sup_{x\in[0,\infty)} \mu_0(x) < \infty$ and $\overline{\mu_1} = \sup_{x\in[0,\infty)} \mu_1(x) < \infty$, then $A + U + E$ generates a positive contraction C_0- semigroup $T(t)$.*

A detailed proof of the Theorem 2.1 can be found in the Appendix. It is not difficult to verify that X^*, the dual space of X, is as follows.

$$X^* = \left\{ (q_0^*, q_1^*) \,\middle|\, \begin{array}{l} q_0^* \in Y_1^*, \quad q_1^* \in Y_2^*, \\ |||(q_0^*, q_1^*)||| = \sup\{|||q_0^*|||_{Y_1^*}, |||q_1^*|||_{Y_2^*}\} \end{array} \right\},$$

here

$$Y_1^* = \left\{ q_0^* \,\middle|\, \begin{array}{l} q_0^*(x) = (q_{0,0}^*, q_{1,0}^*(x), q_{2,0}^*(x), q_{3,0}^*(x), \cdots), \\ |||q_0^*||| = \sup\left\{ |q_{0,0}^*|, \sup\limits_{n\geq 1} \|q_{n,0}^*\|_{L^\infty[0,\infty)} \right\} < \infty \end{array} \right\},$$

$$Y_2^* = \left\{ q_1^* \,\middle|\, \begin{array}{l} q_1^*(x) = (q_{0,1}^*, q_{1,1}^*(x), q_{2,1}^*(x), q_{3,1}^*(x), \cdots) \\ |||q_1^*||| = \sup\limits_{n\geq 1} \|q_{n,1}^*\|_{L^\infty[0,\infty)} < \infty \end{array} \right\}.$$

It is obvious that X^* is a Banach space. If we take a set S in X as

$$S = \left\{ (p_0, p_1) \in X \,\middle|\, p_{0,0} \geq 0,\ p_{n,0}(x) \geq 0,\ p_{n,1}(x) \geq 0,\ \forall n \geq 1,\ x \in [0,\infty) \right\}.$$

Then S is cone in X. for $(p_0, p_1) \in D(A) \cap S$, we take

$$(q_0^*, q_1^*) = \|(p_0, p_1)\| \left(\begin{pmatrix} 1 \\ 1 \\ 1 \\ \vdots \end{pmatrix}, \begin{pmatrix} 1 \\ 1 \\ 1 \\ \vdots \end{pmatrix} \right) \in X^*,$$

For such $(p_0^*, p_1^*) \in X^*$, by using the boundary condition we have

$$\begin{aligned}
\langle (A+U+E)(p_0,p_1), (q_0^*, q_1^*) \rangle &= \|(p_0,p_1)\| \left\{ -\lambda p_{0,0} + \int_0^\infty \mu_0(x) p_{1,0}(x) dx + \int_0^\infty \mu_1(x) p_{1,1}(x) dx \right\} \\
&+ \int_0^\infty \|(p_0,p_1)\| \left\{ -\frac{dp_{1,0}(x)}{dx} - (\lambda + \theta + \mu_0(x)) p_{1,0}(x) \right\} dx \\
&+ \sum_{n=2}^\infty \int_0^\infty \|(p_0,p_1)\| \left\{ -\frac{dp_{n,0}(x)}{dx} - (\lambda + \theta + \mu_0(x)) p_{n,0}(x) + \lambda p_{n-1,0}(x) \right\} \\
&+ \int_0^\infty \|(p_0,p_1)\| \left\{ -\frac{dp_{1,1}(x)}{dx} - (\lambda + \mu_1(x)) p_{1,1}(x) \right\} dx \\
&+ \sum_{n=2}^\infty \int_0^\infty \|(p_0,p_1)\| \left\{ -\frac{dp_{n,1}(x)}{dx} - (\lambda + \mu_1(x)) p_{n,1}(x) + \lambda p_{n-1,1}(x) \right\} \\
&= \|(p_0,p_1)\| \left\{ -\lambda p_{0,0} + \int_0^\infty \mu_0(x) p_{1,0}(x) dx + \int_0^\infty \mu_1(x) p_{1,1}(x) dx \right. \\
&- \sum_{n=1}^\infty \int_0^\infty dp_{n,0}(x) - \sum_{n=1}^\infty \int_0^\infty (\lambda + \theta + \mu_0(x)) p_{n,0}(x) dx + \lambda \sum_{n=1}^\infty \int_0^\infty p_{n,0}(x) dx \\
&\left. - \sum_{n=1}^\infty \int_0^\infty dp_{n,1}(x) - \sum_{n=1}^\infty \int_0^\infty (\lambda + \mu_1(x)) p_{n,1}(x) dx + \lambda \sum_{n=1}^\infty \int_0^\infty p_{n,1}(x) dx \right\} \\
&= \|(p_0,p_1)\| \left\{ -\lambda p_{0,0} + \int_0^\infty \mu_0(x) p_{1,0}(x) dx + \int_0^\infty \mu_1(x) p_{1,1}(x) dx \right. \\
&+ \sum_{n=1}^\infty p_{n,0}(0) - \sum_{n=1}^\infty \int_0^\infty (\theta + \mu_0(x)) p_{n,0}(x) dx \\
&\left. + \sum_{n=1}^\infty p_{n,1}(0) - \sum_{n=1}^\infty \int_0^\infty \mu_1(x) p_{n,1}(x) dx \right\} \\
&= \|(p_0,p_1)\| \left\{ -\lambda p_{0,0} + \int_0^\infty \mu_0(x) p_{1,0}(x) dx + \int_0^\infty \mu_1(x) p_{1,1}(x) dx \right. \\
&+ \lambda p_{0,0} - \sum_{n=1}^\infty \int_0^\infty (\theta + \mu_0(x)) p_{n,0}(x) dx \\
&+ \sum_{n=1}^\infty \int_0^\infty \mu_1(x) p_{n+1,1}(x) dx + \theta \sum_{n=1}^\infty \int_0^\infty p_{n,0}(x) dx
\end{aligned}$$

$$+\sum_{n=1}^{\infty}\int_0^{\infty}\mu_0(x)p_{n+1,0}(x)dx-\sum_{n=1}^{\infty}\int_0^{\infty}\mu_1(x)p_{n,1}(x)dx\Bigg\}$$
$$=0$$

which shows that $A+U+E$ is a conservative operator. Since the initial value $(p_0, p_1)(0)) \in D(A^2) \cap S$, by using the Fattorini theorem (Fattorini, 1983) we obtain the following result. see (Gupur, Li, & Zhu, 2001) for detail proof.

Theorem 2.2 *$T(t)$ is isometric for the initial value of the system (1.4), that is,*

$$\|T(t)(p_0, p_1)(0)\| = \|(p_0, p_1)(0)\|, \quad t \in [0, \infty). \tag{2.1}$$

From Theorem 2.1 and Theorem 2.2 we obtain well-posedness of the system (1.4).

Theorem 2.3 *If $\mu_0(x)$ and $\mu_1(x)$ are satisfy $\overline{\mu_0} = \sup_{x\in[0,\infty)} \mu_0(x) < \infty$ and $\overline{\mu_1} = \sup_{x\in[0,\infty)} \mu_1(x) < \infty$, then the system (1.4) has a unique positive time-dependent solution $(p_0, p_1)(x,t)$ satisfying*

$$\|(p_0, p_1)(\cdot, t)\| = 1, \quad \forall t \in [0, \infty).$$

Proof. Since $(p_0, p_1)(0) \in D(A^2) \cap S$, From Theorem 2.1 and (Gupur, Li, & Zhu, 2001) we know that the system (1.4) has a unique positive time-dependent solution $(p_0, p_1)(x,t)$ which can be expressed as

$$(p_0, p_1)(x,t) = T(t)(p_0, p_1)(0), \qquad t \in [0, \infty).$$

From which together with Theorem 2.2 (i.e., (2.1)) we derive

$$\|(p_0, p_1)(.,t)\| = \|T(t)(p_0, p_1)(0)\| = \|(p_0, p_1)(0)\| = 1, \qquad t \in [0, \infty). \tag{2.2}$$

□

(2.2) just reflects the physical meaning of $(p_0, p_1)(x,t)$.

3. Asymptotic Behavior of the Time-Dependent Solution of the System (1.4) when $\mu_0(x) = \mu_0$ and $\mu_1(x) = \mu_1$

Lemma 3.1 *$(A+U+E)^*$, the adjoint operator of $A+U+E$, is as follows.*

$$(A+U+E)^*(q_0^*, q_1^*) = (\mathcal{L}+\mathcal{N}+\mathcal{R}+\mathcal{J})(q_0^*, q_1^*),$$
$$(q_0^*, q_1^*) \in D(A+U+E)^*,$$

where

$$\mathcal{L}(q_0^*, q_1^*)(x) = \left(\begin{pmatrix} -\lambda & 0 & 0 & \cdots \\ 0 & \frac{d}{dx}-(\lambda+\theta+\mu_0) & 0 & \cdots \\ 0 & 0 & \frac{d}{dx}-(\lambda+\theta+\mu_0) & \cdots \\ \vdots & \vdots & \vdots & \ddots \end{pmatrix}\begin{pmatrix} q_{0,0}^* \\ q_{1,0}^*(x) \\ q_{2,0}^*(x) \\ \vdots \end{pmatrix},\right.$$
$$\left.\begin{pmatrix} \frac{d}{dx}-(\lambda+\mu_1) & 0 & 0 & \cdots \\ 0 & \frac{d}{dx}-(\lambda+\mu_1) & 0 & \cdots \\ 0 & 0 & \frac{d}{dx}-(\lambda+\mu_1) & \cdots \\ \vdots & \vdots & \vdots & \ddots \end{pmatrix}\begin{pmatrix} q_{1,1}^*(x) \\ q_{2,1}^*(x) \\ q_{3,1}^*(x) \\ \vdots \end{pmatrix}\right),$$

$$\mathcal{N}(q_0^*,q_1^*)(x)=\left(\begin{pmatrix}0&0&0&0&\cdots\\0&0&\lambda&0&\cdots\\0&0&0&\lambda&\cdots\\\vdots&\vdots&\vdots&\vdots&\ddots\end{pmatrix}\begin{pmatrix}q_{0,0}^*\\q_{1,0}^*(x)\\q_{2,0}^*(x)\\\vdots\end{pmatrix}+\begin{pmatrix}0&\lambda&0&\cdots\\0&0&0&\cdots\\0&0&0&\cdots\\\vdots&\vdots&\vdots&\ddots\end{pmatrix}\begin{pmatrix}q_{0,0}^*\\q_{1,0}^*(0)\\q_{2,0}^*(0)\\\vdots\end{pmatrix},\right.$$

$$\left.\begin{pmatrix}0&\lambda&0&0&\cdots\\0&0&\lambda&0&\cdots\\0&0&0&\lambda&\cdots\\\vdots&\vdots&\vdots&\vdots&\ddots\end{pmatrix}\begin{pmatrix}q_{1,1}^*(x)\\q_{2,1}^*(x)\\q_{3,1}^*(x)\\\vdots\end{pmatrix}\right),$$

$$\mathcal{R}(q_0^*,q_1^*)(x)=\left(\begin{pmatrix}0&0&0&\cdots\\\mu_0&0&0&\cdots\\0&0&0&\cdots\\0&0&0&\cdots\\\vdots&\vdots&\vdots&\ddots\end{pmatrix}\begin{pmatrix}q_{0,0}^*\\q_{1,0}^*(0)\\q_{2,0}^*(0)\\q_{3,0}^*(0)\\\vdots\end{pmatrix}+\begin{pmatrix}0&0&0&0&\cdots\\0&0&0&0&\cdots\\\mu_0&0&0&0&\cdots\\0&\mu_0&0&0&\cdots\\\vdots&\vdots&\vdots&\vdots&\ddots\end{pmatrix}\begin{pmatrix}q_{1,1}^*(0)\\q_{2,1}^*(0)\\q_{3,1}^*(0)\\q_{4,1}^*(0)\\\vdots\end{pmatrix},\right.$$

$$\left.\begin{pmatrix}0&0&0&0&\cdots\\\mu_1&0&0&0&\cdots\\0&\mu_1&0&0&\cdots\\0&0&\mu_1&0&\cdots\\\vdots&\vdots&\vdots&\vdots&\ddots\end{pmatrix}\begin{pmatrix}q_{1,1}^*(0)\\q_{2,1}^*(0)\\q_{3,1}^*(0)\\q_{4,1}^*(0)\\\vdots\end{pmatrix}\right),$$

$$\mathcal{J}(q_0^*,q_1^*)(x)=\left(\begin{pmatrix}0&0&0&\cdots\\\theta&0&0&\cdots\\0&\theta&0&\cdots\\0&0&\theta&\cdots\\\vdots&\vdots&\vdots&\ddots\end{pmatrix}\begin{pmatrix}q_{1,1}^*(0)\\q_{2,1}^*(0)\\q_{3,1}^*(0)\\q_{4,1}^*(0)\\\vdots\end{pmatrix},\begin{pmatrix}\mu_1&0&0&\cdots\\0&0&0&\cdots\\0&0&0&\cdots\\\vdots&\vdots&\vdots&\ddots\end{pmatrix}\begin{pmatrix}q_{0,0}^*\\q_{1,0}^*(0)\\q_{2,0}^*(0)\\\vdots\end{pmatrix}\right),$$

$$D((A+U+E)^*)=\left\{(q_0^*,q_1^*)\in X^*\;\middle|\;\begin{array}{l}\frac{dq_{0,n}^*(x)}{dx}\ and\ \frac{dq_{1,n}^*(x)}{dx}\ exist\ and\\ q_{n,0}^*(\infty)=q_{n,1}^*(\infty)=\alpha,\quad n\geq 1\end{array}\right\},$$

here α in $D((A+U+E)^*)$ is a constant which is irrelevant to n.

The proof is easy computaion, and we omit the detail proof.

Lemma 3.2

$$\left\{\gamma\in C\;\middle|\;\begin{array}{l}Re\gamma+\lambda+\theta+\mu_0>0,\ Re\gamma+\lambda+\mu_1>0,\\ \sup\left\{\frac{\lambda}{|\gamma+\lambda|},\frac{\lambda|\gamma+\lambda+\mu_1|}{(Re\gamma+\lambda+\mu_1)(|\gamma+\lambda+\mu_1|-\mu_1)},\right.\\ \frac{\lambda\mu_0}{|\gamma+\lambda||\gamma+\lambda+\theta+\mu_0|}+\frac{\lambda\theta\mu_1}{|\gamma+\lambda||\gamma+\lambda+\theta+\mu_0||\gamma+\lambda+\mu_1|}+\frac{\lambda\theta}{|\gamma+\lambda+\theta+\mu_0|Re\gamma+\lambda+\mu_1|}+\frac{\lambda}{Re\gamma+\lambda+\theta+\mu_0},\\ \left.\frac{\lambda(\theta+\mu_0)|\gamma+\lambda+\mu_1|}{|\gamma+\lambda+\theta+\mu_0|(Re\gamma+\lambda+\mu_1)(|\gamma+\lambda+\mu_1|-\mu_1)}+\frac{\lambda}{Re\gamma+\lambda+\theta+\mu_0}\right\}<1.\end{array}\right\},$$

belongs to the resolvent set of $(A+U+E)^*$, Especially, all points on the imaginary axis except for zero belong to the resolvent set of $(A+U+E)^*$ and $A+U+E$.

Proof. For any given $(y_0^*,y_1^*)\in X^*$ we consider the equation $[\gamma I-(\mathcal{L}+\mathcal{R}+\mathcal{J})](q_0^*,q_1^*)=(y_0^*,y_1^*)$. that is,

$$(\gamma+\lambda)q_{0,0}^*=y_{0,0}^*, \tag{3.1}$$

$$\frac{dq_{1,0}^*(x)}{dx}=(\gamma+\lambda+\theta+\mu_0)q_{1,0}^*(x)-\mu_0q_{0,0}^*-\theta q_{1,1}^*(0)-y_{0,1}^*(x), \tag{3.2}$$

$$\frac{dq_{n,0}^*(x)}{dx}=(\gamma+\lambda+\theta+\mu_0)q_{n,0}^*(x)-\mu_0q_{n-1,1}^*(0)-\theta q_{n,1}^*(0)-y_{n,0}^*(x),\quad n\geq 2, \tag{3.3}$$

$$\frac{dq_{1,1}^*(x)}{dx}=(\gamma+\lambda+\mu_1)q_{1,1}^*(x)-\mu_1q_{0,0}^*-y_{1,1}^*(x), \tag{3.4}$$

$$\frac{dq_{n,1}^*(x)}{dx}=(\gamma+\lambda+\mu_1)q_{n,1}^*(x)-\mu_1q_{n-1,1}^*(0)-y_{n,1}^*(x),\quad n\geq 2, \tag{3.5}$$

$$q_{n,0}^*(\infty)=q_{n,1}^*(\infty)=\alpha,\quad n\geq 1. \tag{3.6}$$

By solving (3.1)-(3.5) we have

$$q_{0,0}^* = \frac{1}{\gamma+\lambda} y_{0,0}^*, \tag{3.7}$$

$$q_{1,0}^*(x) = a_1^* e^{(\gamma+\lambda+\theta+\mu_0)x} - e^{(\gamma+\lambda+\theta+\mu_0)x} \int_0^x \left[\mu_0 q_{0,0}^* + \theta q_{1,1}^*(0)\right] e^{-(\gamma+\lambda+\theta+\mu_0)\tau} d\tau$$
$$- e^{(\gamma+\lambda+\theta+\mu_0)x} \int_0^x y_{1,0}^*(\tau)\, e^{-(\gamma+\lambda+\theta+\mu_0)\tau} d\tau, \tag{3.8}$$

$$q_{n,0}^*(x) = a_n^* e^{(\gamma+\lambda+\theta+\mu_0)x} - e^{(\gamma+\lambda+\theta+\mu_0)x} \int_0^x \left[\mu_0 q_{n-1,1}^*(0) + \theta q_{n,1}^*(0)\right] e^{-(\gamma+\lambda+\theta+\mu_0)\tau} d\tau$$
$$- e^{(\gamma+\lambda+\theta+\mu_0)x} \int_0^x y_{n,0}^*(\tau)\, e^{-(\gamma+\lambda+\theta+\mu_0)\tau} d\tau, \quad n \geq 2, \tag{3.9}$$

$$q_{1,1}^*(x) = b_1^* e^{(\gamma+\lambda+\mu_1)x} - e^{(\gamma+\lambda+\mu_1)x} \int_0^x \mu_1 q_{0,0}^*\, e^{-(\gamma+\lambda+\mu_1)\tau} d\tau$$
$$- e^{(\gamma+\lambda+\mu_1)x} \int_0^x y_{1,1}^*(\tau) e^{-(\gamma+\lambda+\mu_1)\tau} d\tau, \tag{3.10}$$

$$q_{n,1}^*(x) = b_n^* e^{(\gamma+\lambda+\mu_1)x} - e^{(\gamma+\lambda+\mu_1)x} \int_0^x \mu_1 q_{n-1,1}^*(0)\, e^{-(\gamma+\lambda+\mu_1)\tau} d\tau$$
$$- e^{(\gamma+\lambda+\mu_1)x} \int_0^x y_{n,1}^*(\tau)\, e^{-(\gamma+\lambda+\mu_1)\tau} d\tau, \quad n \geq 2, \tag{3.11}$$

Multiplying the both side of (3.8), (3.9) by $e^{-(\gamma+\lambda+\theta+\mu_0)x}$, the both side of (3.10), (3.11) by $e^{-(\gamma+\lambda+\mu_1)x}$, and taking the limit $x \to \infty$ as well as using (3.6) it gives

$$a_1^* = \int_0^\infty \left[\mu_0 q_{0,0}^* + \theta q_{1,1}^*(0)\right] e^{-(\gamma+\lambda+\theta+\mu_0)\tau} d\tau + \int_0^\infty y_{1,0}^*(\tau)\, e^{-(\gamma+\lambda+\theta+\mu_0)\tau} d\tau, \tag{3.12}$$

$$a_n^* = \int_0^\infty \left[\mu_0 q_{n-1,1}^*(0) + \theta q_{n,1}^*(0)\right] e^{-(\gamma+\lambda+\theta+\mu_0)\tau} d\tau + \int_0^\infty y_{n,0}^*(\tau)\, e^{-(\gamma+\lambda+\theta+\mu_0)\tau} d\tau, \quad n \geq 2, \tag{3.13}$$

$$b_1^* = \int_0^\infty \mu_1 q_{0,0}^*\, e^{-(\gamma+\lambda+\mu_1)\tau} d\tau + \int_0^\infty y_{1,1}^*(\tau)\, e^{-(\gamma+\lambda+\mu_1)\tau} d\tau, \tag{3.14}$$

$$b_n^* = \int_0^\infty \mu_1 q_{n-1,1}^*(0)\, e^{-(\gamma+\lambda+\mu_1)\tau} d\tau + \int_0^\infty y_{n,1}^*(\tau)\, e^{-(\gamma+\lambda+\mu_1)\tau} d\tau, \quad n \geq 2, \tag{3.15}$$

By inserting (3.12)-(3.15) into (3.8)-(3.11) we deduce (without lose of generality assume $\mathrm{Re}\gamma+\lambda+\theta+\mu_0 > 0, \mathrm{Re}\gamma+\lambda+\mu_1 > 0$)

$$q_{1,0}^*(x) = \frac{1}{\gamma+\lambda+\theta+\mu_0} \left[\mu_0 q_{0,0}^* + \theta q_{1,1}^*(0)\right] + e^{(\gamma+\lambda+\theta+\mu_0)x} \int_x^\infty y_{1,0}^*(\tau) e^{-(\gamma+\lambda+\theta+\mu_0)\tau} d\tau, \tag{3.16}$$

$$q_{n,0}^*(x) = \frac{1}{\gamma+\lambda+\theta+\mu_0} \left[\mu_0 q_{n-1,1}^*(0) + \theta q_{n,1}^*(0)\right] + e^{(\gamma+\lambda+\theta+\mu_0)x} \int_x^\infty y_{n,0}^*(\tau) e^{-(\gamma+\lambda+\theta+\mu_0)\tau} d\tau, \quad n \geq 2, \tag{3.17}$$

$$q_{1,1}^*(x) = \frac{\mu_1}{\gamma+\lambda+\mu_1} q_{0,0}^* + e^{(\gamma+\lambda+\mu_1)x} \int_x^\infty y_{1,1}^*(\tau) e^{-(\gamma+\lambda+\mu_1)\tau} d\tau, \tag{3.18}$$

$$q_{n,1}^*(x) = \frac{\mu_1}{\gamma+\lambda+\mu_1} q_{n-1,1}^*(0) + e^{(\gamma+\lambda+\mu_1)x} \int_x^\infty y_{n,1}^*(\tau) e^{-(\gamma+\lambda+\mu_1)\tau} d\tau, \quad n \geq 2, \tag{3.19}$$

By using (3.18), (3.19) repeatedly we obtain that, by induction,

$$q_{k,1}^*(x) = \left(\frac{\mu_1}{\gamma+\lambda+\mu_1}\right)^k q_{0,0}^* + \left(\frac{\mu_1}{\gamma+\lambda+\mu_1}\right)^{k-1} e^{(\gamma+\lambda+\mu_1)x} \int_0^\infty y_{1,1}^*(\tau) e^{-(\gamma+\lambda+\mu_1)\tau} d\tau$$
$$+ \left(\frac{\mu_1}{\gamma+\lambda+\mu_1}\right)^{k-2} e^{(\gamma+\lambda+\mu_1)x} \int_0^\infty y_{2,1}^*(\tau) e^{-(\gamma+\lambda+\mu_1)\tau} d\tau + \cdots$$
$$+ \frac{\mu_1}{\gamma+\lambda+\mu_1} e^{(\gamma+\lambda+\mu_1)x} \int_0^\infty y_{k-1,1}^*(\tau) e^{-(\gamma+\lambda+\mu_1)\tau} d\tau$$
$$+ e^{(\gamma+\lambda+\mu_1)x} \int_x^\infty y_{k,1}^*(\tau) e^{-(\gamma+\lambda+\mu_1)\tau} d\tau, \quad k \geq 1, \tag{3.20}$$

From (3.20) and (3.7) we estimate that

$$\begin{aligned}
\|q^*_{k,1}\|_{L^\infty[0,\infty)} \le & \frac{1}{|\gamma+\lambda|}\left(\frac{\mu_1}{|\gamma+\lambda+\mu_1|}\right)^k |y^*_{0,0}| + \frac{1}{\mathrm{Re}\gamma+\lambda+\mu_1}\left(\frac{\mu_1}{|\gamma+\lambda+\mu_1|}\right)^{k-1}\|y^*_{1,1}\|_{L^\infty[0,\infty)} \\
& + \frac{1}{\mathrm{Re}\gamma+\lambda+\mu_1}\left(\frac{\mu_1}{|\gamma+\lambda+\mu_1|}\right)^{k-2}\|y^*_{2,1}\|_{L^\infty[0,\infty)} + \cdots \\
& + \frac{1}{\mathrm{Re}\gamma+\lambda+\mu_1}\frac{\mu_1}{|\gamma+\lambda+\mu_1|}\|y^*_{k-1,1}\|_{L^\infty[0,\infty)} + \frac{1}{\mathrm{Re}\gamma+\lambda+\mu_1}\|y^*_{k,1}\|_{L^\infty[0,\infty)} \\
\le & \left\{\frac{1}{|\gamma+\lambda|}\left(\frac{\mu_1}{|\gamma+\lambda+\mu_1|}\right)^k + \frac{1}{\mathrm{Re}\gamma+\lambda+\mu_1}\sum_{j=0}^{k-1}\left(\frac{\mu_1}{|\gamma+\lambda+\mu_1|}\right)^j\right\}\|(y^*_0,y^*_1)\| \\
= & \left\{\frac{1}{|\gamma+\lambda|}\left(\frac{\mu_1}{|\gamma+\lambda+\mu_1|}\right)^k + \frac{|\gamma+\lambda+\mu_1|}{(\mathrm{Re}\gamma+\lambda+\mu_1)(|\gamma+\lambda+\mu_1|-\mu_1)}\right. \\
& \left. - \frac{|\gamma+\lambda+\mu_1|}{(\mathrm{Re}\gamma+\lambda+\mu_1)(|\gamma+\lambda+\mu_1|-\mu_1)}\left(\frac{\mu_1}{|\gamma+\lambda+\mu_1|}\right)^{k-1}\right\}\|(y^*_0,y^*_1)\| \\
\le & \left\{\frac{\mu_1}{|\gamma+\lambda|(\mathrm{Re}\gamma+\lambda+\mu_1)}\left(\frac{\mu_1}{|\gamma+\lambda+\mu_1|}\right)^{k-1} + \frac{|\gamma+\lambda+\mu_1|}{(\mathrm{Re}\gamma+\lambda+\mu_1)(|\gamma+\lambda+\mu_1|-\mu_1)}\right. \\
& \left. - \frac{|\gamma+\lambda+\mu_1|}{(\mathrm{Re}\gamma+\lambda+\mu_1)(|\gamma+\lambda+\mu_1|-\mu_1)}\left(\frac{\mu_1}{|\gamma+\lambda+\mu_1|}\right)^{k-1}\right\}\|(y^*_0,y^*_1)\|, \quad k \ge 1.
\end{aligned} \tag{3.21}$$

Note that the following inequality holds.

$$|\gamma+\lambda| \ge |\gamma+\lambda+\mu_1| - \mu_1$$

which implies

$$\begin{aligned}
& \frac{\mu_1}{|\gamma+\lambda|(\mathrm{Re}\gamma+\lambda+\mu_1)}\left(\frac{\mu_1}{|\gamma+\lambda+\mu_1|}\right)^{k-1} - \frac{|\gamma+\lambda+\mu_1|}{(\mathrm{Re}\gamma+\lambda+\mu_1)(|\gamma+\lambda+\mu_1|-\mu_1)}\left(\frac{\mu_1}{|\gamma+\lambda+\mu_1|}\right)^{k-1} \\
\le & \frac{\mu_1}{|\gamma+\lambda|(\mathrm{Re}\gamma+\lambda+\mu_1)}\left(\frac{\mu_1}{|\gamma+\lambda+\mu_1|}\right)^{k} - \frac{|\gamma+\lambda+\mu_1|}{(\mathrm{Re}\gamma+\lambda+\mu_1)(|\gamma+\lambda+\mu_1|-\mu_1)}\left(\frac{\mu_1}{|\gamma+\lambda+\mu_1|}\right)^{k}
\end{aligned}$$

From which together with (3.21) it follows that

$$\begin{aligned}
\sup_{k\ge1}\|q^*_{k,1}\|_{L^\infty[0,\infty)} \le & \lim_{k\to\infty}\left\{\frac{\mu_1}{|\gamma+\lambda|(\mathrm{Re}\gamma+\lambda+\mu_1)}\left(\frac{\mu_1}{|\gamma+\lambda+\mu_1|}\right)^{k-1} + \frac{|\gamma+\lambda+\mu_1|}{(\mathrm{Re}\gamma+\lambda+\mu_1)(|\gamma+\lambda+\mu_1|-\mu_1}\right. \\
& \left. - \frac{|\gamma+\lambda+\mu_1|}{(\mathrm{Re}\gamma+\lambda+\mu_1)(|\gamma+\lambda+\mu_1|-\mu_1}\left(\frac{\mu_1}{|\gamma+\lambda+\mu_1|}\right)^{k-1}\right\}\|(y^*_0,y^*_1)\| \\
\le & \frac{|\gamma+\lambda+\mu_1|}{(\mathrm{Re}\gamma+\lambda+\mu_1)(|\gamma+\lambda+\mu_1|-\mu_1)}\|(y^*_0,y^*_1)\|,
\end{aligned} \tag{3.22}$$

Substituting (3.20) into (3.17) we conclude similarly that

$$\begin{aligned}
\|q^*_{k,0}\|_{L^\infty[0,\infty)} \le & \left\{\frac{\mu_0}{|\gamma+\lambda+\theta+\mu_0||\gamma+\lambda|}\left(\frac{\mu_1}{|\gamma+\lambda+\mu_1|}\right)^{k-1}\right. \\
& + \frac{(\theta+\mu_0)|\gamma+\lambda+\mu_1|}{|\gamma+\lambda+\theta+\mu_0|(\mathrm{Re}\gamma+\lambda+\mu_1)(|\gamma+\lambda+\mu_1|-\mu_1)} \\
& + \frac{\mu_0}{|\gamma+\lambda+\theta+\mu_0|(\mathrm{Re}\gamma+\lambda+\mu_1)(|\gamma+\lambda+\mu_1|-\mu_1)}\left(\frac{\mu_1}{|\gamma+\lambda+\mu_1|}\right)^{k-2} \\
& + \frac{\theta}{|\gamma+\lambda+\theta+\mu_0||\gamma+\lambda|}\left(\frac{\mu_1}{\gamma+\lambda+\mu_1}\right)^{k} \\
& + \frac{\theta}{|\gamma+\lambda+\theta+\mu_0|(\mathrm{Re}\gamma+\lambda+\mu_1)(|\gamma+\lambda+\mu_1|-\mu_1)}\left(\frac{\mu_1}{|\gamma+\lambda+\mu_1|}\right)^{k-1} \\
& \left. + \frac{1}{\mathrm{Re}\gamma+\lambda+\theta+\mu_0}\right\}\|(y^*_0,y^*_1)\|_{L^\infty[0,\infty)}, \quad k \ge 2.
\end{aligned} \tag{3.23}$$

We can easy to check that the following inequality holds.

$$\begin{aligned}
&\frac{\mu_0}{|\gamma+\lambda+\theta+\mu_0||\gamma+\lambda|}\left(\frac{\mu_1}{|\gamma+\lambda+\mu_1|}\right)^{k-1}\\
&+\frac{\mu_0}{|\gamma+\lambda+\theta+\mu_0|(\mathrm{Re}\gamma+\lambda+\mu_1)(|\gamma+\lambda+\mu_1|-\mu_1)}\left(\frac{\mu_1}{|\gamma+\lambda+\mu_1|}\right)^{k-2}\\
&+\frac{\theta}{|\gamma+\lambda+\theta+\mu_0||\gamma+\lambda|}\left(\frac{\mu_1}{\gamma+\lambda+\mu_1}\right)^{k}\\
&+\frac{\theta}{|\gamma+\lambda+\theta+\mu_0|(\mathrm{Re}\gamma+\lambda+\mu_1)(|\gamma+\lambda+\mu_1|-\mu_1)}\left(\frac{\mu_1}{|\gamma+\lambda+\mu_1|}\right)^{k-1}\\
\leq&\frac{\mu_0}{|\gamma+\lambda+\theta+\mu_0||\gamma+\lambda|}\left(\frac{\mu_1}{|\gamma+\lambda+\mu_1|}\right)^{k}\\
&+\frac{\mu_0}{|\gamma+\lambda+\theta+\mu_0|(\mathrm{Re}\gamma+\lambda+\mu_1)(|\gamma+\lambda+\mu_1|-\mu_1)}\left(\frac{\mu_1}{|\gamma+\lambda+\mu_1|}\right)^{k-1}\\
&+\frac{\theta}{|\gamma+\lambda+\theta+\mu_0||\gamma+\lambda|}\left(\frac{\mu_1}{\gamma+\lambda+\mu_1}\right)^{k+1}\\
&+\frac{\theta}{|\gamma+\lambda+\theta+\mu_0|(\mathrm{Re}\gamma+\lambda+\mu_1)(|\gamma+\lambda+\mu_1|-\mu_1)}\left(\frac{\mu_1}{|\gamma+\lambda+\mu_1|}\right)^{k},k\geq 2.
\end{aligned}$$

From which together with (3.23) we know that

$$\begin{aligned}
\sup_{k\geq 2}\|q^*_{k,0}\|_{L^\infty[0,\infty)}\leq&\lim_{n\to\infty}\left\{\frac{\mu_0}{|\gamma+\lambda+\theta+\mu_0||\gamma+\lambda|}\left(\frac{\mu_1}{|\gamma+\lambda+\mu_1|}\right)^{k-1}\right.\\
&+\frac{(\theta+\mu_0)|\gamma+\lambda+\mu_1|}{|\gamma+\lambda+\theta+\mu_0|(\mathrm{Re}\gamma+\lambda+\mu_1)(|\gamma+\lambda+\mu_1|-\mu_1)}\\
&+\frac{\mu_0}{|\gamma+\lambda+\theta+\mu_0|(\mathrm{Re}\gamma+\lambda+\mu_1)(|\gamma+\lambda+\mu_1|-\mu_1)}\left(\frac{\mu_1}{|\gamma+\lambda+\mu_1|}\right)^{k-2}\\
&+\frac{\theta}{|\gamma+\lambda+\theta+\mu_0||\gamma+\lambda|}\left(\frac{\mu_1}{\gamma+\lambda+\mu_1}\right)^{k}\\
&+\frac{\theta}{|\gamma+\lambda+\theta+\mu_0|(\mathrm{Re}\gamma+\lambda+\mu_1)(|\gamma+\lambda+\mu_1|-\mu_1)}\left(\frac{\mu_1}{|\gamma+\lambda+\mu_1|}\right)^{k-1}\\
&\left.+\frac{1}{\mathrm{Re}\gamma+\lambda+\theta+\mu_0}\right\}\|(y_0^*,y_1^*)\|_{L^\infty[0,\infty)}\\
=&\left\{\frac{(\theta+\mu_0)|\gamma+\lambda+\mu_1|}{|\gamma+\lambda+\theta+\mu_0|(\mathrm{Re}\gamma+\lambda+\mu_1)(|\gamma+\lambda+\mu_1|-\mu_1)}\right.\\
&\left.+\frac{1}{\mathrm{Re}\gamma+\lambda+\theta+\mu_0}\right\}\|(y_0^*,y_1^*)\|_{L^\infty[0,\infty)}.
\end{aligned}\tag{3.24}$$

Combining (3.7), (3.16), (3.22) with (3.24) we estimate

$$\begin{aligned}
\|(q_0^*,q_1^*)\|=&\max\left\{|p_{0,0}|,\sup_{n\geq 1}\|q^*_{n,0}\|_{L^\infty[0,\infty)},\sup_{n\geq 1}\|q^*_{n,1}\|_{L^\infty[0,\infty)}\right\}\\
=&\max\left\{\frac{1}{|\gamma+\lambda|},\frac{|\gamma+\lambda+\mu_1|}{(\mathrm{Re}\gamma+\lambda+\mu_1)(|\gamma+\lambda+\mu_1|-\mu_1)},\right.\\
&\frac{\mu_0}{|\gamma+\lambda||\gamma+\lambda+\theta+\mu_0|}+\frac{\theta\mu_0}{|\gamma+\lambda||\gamma+\lambda+\theta+\mu_0||\gamma+\lambda+\mu_1|}\\
&+\frac{\theta}{|\gamma+\lambda+\theta+\mu_0||\mathrm{Re}\gamma+\lambda+\mu_1|}+\frac{1}{\mathrm{Re}\gamma+\lambda+\theta+\mu_0}\\
&\frac{(\theta+\mu_0)|\gamma+\lambda+\mu_1|}{|\gamma+\lambda+\theta+\mu_0|(\mathrm{Re}\gamma+\lambda+\mu_1)(|\gamma+\lambda+\mu_1|-\mu_1)}\\
&\left.+\frac{1}{\mathrm{Re}\gamma+\lambda+\theta+\mu_0}\right\}\|(y_0^*,y_1^*)\|_{L^\infty[0,\infty)}.
\end{aligned}\tag{3.25}$$

(3.25) shows that

$$\begin{aligned}
\|(\gamma I-\mathcal{L}-\mathcal{R}-\mathcal{J})^{-1}\| \le \max\Big\{&\frac{1}{|\gamma+\lambda|}, \frac{|\gamma+\lambda+\mu_1|}{(\mathrm{Re}\gamma+\lambda+\mu_1)(|\gamma+\lambda+\mu_1|-\mu_1}, \\
&\frac{\mu_0}{|\gamma+\lambda||\gamma+\lambda+\theta+\mu_0|}+\frac{\theta\mu_1}{|\gamma+\lambda||\gamma+\lambda+\theta+\mu_0||\gamma+\lambda+\mu_1|} \\
&+\frac{\theta}{|\gamma+\lambda+\theta+\mu_0||\mathrm{Re}\gamma+\lambda+\mu_1|}+\frac{1}{\mathrm{Re}\gamma+\lambda+\theta+\mu_0} \\
&\frac{(\theta+\mu_0)|\gamma+\lambda+\mu_1|}{|\gamma+\lambda+\theta+\mu_0|(\mathrm{Re}\gamma+\lambda+\mu_1)(|\gamma+\lambda+\mu_1|-\mu_1)} \\
&+\frac{1}{\mathrm{Re}\gamma+\lambda+\theta+\mu_0}\Big\}. \qquad (3.26)
\end{aligned}$$

Together with the fact $\|\mathcal{N}\|=\lambda$ we conclude that, when

$$\begin{aligned}
\|(\gamma I-\mathcal{L}-\mathcal{R}-\mathcal{J})^{-1}\mathcal{N}\| &\le \|(\gamma I-\mathcal{L}-\mathcal{R}-\mathcal{J})^{-1}\|\|\mathcal{N}\| \\
&\le \max\Big\{\frac{\lambda}{|\gamma+\lambda|}, \frac{\lambda|\gamma+\lambda+\mu_1|}{(\mathrm{Re}\gamma+\lambda+\mu_1)(|\gamma+\lambda+\mu_1|-\mu_1}, \\
&\frac{\lambda\mu_0}{|\gamma+\lambda||\gamma+\lambda+\theta+\mu_0|}+\frac{\lambda\theta\mu_1}{|\gamma+\lambda||\gamma+\lambda+\theta+\mu_0||\gamma+\lambda+\mu_1|} \\
&+\frac{\lambda\theta}{|\gamma+\lambda+\theta+\mu_0||\mathrm{Re}\gamma+\lambda+\mu_1|}+\frac{\lambda}{\mathrm{Re}\gamma+\lambda+\theta+\mu_0} \\
&\frac{\lambda(\theta+\mu_0)|\gamma+\lambda+\mu_1|}{|\gamma+\lambda+\theta+\mu_0|(\mathrm{Re}\gamma+\lambda+\mu_1)(|\gamma+\lambda+\mu_1|-\mu_1)} \\
&+\frac{\lambda}{\mathrm{Re}\gamma+\lambda+\theta+\mu_0}\Big\} \\
&<1, \qquad (3.27)
\end{aligned}$$

$[(I-(\gamma I-\mathcal{L}-\mathcal{R}-\mathcal{J})^{-1}\mathcal{N}]^{-1}$ exists and is bounded. By noting

$$(\gamma I-\mathcal{L}-\mathcal{R}-\mathcal{J}-\mathcal{N})^{-1}=[(I-(\gamma I-\mathcal{L}-\mathcal{R}-\mathcal{J})^{-1}\mathcal{N}]^{-1}(\gamma I-\mathcal{L}-\mathcal{R}-\mathcal{J})^{-1}$$

we know that $(\gamma I-\mathcal{L}-\mathcal{R}-\mathcal{J}-\mathcal{N})^{-1}$ exists and is bounded when (3.27) holds. that is to say, (29) belongs to the resolvent set of $(A+U+E)^*$(see Gupur, Li, & Zhu, 2001). In particular, if $\gamma=i\omega,\ \omega\in\mathbb{R}\backslash\{0\},\ i^2=-1$, then all the γ naturally belong to (3.27). In fact, by simple calculation, we have

$$\frac{\lambda}{\sqrt{\omega^2+(\lambda)^2}}<\frac{\lambda}{\lambda}=1,$$

$$\frac{\lambda}{\lambda+\mu_1}\frac{\sqrt{\omega^2+(\lambda+\mu_1)^2}}{\sqrt{\omega^2+(\lambda+\mu_1)^2}-\mu_1}<1,$$

$$\begin{aligned}
&\frac{\lambda\mu_0}{\sqrt{\omega^2+\lambda^2}\sqrt{\omega^2+(\lambda+\theta+\mu_0)^2}}+\frac{\lambda\theta\mu_1}{\sqrt{\omega^2+\lambda^2}\sqrt{\omega^2+(\lambda+\theta+\mu_0)^2}\sqrt{\omega^2+(\lambda+\mu_1)^2}} \\
&+\frac{\lambda\theta}{\sqrt{\omega^2+(\lambda+\theta+\mu_0)^2}(\lambda+\mu_1)}+\frac{\lambda}{\lambda+\theta+\mu_0}<1,
\end{aligned}$$

$$\frac{\lambda(\theta+\mu_0)\sqrt{a^2+(\lambda+\mu_1)^2}}{\sqrt{\omega^2+(\lambda+\theta+\mu_0)^2}(\lambda+\mu_1)(\sqrt{\omega^2+(\lambda+\mu_1)^2}-\mu_1)}+\frac{\lambda}{\lambda+\theta+\mu_0}<1.$$

The above inequalities show that the resolvent set of $(A+U+E)^*$ contain all points on the imaginary axis except zero so as $(A+U+E)$. □

Lemma 3.3 *If $\lambda<\mu_1$, then 0 is an eigenvalue of $A+U+E$ with geometric multiplicity one.*

Proof. We consider the equation $(A + U + E)(p_0, p_1) = 0$, which is equivalent to

$$\lambda p_{0,0} = \mu_0 \int_0^\infty p_{1,0}(x)dx + \mu_1 \int_0^\infty p_{1,1}(x)dx, \tag{3.28}$$

$$\frac{dp_{1,0}(x)}{dx} = -(\lambda + \theta + \mu_0)p_{1,0}(x), \tag{3.29}$$

$$\frac{dp_{n,0}(x)}{dx} = -(\lambda + \theta + \mu_0)p_{n,0}(x) + \lambda p_{n-1,0}(x), \quad n \geq 2, \tag{3.30}$$

$$\frac{dp_{1,1}(x)}{dx} = -(\lambda + \mu_1)p_{1,1}(x), \tag{3.31}$$

$$\frac{dp_{n,1}(x)}{dx} = -(\lambda + \mu_1)p_{n,1}(x) + \lambda p_{n-1,1}(x), \quad n \geq 2, \tag{3.32}$$

$$p_{1,0}(0) = \lambda p_{0,0}, \tag{3.33}$$

$$p_{n,0}(0) = 0, \quad n \geq 2, \tag{3.34}$$

$$p_{n,1}(0) = \theta \int_0^\infty p_{n,0}(x)dx + \mu_0 \int_0^\infty p_{n+1,0}(x)dx + \mu_1 \int_0^\infty p_{n+1,1}(x)dx, \quad n \geq 1, \tag{3.35}$$

It is difficult to determine the expression of all $p_{n,0}$ and $p_{n,1}$ for $n \geq 1$ and to verify $(p_0, p_1) \in D(A + U + E)$. Hence, we use an indirect method. We define the probability generating functions for $|z| < 1$

$$P_0(x, z) = \sum_{n=1}^\infty p_{n,0}(x)z^n, \quad P_1(x, z) = \sum_{n=1}^\infty p_{n,1}(x)z^n$$

then Theorem 2.3 ensures that $P_0(x, z)$ and $P_1(x, z)$ are well-defined. (3.29) and (3.30) gives

$$\begin{aligned}
\frac{\partial \sum_{n=1}^\infty p_{n,0}(x)z^n}{\partial x} &= -(\lambda + \theta + \mu_0) \sum_{n=1}^\infty p_{n,0}(x)z^n + \lambda \sum_{n=2}^\infty p_{n-1,0}(x)z^n \\
&\Longrightarrow \\
\frac{\partial P_0(x, z)}{\partial x} &= -(\lambda + \theta + \mu_0)P_0(x, z) + \lambda z P_0(x, z) \\
&= (\lambda z - \lambda - \theta - \mu_0)P_0(x, z) \\
&\Longrightarrow \\
P_0(x, z) &= P_0(0, z)e^{(\lambda z - \lambda - \theta - \mu_0)x},
\end{aligned} \tag{3.36}$$

(3.31) and (3.32) imply

$$\begin{aligned}
\frac{\partial \sum_{n=1}^\infty p_{n,1}(x)z^n}{\partial x} &= -(\lambda + \mu_1) \sum_{n=1}^\infty p_{n,1}(x)z^n + \lambda \sum_{n=2}^\infty p_{n-1,1}(x)z^n \\
&\Longrightarrow \\
\frac{\partial P_1(x, z)}{\partial x} &= -(\lambda + \mu_1)P_1(x, z) + \lambda z P_1(x, z) \\
&= (\lambda z - \lambda - \mu_1)P_1(x, z) \\
&\Longrightarrow \\
P_1(x, z) &= P_1(0, z)e^{(\lambda z - \lambda - \mu_1)x},
\end{aligned} \tag{3.37}$$

From (3.33)-(3.35) and (3.28), we deduce

$$P_0(0,z) = \lambda z p_{0,0}, \tag{3.38}$$

$$\begin{aligned}
P_1(0,z) &= \sum_{n=1}^{\infty} P_1(0) z^n \\
&= \theta \int_0^{\infty} \sum_{n=1}^{\infty} p_{n,0}(x) z^n dx + \mu_0 \int_0^{\infty} \sum_{n=1}^{\infty} p_{n+1,0}(x) z^n dx + \mu_1 \sum_{n=1}^{\infty} \int_0^{\infty} p_{n+1,1}(x) z^n dx \\
&= \theta \int_0^{\infty} P_0(x,z) dx + \mu_0 \int_0^{\infty} \frac{1}{z}\Big[\sum_{n=1}^{\infty} p_{n,0}(x) z^n - p_{1,0}(x) z\Big] dx \\
&\quad + \mu_1 \int_0^{\infty} \frac{1}{z}\Big[\sum_{n=1}^{\infty} p_{n,1}(x) z^n - p_{1,1}(x) z\Big] dx \\
&= \theta \int_0^{\infty} P_0(x,z) dx + \frac{\mu_0}{z} \int_0^{\infty} P_0(x,z) dx - \mu_0 \int_0^{\infty} p_{1,0}(x) dx \\
&\quad + \frac{\mu_1}{z} \int_0^{\infty} P_1(x,z) dx - \mu_1 \int_0^{\infty} p_{1,1}(x) dx \\
&= \theta \int_0^{\infty} P_0(x,z) dx + \frac{\mu_0}{z} \int_0^{\infty} P_0(x,z) dx + \frac{\mu_1}{z} \int_0^{\infty} P_1(x,z) dx - \lambda p_{0,0},
\end{aligned} \tag{3.39}$$

By inserting (3.36), (3.37) and (3.38) into (3.39), we determine

$$\begin{aligned}
P_1(0,z) &= \theta \int_0^{\infty} P_0(0,z) e^{(\lambda z - \lambda - \theta - \mu_0)x} dx + \frac{\mu_0}{z} \int_0^{\infty} P_0(0,z) e^{(\lambda z - \lambda - \theta - \mu_0)x} dx \\
&\quad + \frac{\mu_1}{z} \int_0^{\infty} P_1(0,z) e^{(\lambda z - \lambda - \mu_1)x} dx - \lambda p_{0,0} \\
&= \theta \int_0^{\infty} \lambda z p_{0,0} e^{(\lambda z - \lambda - \theta - \mu_0)x} dx + \frac{\mu_0}{z} \int_0^{\infty} \lambda z p_{0,0} e^{(\lambda z - \lambda - \theta - \mu_0)x} dx \\
&\quad + \frac{\mu_1}{z} \int_0^{\infty} P_1(0,z) e^{(\lambda z - \lambda - \mu_1)x} dx - \lambda p_{0,0} \\
&= \frac{\theta z}{\lambda z - \lambda - \theta - \mu_0} \lambda p_{0,0} + \frac{\mu_0}{\lambda z - \lambda - \theta - \mu_0} \lambda p_{0,0} \\
&\quad + \frac{\mu_1}{z(\lambda z - \lambda - \mu_1)} P_1(0,z) - \lambda p_{0,0} \\
&\Longrightarrow
\end{aligned}$$

$$P_1(0,z) = \frac{\frac{\theta z + \mu_0}{\lambda z - \lambda - \theta - \mu_0} - 1}{1 - \frac{\mu_1}{z(\lambda z - \lambda - \mu_1)}} \lambda p_{0,0}. \tag{3.40}$$

By (3.40) and the L'Hospital rule we calculate

$$\begin{aligned}
\lim_{z\to 1} P_1(0,z) &= \lim_{z\to 1} \frac{\dfrac{\theta z + \mu_0}{\lambda z - \lambda - \theta - \mu_0} - 1}{1 - \dfrac{\mu_1}{z(\lambda z - \lambda - \mu_1)}} \lambda p_{0,0} = \lim_{z\to 1} \frac{\dfrac{\theta(\lambda z - \lambda - \theta - \mu_0) - \lambda(\theta z + \mu_0)}{(\lambda z - \lambda - \theta - \mu_0)^2}}{\dfrac{\mu_1[(\lambda z - \lambda - \mu_1) + \lambda z]}{z^2(\lambda z - \lambda - \mu_1)^2}} \lambda p_{0,0} \\
&= \frac{\dfrac{\theta(-\theta - \mu_0) - \lambda(\theta + \mu_0)}{(-\theta - \mu_0)^2}}{\dfrac{-\mu_1 + \lambda}{\mu_1}} \lambda p_{0,0} = \frac{(\lambda + \theta)\mu_1}{(\theta + \mu_0)(\mu_1 - \lambda)} \lambda p_{0,0} < \infty.
\end{aligned} \tag{3.41}$$

By combining (3.36), (3.37), (3.38) and (3.41), we have

$$\sum_{n=1}^{\infty} p_{n,0}(x) = \lim_{z\to 1} P_0(x,z) = \lim_{z\to 1} P_0(0,z)e^{(\lambda z-\lambda-\theta-\mu_0)x}$$
$$= \lambda p_{0,0}e^{-(\theta+\mu_0)x}$$
$$\Longrightarrow$$
$$\sum_{n=1}^{\infty}\int_0^{\infty} p_{n,0}(x)dx = \frac{\lambda}{\theta+\mu_0}p_{0,0} < \infty, \tag{3.42}$$
$$\sum_{n=1}^{\infty} p_{n,1}(x) = \lim_{z\to 1} P_1(x,z) = \lim_{z\to 1} P_1(0,z)e^{(\lambda z-\lambda-\mu_1)x}$$
$$= \frac{(\lambda+\theta)\mu_1}{(\theta+\mu_0)(\mu_1-\lambda)}\lambda p_{0,0}e^{-\mu_1 x}$$
$$\Longrightarrow$$
$$\sum_{n=1}^{\infty}\int_0^{\infty} p_{n,1}(x)dx = \frac{(\lambda+\theta)\lambda}{(\theta+\mu_0)(\mu_1-\lambda)}p_{0,0} < \infty. \tag{3.43}$$

(3.42) and (3.43) imply

$$\|(p_0,p_1)\| = \|p_0\| + \|p_1\| < \infty.$$

This shows that 0 is an eigenvalue of $A + U + E$. Moreover, by solving (3.29) - (3.31) we have

$$p_{n,0}(x) = e^{-(\lambda+\theta+\mu_0)x}\sum_{k=1}^{n} p_{k,0}(0)\frac{(\lambda x)^{n-k}}{(n-k)!}, \quad n \geq 1,$$
$$p_{n,1}(x) = e^{-(\lambda+\mu_1)x}\sum_{k=1}^{n} p_{k,1}(0)\frac{(\lambda x)^{n-k}}{(n-k)!}, \quad n \geq 1,$$
$$p_{1,0}(0) = \lambda p_{0,0},$$
$$p_{n,0}(0) = 0, \quad n \geq 2,$$
$$p_{n,1}(0) = \theta\int_0^{\infty} p_{n,0}(x)dx + \mu_0\int_0^{\infty} p_{n+1,0}(x)dx + \mu_1\int_0^{\infty} p_{n+1,1}(x)dx, \quad n \geq 1,$$

From the above, we know that the eigenvectors corresponding to zero span one dimensional linear space, that is to say, the geometric multiplicity of 0 is one. □

From Theorem 2.3 and Lemma 3.2 we know that 0 is an eigenvalue of $(A+U+E)^*$. Furthermore, we deduce the following result.

Lemma 3.4 *If $\lambda < \mu_1$, then 0 is an eigenvalue of $(A + U + E)^*$ with geometric multiplicity one.*

Since Theorem 2.3, Lemma 3.2, Lemma 3.3 and Lemma 3.4 are just the conditions of Theorem 14 in (Gupur, Li, & Zhu, 2001), we conclude the following result.

Theorem 1 *If $\lambda < \mu_1$, then the time-dependent solution of the system (1.4) strongly converges to its steady-state solution, that is,*

$$\lim_{t\to\infty} \|(p_0,p_1)(\cdot,t) - \beta(p_0,p_1)(\cdot)\| = 0,$$

here $(p_0,p_1)(x)$ is the eigenvector in Lemma 3.3 and β is decided by the eigenvector in Lemma 3.4 and initial value $(p_0,p_1)(0)$.

4. Conclusion

In this paper, we do dynamic analysis for the M/G/1 queueing model with single working vacation and vacation interruption by using the functional analysis method. we prove the existence of a unique nonnegative time-dependent solution of the model, and when the service completion rate are constant we obtain that the time-dependent solution of the model strongly converges to its steady-state solution. These results confirmed all two hypotheses which stated in the introduction.

In addition, from Theorem 3.1 we can prove the time-dependent queueing size at the departure point converges a positive number, the time-dependent queueing length and the time-dependent waiting time also converges to the corresponding steady-state queuing length and steady-state waiting time.

Our studies in this field, see (Kasim & Gupur, 2011) and (Gupur, 2011; Gupur, 2014) for instance, suggest that there are infinitely many eigenvalues of $A + U + E$ on the left half complex plane, that is to say, it is impossible that the time-dependent solution of the system (1.4) exponentially converges to its steady-state solution. Of course, it needs to verify. that is our next research work.

Acknowledgements

The author would like to express his sincere thanks to the anonymous referees and associated editor for his/her careful reading of the manuscript. The author' research work was supported by the Tian Yuan Special Funds of the National Natural Science Foundation of China (No:11526175),the Natural Science Foundation of Xinjiang University(No:BS130104).

Appendix

Proof of Theorem 2.1 The proof will be divided into four steps. Let us first prove $(\gamma I - A)^{-1}$ exists and is bounded for some γ. For any given $(y_0, y_1) \in X$, we consider the equation $(\gamma I - A)(p_0, p_1) = (y_0, y_1)$, that is,

$$(\gamma + \lambda)p_{0,0} = y_{0,0}, \tag{A.1}$$
$$\frac{dp_{n,j}(x)}{dx} = -\gamma p_{n,j}(x) + y_{n,j}(x), \quad n \geq 1, \quad j = 0, 1, \tag{A.2}$$
$$p_{1,0}(0) = \lambda p_{0,0}, \quad p_{n,0}(0) = 0, \quad n \geq 2, \tag{A.3}$$
$$p_{n,1}(0) = \theta \int_0^\infty p_{n,0}(x)dx + \int_0^\infty \mu_0(x)p_{n+1,0}(x)dx + \int_0^\infty \mu_1(x)p_{n+1,1}(x)dx, \quad n \geq 1. \tag{A.4}$$

By solving (A.1)-(A.2), we have

$$p_{0,0} = \frac{1}{\gamma + \lambda}y_0, \tag{A.5}$$
$$p_{n,j}(x) = a_{n,j}\, e^{-\gamma x} + e^{-\gamma x}\int_0^x y_{n,j}(\tau)e^{\gamma\tau}d\tau, \quad n \geq 1, \quad j = 0, 1, \tag{A.6}$$

(A.3)-(A.4) together with (A.5)-(A.6) we get that

$$a_{1,0} = p_{1,0}(0) = \lambda p_{0,0} = \frac{\lambda}{\gamma + \lambda}y_{0,0}, \tag{A.7}$$
$$a_{n,0} = p_{n,0}(0) = 0, \quad n \geq 2, \tag{A.8}$$
$$\begin{aligned}
a_{1,1} =& p_{1,1}(0)\\
=&\theta \int_0^\infty p_{1,0}(x)dx + \int_0^\infty \mu_0(x)p_{2,0}(x)dx + \int_0^\infty \mu_1(x)p_{2,1}(x)dx\\
=&\theta \int_0^\infty \left[a_{1,0}e^{-\gamma x} + e^{-\gamma x}\int_0^x y_{1,0}(\tau)e^{\gamma\tau}d\tau\right]dx\\
&+ \int_0^\infty \mu_0(x)\left[a_{2,0}e^{-\gamma x} + e^{-\gamma x}\int_0^x y_{2,0}(\tau)e^{\gamma\tau}d\tau\right]dx\\
&+ \int_0^\infty \mu_1(x)\left[a_{2,1}e^{-\gamma x} + e^{-\gamma x}\int_0^x y_{2,1}(\tau)e^{\gamma\tau}d\tau\right]dx\\
=&\frac{\theta}{\gamma}a_{1,0} + \theta\int_0^\infty e^{-\gamma x}\int_0^x y_{1,0}(\tau)e^{\gamma\tau}d\tau dx + \int_0^\infty \mu_0(x)e^{-\gamma x}\int_0^x y_{2,0}(\tau)e^{\gamma\tau}d\tau dx\\
&+ a_{2,1}\int_0^\infty \mu_1(x)e^{-\gamma x}dx + \int_0^\infty e^{-\gamma x}\int_0^x y_{2,1}(\tau)e^{\gamma\tau}d\tau dx,
\end{aligned} \tag{A.9}$$

$$
\begin{aligned}
a_{n,1} =& p_{n,1}(0) \\
=& \theta \int_0^\infty p_{n,0}(x)dx + \int_0^\infty \mu_0(x)p_{n+1,0}(x)dx + \int_0^\infty \mu_1(x)p_{n+1,1}(x)dx \\
=& \theta \int_0^\infty \left[a_{n,0}e^{-\gamma x} + e^{-\gamma x}\int_0^x y_{n,0}(\tau)e^{\gamma\tau}d\tau\right]dx \\
&+ \int_0^\infty \mu_0(x)\left[a_{n+1,0}e^{-\gamma x} + e^{-\gamma x}\int_0^x y_{n+1,0}(\tau)e^{\gamma\tau}d\tau\right]dx \\
&+ \int_0^\infty \mu_1(x)\left[a_{n+1,1}e^{-\gamma x} + e^{-\gamma x}\int_0^x y_{n+1,1}(\tau)e^{\gamma\tau}d\tau\right]dx \\
=& \theta \int_0^\infty e^{-\gamma x}\int_0^x y_{n,0}(\tau)e^{\gamma\tau}d\tau dx + \int_0^\infty \mu_0(x)e^{-\gamma x}\int_0^x y_{n+1,0}(\tau)e^{\gamma\tau}d\tau dx \\
&+ a_{n+1,1}\int_0^\infty \mu_1(x)e^{-\gamma x}dx + \int_0^\infty \mu_1(x)e^{-\gamma x}\int_0^x y_{n+1,1}(\tau)e^{\gamma\tau}d\tau dx, \quad n \geq 2.
\end{aligned} \tag{A.10}
$$

If we set

$$
C = \begin{pmatrix}
1 & -\int_0^\infty \mu_1(x)e^{-\gamma x}dx & 0 & 0 & \cdots \\
0 & 1 & -\int_0^\infty \mu_1(x)e^{-\gamma x}dx & 0 & \cdots \\
0 & 0 & 1 & -\int_0^\infty \mu_1(x)e^{-\gamma x}dx & \cdots \\
\vdots & \vdots & \vdots & \vdots & \ddots
\end{pmatrix},
$$

$$
\vec{a} = (a_{1,1}, a_{2,1}, a_{3,1}, \cdots)^T,
$$

then (A.9)-(A.10) give

$$
C\vec{a} = \begin{pmatrix}
\frac{\theta}{\gamma}a_{1,0} + \theta\int_0^\infty e^{-\gamma x}\int_0^x y_{1,0}(\tau)e^{\gamma\tau}d\tau dx + \int_0^\infty \mu_0(x)e^{-\gamma x}\int_0^x y_{2,0}(\tau)e^{\gamma\tau}d\tau dx + \int_0^\infty \mu_1(x)e^{-\gamma x}\int_0^x y_{2,1}(\tau)e^{\gamma\tau}d\tau dx \\
\theta\int_0^\infty e^{-\gamma x}\int_0^x y_{2,0}(\tau)e^{\gamma\tau}d\tau dx + \int_0^\infty \mu_0(x)e^{-\gamma x}\int_0^x y_{3,0}(\tau)e^{\gamma\tau}d\tau dx + \int_0^\infty \mu_1(x)e^{-\gamma x}\int_0^x y_{3,1}(\tau)e^{\gamma\tau}d\tau dx \\
\theta\int_0^\infty e^{-\gamma x}\int_0^x y_{3,0}(\tau)e^{\gamma\tau}d\tau dx + \int_0^\infty \mu_0(x)e^{-\gamma x}\int_0^x y_{4,0}(\tau)e^{\gamma\tau}d\tau dx + \int_0^\infty \mu_1(x)e^{-\gamma x}\int_0^x y_{4,1}(\tau)e^{\gamma\tau}d\tau dx \\
\theta\int_0^\infty e^{-\gamma x}\int_0^x y_{4,0}(\tau)e^{\gamma\tau}d\tau dx + \int_0^\infty \mu_0(x)e^{-\gamma x}\int_0^x y_{5,0}(\tau)e^{\gamma\tau}d\tau dx + \int_0^\infty \mu_1(x)e^{-\gamma x}\int_0^x y_{5,1}(\tau)e^{\gamma\tau}d\tau dx \\
\vdots
\end{pmatrix}, \tag{A.11}
$$

It is easy to calculate the inverse of C as follows:

$$
C^{-1} = \begin{pmatrix}
1 & \int_0^\infty \mu_1(x)e^{-\gamma x}dx & (\int_0^\infty \mu_1(x)e^{-\gamma x}dx)^2 & (\int_0^\infty \mu_1(x)e^{-\gamma x}dx)^3 & \cdots \\
0 & 1 & \int_0^\infty \mu_1(x)e^{-\gamma x}dx & (\int_0^\infty \mu_1(x)e^{-\gamma x}dx)^2 & \cdots \\
0 & 0 & 1 & \int_0^\infty \mu_1(x)e^{-\gamma x}dx & \cdots \\
0 & 0 & 0 & 1 & \cdots \\
\vdots & \vdots & \vdots & \vdots & \ddots
\end{pmatrix},
$$

which together with (A.11) we obtain that

$$
\begin{aligned}
a_{1,1} =& \frac{\lambda\theta}{\gamma(\gamma+\lambda)}y_{0,0} + \sum_{k=0}^\infty \left(\int_0^\infty \mu_1(x)e^{-\gamma x}dx\right)^k \left[\theta\int_0^\infty e^{-\gamma x}\int_0^x y_{k+1,0}(\tau)e^{\gamma\tau}d\tau dx\right. \\
&\left. + \int_0^\infty \mu_0(x)e^{-\gamma x}\int_0^x y_{k+2,0}(\tau)e^{\gamma\tau}d\tau dx + \int_0^\infty \mu_1(x)e^{-\gamma x}\int_0^x y_{k+2,1}(\tau)e^{\gamma\tau}d\tau dx\right],
\end{aligned} \tag{A.12}
$$

$$
\begin{aligned}
a_{n,1} =& \sum_{k=0}^\infty \left(\int_0^\infty \mu_1(x)e^{-\gamma x}dx\right)^k \left[\theta\int_0^\infty e^{-\gamma x}\int_0^x y_{k+n,0}(\tau)e^{\gamma\tau}d\tau dx\right. \\
&\left. + \int_0^\infty \mu_0(x)e^{-\gamma x}\int_0^x y_{k+n+1,0}(\tau)e^{\gamma\tau}d\tau dx + \int_0^\infty \mu_1(x)e^{-\gamma x}\int_0^x y_{k+n+1,1}(\tau)e^{\gamma\tau}d\tau dx\right], \quad n \geq 2.
\end{aligned} \tag{A.13}
$$

Now, from (A.5)-(A.6) and using the Fubini theorem we calculate (assume $\gamma > 0$)

$$\begin{aligned}
\|p_{n,j}\|_{L^1[0,\infty)} &= \int_0^\infty |p_{n,j}(x)|dx \\
&= \int_0^\infty \left| a_{n,j}e^{-\gamma x} + e^{-\gamma x}\int_0^x y_{n,j}(\tau)e^{\gamma\tau}d\tau\right| dx \\
&\le |a_{n,j}|\int_0^\infty e^{-\gamma x}dx + \int_0^\infty e^{-\gamma x}\int_0^x |y_{n,j}(\tau)|e^{\gamma\tau}d\tau dx \\
&= \frac{1}{\gamma}|a_{n,j}| + \int_0^\infty y_{n,j}(\tau)e^{\gamma\tau}\int_\tau^\infty e^{-\gamma x}dxd\tau \\
&= \frac{1}{\gamma}|a_{n,j}| + \frac{1}{\gamma}\|y_{n,j}\|_{L^1[0,\infty)}, \quad j = 0, 1, \\
&\Longrightarrow \\
\|(p_0, p_1)\| &= \|p_0\| + \|p_1\| \\
&= |p_{0,0}| + \sum_{n=1}^\infty \|p_{n,0}\|_{L^1[0,\infty)} + \sum_{n=1}^\infty \|p_{n,1}\|_{L^1[0,\infty)} \\
&\le \frac{1}{\gamma}|y_{0,0}| + \frac{1}{\gamma}\sum_{n=1}^\infty \|y_{n,0}\|_{L^1[0,\infty)} \\
&\quad + \frac{1}{\gamma}\sum_{n=1}^\infty |a_{n,1}| + \frac{1}{\gamma}\sum_{n=1}^\infty \|y_{n,1}\|_{L^1[0,\infty)},
\end{aligned} \tag{A.14}$$

Combining (A.12) and (A.13) with the Fubini theorem we estimate

$$\begin{aligned}
\sum_{n=1}^\infty |a_{n,1}| \le& \frac{\lambda\theta}{\gamma(\gamma+\lambda)}|y_{0,0}| \\
&+ \theta\sum_{n=1}^\infty\sum_{k=0}^\infty\left(\int_0^\infty \mu_1(x)e^{-\gamma x}dx\right)^k \int_0^\infty e^{-\gamma x}\int_0^x |y_{k+n,0}(\tau)|e^{\gamma\tau}d\tau dx \\
&+ \sum_{n=1}^\infty\sum_{k=0}^\infty\left(\int_0^\infty \mu_1(x)e^{-\gamma x}dx\right)^k \int_0^\infty \mu_0(x)e^{-\gamma x}\int_0^x |y_{k+n+1,0}(\tau)|e^{\gamma\tau}d\tau dx \\
&+ \sum_{n=1}^\infty\sum_{k=0}^\infty\left(\int_0^\infty \mu_1(x)e^{-\gamma x}dx\right)^k \int_0^\infty \mu_1(x)e^{-\gamma x}\int_0^x |y_{k+n+1,1}(\tau)|e^{\gamma\tau}d\tau dx \\
=& \frac{\lambda\theta}{\gamma(\gamma+\lambda)}|y_{0,0}| \\
&+ \theta\sum_{n=1}^\infty\sum_{k=0}^\infty\left(\int_0^\infty \mu_1(x)e^{-\gamma x}dx\right)^k \int_0^\infty |y_{k+n,0}(\tau)|e^{\gamma\tau}\int_\tau^\infty e^{-\gamma x}dxd\tau \\
&+ \sum_{n=1}^\infty\sum_{k=0}^\infty\left(\int_0^\infty \mu_1(x)e^{-\gamma x}dx\right)^k \int_0^\infty |y_{k+n+1,0}(\tau)|e^{\gamma\tau}\int_\tau^\infty \mu_0(x)e^{-\gamma x}dxd\tau \\
&+ \sum_{n=1}^\infty\sum_{k=0}^\infty\left(\int_0^\infty \mu_1(x)e^{-\gamma x}dx\right)^k \int_0^\infty |y_{k+n+1,1}(\tau)|e^{\gamma\tau}\int_\tau^\infty \mu_1(x)e^{-\gamma x}dxd\tau \\
\le& \frac{\lambda\theta}{\gamma(\gamma+\lambda)}|y_{0,0}| \\
&+ \theta\sum_{n=1}^\infty\sum_{k=0}^\infty\left(\int_0^\infty \overline{\mu_1}e^{-\gamma x}dx\right)^k \int_0^\infty |y_{k+n,0}(\tau)|e^{\gamma\tau}\int_\tau^\infty e^{-\gamma x}dxd\tau \\
&+ \sum_{n=1}^\infty\sum_{k=0}^\infty\left(\int_0^\infty \overline{\mu_1}e^{-\gamma x}dx\right)^k \int_0^\infty |y_{k+n+1,0}(\tau)|e^{\gamma\tau}\int_\tau^\infty \overline{\mu_0}e^{-\gamma x}dxd\tau
\end{aligned}$$

$$
\begin{aligned}
&+\sum_{n=1}^{\infty}\sum_{k=0}^{\infty}\left(\int_0^{\infty}\overline{\mu_1}e^{-\gamma x}dx\right)^k\int_0^{\infty}|y_{k+n+1,1}(\tau)|e^{\gamma\tau}\int_{\tau}^{\infty}\overline{\mu_1}e^{-\gamma x}dxd\tau\\
\le&\frac{\lambda\theta}{\gamma(\gamma+\lambda)}|y_{0,0}|+\frac{\theta}{\gamma}\sum_{n=1}^{\infty}\sum_{k=0}^{\infty}\left(\frac{\overline{\mu_1}}{\gamma}\right)^k\int_0^{\infty}|y_{k+n,0}(\tau)|d\tau\\
&+\frac{\overline{\mu_0}}{\gamma}\sum_{n=1}^{\infty}\sum_{k=0}^{\infty}\left(\frac{\overline{\mu_1}}{\gamma}\right)^k\int_0^{\infty}|y_{k+n+1,0}(\tau)|d\tau\\
&+\sum_{n=1}^{\infty}\sum_{k=0}^{\infty}\left(\frac{\overline{\mu_1}}{\gamma}\right)^{k+1}\int_0^{\infty}|y_{k+n+1,1}(\tau)|d\tau\\
\le&\frac{\lambda\theta}{\gamma(\gamma+\lambda)}|y_{0,0}|+\frac{\theta}{\gamma}\sum_{k=0}^{\infty}\left(\frac{\overline{\mu_1}}{\gamma}\right)^k\sum_{n=1}^{\infty}\|y_{n,0}\|_{L^1[0,\infty)}\\
&+\frac{\overline{\mu_0}}{\gamma}\sum_{k=0}^{\infty}\left(\frac{\overline{\mu_1}}{\gamma}\right)^k\sum_{n=1}^{\infty}\|y_{n,0}\|_{L^1[0,\infty)}+\sum_{k=0}^{\infty}\left(\frac{\overline{\mu_1}}{\gamma}\right)^{k+1}\sum_{n=1}^{\infty}\|y_{n,1}\|_{L^1[0,\infty)}\\
=&\frac{\lambda\theta}{\gamma(\gamma+\lambda)}|y_{0,0}|+\frac{\theta}{\gamma}\frac{\gamma}{\gamma-\overline{\mu_1}}\sum_{n=1}^{\infty}\|y_{n,0}\|_{L^1[0,\infty)}\\
&+\frac{\overline{\mu_0}}{\gamma}\frac{\gamma}{\gamma-\overline{\mu_1}}\sum_{n=1}^{\infty}\|y_{n,0}\|_{L^1[0,\infty)}+\frac{\overline{\mu_1}}{\gamma-\overline{\mu_1}}\sum_{n=1}^{\infty}\|y_{n,1}\|_{L^1[0,\infty)}\\
=&\frac{\lambda\theta}{\gamma(\gamma+\lambda)}|y_{0,0}|+\frac{\overline{\mu_0}+\theta}{\gamma-\overline{\mu_1}}\sum_{n=1}^{\infty}\|y_{n,0}\|_{L^1[0,\infty)}\\
&+\frac{\overline{\mu_1}}{\gamma-\overline{\mu_1}}\sum_{n=1}^{\infty}\|y_{n,1}\|_{L^1[0,\infty)}.
\end{aligned}
\tag{A.15}
$$

By substituting the (A.15) into (A.14) we estimate (assume $\overline{\mu_1} > \overline{\mu_0} + \theta$ and $\gamma > \overline{\mu_1}$)

$$
\begin{aligned}
\|(p_0,p_1)\|\le&\frac{1}{\gamma}|y_{0,0}|+\frac{1}{\gamma}\sum_{n=1}^{\infty}\|y_{n,0}\|_{L^1[0,\infty)}\\
&+\frac{1}{\gamma}\left\{\frac{\lambda\theta}{\gamma(\gamma+\lambda)}|y_{0,0}|+\frac{\overline{\mu_0}+\theta}{\gamma-\overline{\mu_1}}\sum_{n=1}^{\infty}\|y_{n,0}\|_{L^1[0,\infty)}\right.\\
&\left.+\frac{\overline{\mu_1}}{\gamma-\overline{\mu_1}}\sum_{n=1}^{\infty}\|y_{n,1}\|_{L^1[0,\infty)}\right\}+\frac{1}{\gamma}\sum_{n=1}^{\infty}\|y_{n,1}\|_{L^1[0,\infty)}\\
\le&\frac{\gamma^2+\lambda\gamma+\lambda\theta}{\gamma^2(\gamma+\lambda)}|y_{0,0}|+\frac{\gamma-\overline{\mu_1}+\overline{\mu_0}+\theta}{\gamma(\gamma-\overline{\mu_1})}\sum_{n=1}^{\infty}\|y_{n,0}\|_{L^1[0,\infty)}\\
&+\frac{1}{\gamma-\overline{\mu_1}}\sum_{n=1}^{\infty}\|y_{n,1}\|_{L^1[0,\infty)}\\
\le&\frac{1}{\gamma-\overline{\mu_1}}\left\{|y_{0,0}|+\sum_{n=1}^{\infty}\|y_{n,0}\|_{L^1[0,\infty)}+\sum_{n=1}^{\infty}\|y_{n,1}\|_{L^1[0,\infty)}\right\}\\
=&\frac{1}{\gamma-\overline{\mu_1}}\|(y,z)\|.
\end{aligned}
\tag{A.16}
$$

(A.16) shows that $(\gamma I - A)^{-1}$ exist for $\gamma > \overline{\mu_1}$, and

$$(\gamma I - A)^{-1} : X \to D(A), \quad \|(\gamma I - A)^{-1}\| \le \frac{1}{\gamma-\overline{\mu_1}}.$$

In the following, we will prove that $D(A)$ is dense in X. Since $\forall (p_0, p_1) \in X$ implies

$$|p_{0,0}| + \sum_{n=1}^{\infty}\|p_{n,0}\|_{L^1[0,\infty)} + \sum_{n=1}^{\infty}\|p_{n,1}\|_{L^1[0,\infty)} < \infty$$

It follows that, for any $\epsilon > 0$, there exists a positive integer K such that such that

$$\sum_{n=K}^{\infty} \|p_{n,0}\|_{L^1[0,\infty)} + \sum_{n=K}^{\infty} \|p_{n,1}\|_{L^1[0,\infty)} < \epsilon.$$

Which shows that the set

$$L = \left\{ (p_0, p_1) \;\middle|\; \begin{array}{l} p_0(x) = (p_{0,0}, p_{1,0}(x), p_{2,0}(x), \cdots, p_{K,0}(x), 0, 0, \cdots), \\ p_1(x) = (p_{1,1}(x), p_{2,1}(x), \cdots, p_{K,1}(x), 0, 0, \cdots), \\ p_{i,0}(x), p_{i,1}(x) \in L^1[0,\infty), \quad i = 1, 2, \cdots, K, \\ K \text{ is a finite positive integer.} \end{array} \right\},$$

is dense in X. If we set

$$Z = \left\{ (p_0, p_1) \;\middle|\; \begin{array}{l} p_0(x) = (p_{0,0}, p_{1,0}(x), p_{2,0}(x), \cdots, p_{N,0}(x), 0, 0, \cdots), \\ p_1(x) = (p_{1,1}(x), p_{2,1}(x), \cdots, p_{N,1}(x), 0, 0, \cdots), \\ p_{i,0}(x), p_{i,1}(x) \in C_0^{\infty}[0,\infty), \text{ there exists } c_i > 0,\ d_i > 0 \\ \text{such that } p_{i,0}(x) = 0,\ x \in [0, c_i];\ p_{i,1}(x) = 0,\ x \in [0, d_i]; \\ i = 1, 2, \cdots, N. \end{array} \right\},$$

then from (Adams, 1975) we know that Z is dense in L. Therefore, in order to prove that $D(A)$ is dense in X, it is suffices to prove $Z \subset \overline{D(A)}$. In fact, if $Z \subset \overline{D(A)}$, then $X = \overline{L} = \overline{\overline{Z}} = \overline{Z} \subset \overline{\overline{D(A)}} = \overline{D(A)} \subset X$ implies $X = \overline{D(A)}$.

Take any $(p_0, p_1) \in Z$, there are a finite positive integer N and positive numbers $c_i > 0,\ d_i > 0$ such that, for $i = 1, 2, \cdots N$

$$\begin{aligned} p_0(x) =& (p_{0,0}, p_{1,0}(x), p_{2,0}(x), \cdots, p_{N,0}(x), 0, 0, \cdots),\ p_{i,0}(x) = 0,\ x \in [0, c_i], \\ p_1(x) =& (p_{1,1}(x), p_{2,1}(x), \cdots, p_{N,1}(x), 0, 0, \cdots),\ p_{i,1}(x) = 0,\ x \in [0, d_i], \end{aligned}$$

This implies $p_{i,0}(x) = 0,\ p_{i,1}(x) = 0$ for $x \in [0, 2s]$ where $0 < 2s < \min\{c_0, c_1, \cdots, c_N, d_0, d_1, \cdots d_N\}$. Define

$$(f_0^s, f_1^s)(0) = \left(\begin{pmatrix} p_{0,0} \\ \lambda p_{0,0} \\ 0 \\ \vdots \\ 0 \\ 0 \\ 0 \\ \vdots \end{pmatrix}, \begin{pmatrix} \theta \int_{2s}^{\infty} p_{1,0}(x)dx + \int_{2s}^{\infty} \mu_0(x) p_{2,1}(x)dx + \int_{2s}^{\infty} \mu_1(x) p_{2,1}(x)dx \\ \theta \int_{2s}^{\infty} p_{2,0}(x)dx + \int_{2s}^{\infty} \mu_0(x) p_{3,1}(x)dx + \int_{2s}^{\infty} \mu_1(x) p_{3,1}(x)dx \\ \vdots \\ \theta \int_{2s}^{\infty} p_{N-1,0}(x)dx + \int_{2s}^{\infty} \mu_0(x) p_{N,1}(x)dx + \int_{2s}^{\infty} \mu_1(x) p_{N,1}(x)dx \\ 0 \\ 0 \\ \vdots \end{pmatrix} \right),$$

$$(f_0^s, f_1^s)(x) = \left(\begin{pmatrix} p_{0,0} \\ f_{1,0}^s(x) \\ f_{2,0}^s(x) \\ \vdots \\ f_{N-1,0}^s(x) \\ f_{N,0}^s(x) \\ 0 \\ 0 \\ \vdots \end{pmatrix}, \begin{pmatrix} f_{1,1}^s(x) \\ f_{2,1}^s(x) \\ f_{3,1}^s(x) \\ \vdots \\ f_{N-1,1}^s(x) \\ f_{N,1}^s(x) \\ 0 \\ 0 \\ \vdots \end{pmatrix} \right),$$

here

$$f_{i,0}^s(x)=\begin{cases} f_{i,0}^s(0)(1-\frac{x}{s})^2 & x\in[0,s),\\ -u_i(x-s)^2(x-2s)^2 & x\in[s,2s),\\ p_{i,0}(x) & x\in[2s,\infty).\end{cases}\qquad f_{N,0}^s(x)=p_{N,0}(x),$$

$$u_1=\frac{\int_0^s f_{i,0}^s(0)(1-\frac{x}{s})^2dx}{\int_s^{2s}(x-s)^2(x-2s)^2dx},$$

$$u_i=\frac{\int_0^s f_{i,0}^s(0)(1-\frac{x}{s})^2\mu_0(x)dx}{\int_s^{2s}(x-s)^2(x-2s)^2\mu_0(x)dx},\qquad i=2,3,\cdots,N;$$

$$f_{i,1}^s(x)=\begin{cases} f_{i,1}^s(0)(1-\frac{x}{s})^2 & x\in[0,s),\\ -v_i(x-s)^2(x-2s)^2 & x\in[s,2s),\\ p_{i,1}(x) & x\in[2s,\infty).\end{cases}\qquad f_{N,1}^s(x)=p_{N,1}(x),$$

$$\begin{aligned} v_i=&\frac{1}{\int_s^{2s}(x-s)^2(x-2s)^2\mu_1(x)dx}\Big\{\int_0^s f_{i,1}^s(0)(1-\frac{x}{s})^2\mu_1(x)dx\\ &+\int_0^s f_{i-1,0}^s(0)(1-\frac{x}{s})^2\mu_0(x)dx-\theta u_{i-1}\int_s^{2s}(x-s)^2(x-2s)^2\mu_0(x)dx\\ &+\int_0^s f_{i-1,0}^s(0)(1-\frac{x}{s})^2dx-\theta u_{i-1}\int_s^{2s}(x-s)^2(x-2s)^2dx\Big\},\qquad i=2,3\cdots,N.\end{aligned}$$

It is easy to verify that $(f_0^s,f_1^s)\in D(A)$. Moreover

$$\begin{aligned} \|(p_0,p_1)-(f_0^s,f_1^s)\|=&\sum_{n=1}^N\int_0^\infty|p_{n,0}(x)-f_{n,0}^s(x)|dx+\sum_{n=2}^N\int_0^\infty|p_{n,1}(x)-f_{n,1}^s(x)|dx\\ =&\sum_{n=1}^N\int_0^s|f_{n,0}^s(0)|(1-\frac{x}{s})^2dx+\sum_{n=1}^N\int_s^{2s}|u_n|(x-s)^2(x-2s)^2dx\\ &+\sum_{n=2}^N\int_0^s|f_{n,1}^s(0)|(1-\frac{x}{s})^2dx+\sum_{n=2}^N\int_s^{2s}|v_n|(x-s)^2(x-2s)^2dx\\ =&\sum_{n=1}^N|f_{n,0}^s(0)|\frac{s}{3}+\sum_{n=1}^N|u_n|\frac{s^5}{30}+\sum_{n=2}^N|f_{n,1}^s(0)|\frac{s}{3}+\sum_{n=2}^N|v_n|\frac{s^5}{30}\\ &\to0,\quad\text{as } s\to0.\end{aligned}$$

This means that $D(A)$ is dense in X. From the above two steps and Hille-Yosida Theorem we conclude that A generates a C_0- semigroup. see (Gupur, Li, & Zhu, 2001). Now, we verify that U and E are bounded linear operators. From the definition of U and E we have, for $(p_0,p_1)\in X$

$$\begin{aligned} \|U(p_0,p_1)\|\le&\sum_{n=1}^\infty\int_0^\infty|(\lambda+\theta+\mu_0(x))p_{n,0}(x)|dx+\sum_{n=2}^\infty\int_0^\infty|\lambda p_{n,0}(x)|dx\\ &+\sum_{n=1}^\infty\int_0^\infty|(\lambda+\mu_1(x))p_{n,1}(x)|dx+\sum_{n=2}^\infty\int_0^\infty|\lambda p_{n,1}(x)|dx\\ \le&(2\lambda+\theta+\overline{\mu_0})\sum_{n=1}^\infty\int_0^\infty|p_{n,0}(x)|dx+(2\lambda+\overline{\mu_1})\sum_{n=1}^\infty\int_0^\infty|p_{n,1}(x)|dx\\ \le&\max\{2\lambda+\theta+\overline{\mu_0},2\lambda+\overline{\mu_1}\}\|(p_0,p_1)\|,\end{aligned}\tag{A.17}$$

$$\begin{aligned} \|E(p_0,p_1)\|\le&\overline{\mu_0}\int_0^\infty|p_{1,0}(x)|dx+\overline{\mu_1}\int_0^\infty|p_{1,1}(x)|dx\\ \le&\max\{\overline{\mu_0},\overline{\mu_1}\}\|(p_0,p_1)\|.\end{aligned}\tag{A.18}$$

(A.17) and (A.18) show that U and E are bounded linear operators. It is easy to see that U and E are linear operators. Hence, From the perturbation theory of C_0-semigroup we get that $A+U+E$ generates a C_0- semigroup $T(t)$.

Finally, we prove that $A+U+E$ is dispersive operator. For $(p_0, p_1) \in D(A)$, we choose

$$\phi_j(x) = \left(\frac{[p_{0,j}]^+}{p_{0,j}}, \frac{[p_{1,j}(x)]^+}{p_{1,j}(x)}, \frac{[p_{2,j}(x)]^+}{p_{2,j}(x)}, \cdots \right), \quad j = 0, 1.$$

Where

$$[p_{0,0}]^+ = \begin{cases} p_{0,0} & \text{if} \quad p_{0,0} > 0 \\ 0 & \text{if} \quad p_{0,0} \le 0 \end{cases}, \quad [p_{n,j}(x)]^+ = \begin{cases} p_{n,j}(x) & \text{if} \quad p_{n,j}(x) > 0 \\ 0 & \text{if} \quad p_{n,j}(x) \le 0 \end{cases}, \quad n \ge 1, \ j = 0, 1.$$

The boundary condition on $(p_0, p_1) \in D(A)$ imply

$$\sum_{n=1}^{\infty} [p_{n,0}(0)]^+ \le \lambda [p_{0,0}]^+, \tag{A.19}$$

$$\sum_{n=1}^{\infty} [p_{n,1}(0)]^+ \le \theta \sum_{n=1}^{\infty} \int_0^{\infty} [p_{n,0}(x)]^+ dx + \sum_{n=1}^{\infty} \int_0^{\infty} \mu_0(x) [p_{n+1,0}(x)]^+ dx + \sum_{n=1}^{\infty} \int_0^{\infty} \mu_1(x) [p_{n+1,1}(x)]^+ dx. \tag{A.20}$$

If we define $V_{i,j} = \{x \in [0,\infty) \big| p_{i,j}(x) > 0\}$ and $W_{i,j} = \{x \in [0,\infty) \big| p_{i,j}(x) \le 0\}$ for $i \ge 1, \ j = 0, 1$, then we have

$$\begin{aligned} \int_0^{\infty} \frac{dp_{i,j}(x)}{dx} \frac{[p_{i,j}(x)]^+}{p_{i,j}(x)} dx &= \int_{V_{i,j}} \frac{dp_{i,j}(x)}{dx} \frac{[p_{i,j}(x)]^+}{p_{i,j}(x)} dx + \int_{W_{i,j}} \frac{dp_{i,j}(x)}{dx} \frac{[p_{i,j}(x)]^+}{p_{i,j}(x)} dx \\ &= \int_{V_{i,j}} \frac{dp_{i,j}(x)}{dx} \frac{[p_{i,j}(x)]^+}{p_{i,j}(x)} dx = \int_{V_{i,j}} \frac{dp_{i,j}(x)}{dx} dx \\ &= \int_{V_{i,j}} \frac{d[p_{i,j}(x)]^+}{dx} dx = -[p_{i,j}(0)]^+, \quad i \ge 1, \ j = 0, 1. \end{aligned} \tag{A.21}$$

By using boundary condition on $(p_0, p_1) \in D(A)$ and (A.19)-(A.21) for such (ϕ_0, ϕ_1), we derive

$$\begin{aligned} &\langle (A+U+E)(p_0, p_1), (\phi_0, \phi_1) \rangle \\ &= \left\{ -\lambda p_{0,0} + \int_0^{\infty} \mu_0(x) p_{1,0}(x) dx + \int_0^{\infty} \mu_1(x) p_{1,1}(x) dx \right\} \frac{[p_{0,0}]^+}{p_{0,0}} \\ &+ \int_0^{\infty} \left\{ -\frac{dp_{1,0}(x)}{dx} - (\lambda + \theta + \mu_0(x)) p_{1,0}(x) \right\} \frac{[p_{1,0}(x)]^+}{p_{1,0}(x)} dx \\ &+ \sum_{n=2}^{\infty} \int_0^{\infty} \left\{ -\frac{dp_{n,0}(x)}{dx} - (\lambda + \theta + \mu_0(x)) p_{n,0}(x) + \lambda p_{n-1,0}(x) \right\} \frac{[p_{n,0}(x)]^+}{p_{n,0}(x)} dx \\ &+ \int_0^{\infty} \left\{ -\frac{dp_{1,1}(x)}{dx} - (\lambda + \mu_1(x)) p_{1,1}(x) \right\} \frac{[p_{1,1}(x)]^+}{p_{1,1}(x)} dx \\ &+ \sum_{n=2}^{\infty} \int_0^{\infty} \left\{ -\frac{dp_{n,1}(x)}{dx} - (\lambda + \mu_1(x)) p_{n,1}(x) + \lambda p_{n-1,1}(x) \right\} \frac{[p_{n,1}(x)]^+}{p_{n,1}(x)} dx \\ &= -\lambda [p_{0,0}]^+ + \left\{ \int_0^{\infty} \mu_0(x) p_{1,0}(x) dx + \int_0^{\infty} \mu_1(x) p_{1,1}(x) dx \right\} \frac{[p_{0,0}]^+}{p_{0,0}} \\ &- \sum_{n=1}^{\infty} \int_0^{\infty} \frac{dp_{n,0}(x)}{dx} \frac{[p_{n,0}(x)]^+}{p_{n,0}(x)} dx - \sum_{n=1}^{\infty} \int_0^{\infty} (\lambda + \theta + \mu_0(x)) [p_{n,0}(x)]^+ dx \\ &+ \lambda \sum_{n=2}^{\infty} \int_0^{\infty} p_{n-1,0}(x) \frac{[p_{n,0}(x)]^+}{p_{n,0}(x)} dx - \sum_{n=1}^{\infty} \int_0^{\infty} \frac{dp_{n,1}(x)}{dx} \frac{[p_{n,1}(x)]^+}{p_{n,1}(x)} dx \\ &- \sum_{n=1}^{\infty} \int_0^{\infty} (\lambda + \mu_1(x)) [p_{n,1}(x)]^+ dx + \lambda \sum_{n=2}^{\infty} \int_0^{\infty} p_{n-1,1}(x) \frac{[p_{n,1}(x)]^+}{p_{n,1}(x)} dx \end{aligned}$$

$$
\begin{aligned}
&= -\lambda[p_{0,0}]^+ + \left\{\int_0^\infty \mu_0(x)p_{1,0}(x)dx + \int_0^\infty \mu_1(x)p_{1,1}(x)dx\right\}\frac{[p_{0,0}]^+}{p_{0,0}} \\
&\quad + \sum_{n=1}^\infty [p_{n,0}(0)]^+ - \sum_{n=1}^\infty \int_0^\infty (\lambda+\theta+\mu_0(x))[p_{n,0}(x)]^+dx \\
&\quad + \lambda\sum_{n=2}^\infty \int_0^\infty p_{n-1,0}(x)\frac{[p_{n,0}(x)]^+}{p_{n,0}(x)}dx + \sum_{n=1}^\infty [p_{n,1}(0)]^+ \\
&\quad - \sum_{n=1}^\infty \int_0^\infty (\lambda+\mu_1(x))[p_{n,1}(x)]^+dx + \lambda\sum_{n=2}^\infty \int_0^\infty p_{n-1,1}(x)\frac{[p_{n,1}(x)]^+}{p_{n,1}(x)}dx \\
&\le -\lambda[p_{0,0}]^+ + \left\{\int_0^\infty \mu_0(x)[p_{1,0}(x)]^+dx + \int_0^\infty \mu_1(x)[p_{1,1}(x)]^+dx\right\}\frac{[p_{0,0}]^+}{p_{0,0}} \\
&\quad + \lambda[p_{0,0}]^+ - \sum_{n=1}^\infty \int_0^\infty [\lambda+\theta+\mu_0(x)][p_{n,0}(x)]^+dx \\
&\quad + \lambda\sum_{n=2}^\infty \int_0^\infty [p_{n-1,0}(x)]^+\frac{[p_{n,0}(x)]^+}{p_{n,0}(x)}dx + \theta\sum_{n=1}^\infty \int_0^\infty [p_{n,0}(x)]^+dx \\
&\quad + \sum_{n=1}^\infty \int_0^\infty \mu_0(x)[p_{n+1,0}(x)]^+dx + \sum_{n=1}^\infty \int_0^\infty \mu_1(x)[p_{n+1,1}(x)]^+dx \\
&\quad - \sum_{n=1}^\infty \int_0^\infty (\lambda+\mu_1(x))[p_{n,1}(x)]^+dx + \lambda\sum_{n=2}^\infty \int_0^\infty [p_{n-1,1}(x)]^+\frac{[p_{n,1}(x)]^+}{p_{n,1}(x)}dx \\
&= \left\{\int_0^\infty \mu_0(x)[p_{1,0}(x)]^+dx + \int_0^\infty \mu_1(x)[p_{1,1}(x)]^+dx\right\}\frac{[p_{0,0}]^+}{p_{0,0}} \\
&\quad - \int_0^\infty \mu_0(x)[p_{1,0}(x)]^+dx - \lambda\sum_{n=1}^\infty \int_0^\infty [p_{n,0}(x)]^+dx \\
&\quad + \lambda\sum_{n=2}^\infty \int_0^\infty [p_{n-1,0}(x)]^+\frac{[p_{n,0}(x)]^+}{p_{n,0}(x)}dx - \int_0^\infty \mu_1(x)[p_{1,1}(x)]^+dx \\
&\quad - \lambda\sum_{n=1}^\infty \int_0^\infty [p_{n,1}(x)]^+dx + \lambda\sum_{n=2}^\infty \int_0^\infty [p_{n-1,1}(x)]^+\frac{[p_{n,1}(x)]^+}{p_{n,1}(x)}dx \\
&\le \left(\frac{[p_{0,0}]^+}{p_{0,0}} - 1\right)\int_0^\infty \mu_0(x)[p_{1,0}(x)]^+dx \\
&\quad + \frac{[p_{0,0}]^+}{p_{0,0}}\int_0^\infty \mu_1(x)[p_{1,1}(x)]^+dx \\
&\le 0.
\end{aligned}
\tag{A.22}
$$

In the above, we have used the following inequalities:

$$
\begin{aligned}
\int_0^\infty p_{n-1,j}(x)\frac{[p_{n,j}(x)]^+}{p_{n,j}(x)}dx &\le \int_0^\infty [p_{n-1,j}(x)]^+\frac{[p_{n,j}(x)]^+}{p_{n,j}(x)}dx \\
&\le \int_0^\infty [p_{n-1,j}(x)]^+dx, \quad n\ge 1,\ j=0,1.
\end{aligned}
$$

(A.22) shows that $A+U+E$ is a dispersive operator.

From the first step, the second step, the fourth step and Fillips theorem we obtained that $A+U+E$ generates a positive contraction C_0- semigroup. By the uniqueness theorem of the semigroup it follows that this positive contraction C_0-semigroup is just $T(t)$. Thus, we complete the Theorem 2.1. □

References

Adams, R. A. (1975). *Sobolev spaces.* Academic Press, New York.

Doshi, B. T. (1986). Queueing systems with vacations - a survey. *Queueing systems, 1*(1), 29-66. http://dx.doi.org/10.1007/BF01149327

Fattorini, H. O. (1983). *The Cauchy Problem.* Addison-Wesley, Massachusetts.

Gupur, G. (2002). Well-posedness of M/G/1 queueing model with single vacations. *Computers & Mathematics with Applications, 44*(8), 1041-1056. http://dx.doi.org/10.1016/S0898-1221(02)00213-4

Gupur, G. (2010). On the M/M/1 queueing model with compulsory server vacations. *International journal of pure and applied mathematics, 64*(2), 253-304.

Gupur, G. (2011). *Functional analysis methods for reliability models.* Springer, Basel.

Gupur, G. (2014). On eigenvalues of the generator of a C_0-semigroup appearing in queueing theory. *Abstract and Applied Analysis*, Art. ID 896342, 9 pp. http://dx.doi.org/10.1155/2014/896342

Gupur, G., & Guo, B. Z. (2002). Asymptotic property of the solution of exhaustive-service M/M/1 queueing model with single vacations. *International Journal of Differential Equations and Applications, 1*, 29-51.

Gupur, G., Li, X. Z., & Zhu, G. T. (2001). *Functional Analysis Method in Queueing Theory.* Research Information Ltd., Hertfordshire.

Jain, M., & Agrawal, P. K. (2007). $M/E^k/1$ queueing system with working vacation. *Quality Technology & Quantitative Management, 4*(4), 455-470. http://dx.doi.org/10.1080/16843703.2007.11673165

Kim, J. D. Choi, D. W., & Chae, K. C. (2003) *Analysis of queue-length distribution of the M/G/1 queue with working vacations.* in: Hawaii International Conference on Statistics and Related Fields.

Kasim, E., & Gupur, G. (2011). Other eigenvalues of the $M/M/1$ *operator. Acta Analysis Functionalis Applicata, 13*, 45-53.

Li, J. H., Tian, N. S., & Ma, Z. Y. (2007). Performance analysis of GI/M/1 queue with working vacations and vacation interruption. *Applied Mathematical Modelling, 31*, 880-894. http://dx.doi.org/10.1016/j.apm.2007.09.017

Lu, Z. J., & Gupur, G. (2010). Well-posedness of M/G/1 queueing model with compulsory server vacations. *Mathematics in Theory and Practice, 40*(5), 139-148.

Madan, K. C. (1992). An M/G/1 queueing system with compulsory server vacations. *Trabajos de investigación operativa, 7*(1), 105-115. http://dx.doi.org/10.1007/BF02888261

Servi, L. D., & Finn, S. G. (2002). M/M/1 queues with working vacations (M/M/1/WV). *Perform. Eval., 50*(1), 41-52. http://dx.doi.org/10.1016/S0166-5316(02)00057-3

Takagi, H. (1990). Time-dependent analysis of M/G/1 vacation models with exhaustive service. *Queueing Systems, 6*, 369-389. http://dx.doi.org/10.1007/BF02411484

Takagi, H., & Wu, D. (2006). M/G/1 queue with multiple working vacations. *Perform. Eval., 63*, 654-681. http://dx.doi.org/10.1016/j.peva.2005.05.005

Zhang, M., & Hou, Z. (2010). Performance analysis of M/G/1 queue with working vacation and vacation interruption. *Journal of Computational and Applied Mathematics, 234*(10), 2977-2985. http://dx.doi.org/10.1016/j.cam.2010.04.010

5

Mathematical Formulation of Laminated Composite Thick Conical Shells

Mohammad Zannon[1], Hussam Alrabaiah[1]

[1]Department of Mathematics, Tafila Technical University, Tafila, Jordan

Correspondence: Mohammad Zannon, Department of Mathematics, Tafila Technical University, Tafila, Jordan. E-mail: zanno1ms@gmail.com

Abstract

The mathematical formulation of thick conical shells using third order shear deformation of thick shell theory are presented. The equations of motion are obtained using Hamilton's principle. For present analysis, we consider shell's system transverse normal stress, rotary inertia and shear deformation.

Keywords: Conical Shell, free vibrations, third order shear deformation thick shell theory, Equations of Motion, Lame' parameters.

1. Introduction

Many research articles investigated the theory of thick conical shells including (Qatu *et al.*, 2013; Qatu et *al.*, 2010). The vibration behavior of cantilevered laminated composite shallow conical shells were also explored recently (Korjakin et *al.*; Qatu et *al.*, 2010; Mukhopadhyay et *al.*, 2016). (Further, Reddy, 1984) explored vibrations of joined conical shells. Qatu (1994) carried out research about the steady-state torsional oscillations of multilayer truncated cones. The damping of the free vibrations of laminated composite conical shells was investigated by Qatu (1994) and while stiffened conical shells were studies by Reddy, (1984). A variable thickness of composite conical shells was discussed (Asadi & Qatu, 2012; Leissa & Qatu, 2011). Damping in multilayered conical shells was investigated by many researchers (Zannon and Qatu, 2014b). Pre-stressed conical shells were explored thoroughly by Qatu et *al.* (2014, 2015).

Towards this end, in this paper, we propose our contribution towards the mathematical theory of third order shear deformation thick conical shell theory (see Zannon et *al.* (TSDTZ); Qatu *et al.*, 2013) and its stress-strain deformation at the mid thick conical shell surface (Duc & Cong, 2015; Akbari et *al.*, 2015; Jam & Kiani, 2015, Viola et *al.*, 2016).

2. Mathematical Formulation of Conical Shell

The displacement components using the third-order shear deformation shell theory are given in (Asadi & Qatu, 2012; Leissa & Qatu, 2011). Conical shells are one form of engineering solids that are formed by revolving two non-paralleled lines, mostly a line and axis of revolution. We are interested mainly in a particular type of shells which have a circular cross-section (Qatu, 1994).

A closed conical shell with circular sides (Figure 2.1) has a closed shape and the open conical shell can be obtained by cutting the sides of the solid between $\theta_1 \; and \; \theta_2$ (Roh et *al*, 2008). An open conical shell with sides less than the half of the radius of the curvature then the solid is shallow (Qatu et *al.*, 2010; Dung et *al.*, 2014).

A typical fundamental equation of such solid can be written as (with the help of Lame parameters) (Qatu et *al.*, 2010; Dung et *al.*, 2014; Jam & Kiani, 2015; Akbari et *al.*, 2015)

$$(ds)^2 = (d\alpha)^2 + \alpha^2 \sin^2(\varphi) d\theta,$$
$$A = 1; \; B = \alpha \sin(\varphi),$$
$$R_\alpha = \infty; \; R_\beta = \alpha \tan(\varphi).$$

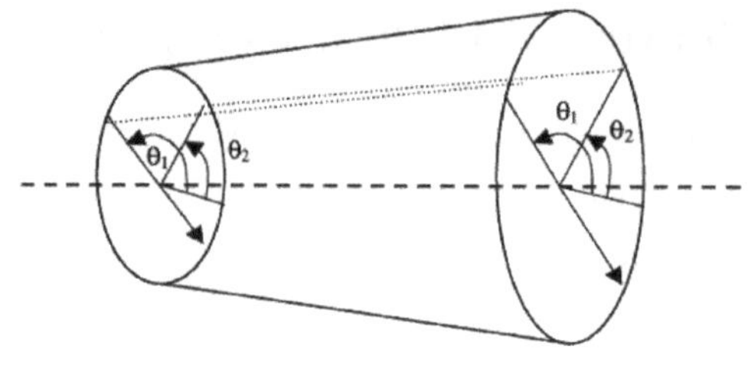

Figure 2.1 A closed coincal shell (Qatu, 1994)

Consider Figure 2.2, which is a side view of the closed conical shell described in Figure 2.1.

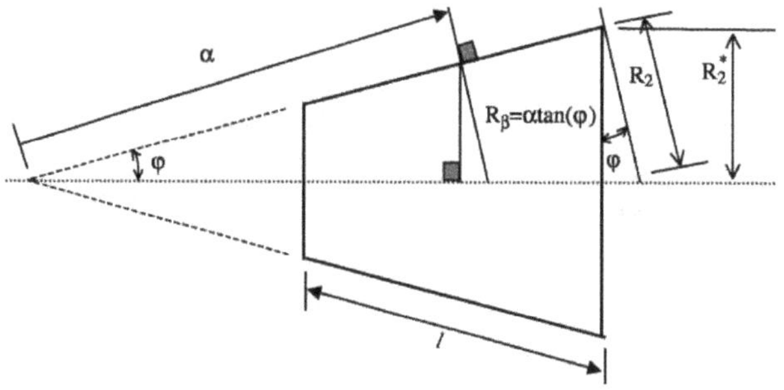

Figure 2.2 A a side view of the closed coincal shell (Qatu, 1994)

3. Equilibrium Equations of Motion

A conical laminated shell with Lame parameters is considered above. The Lame' parameters of middle surface are substituted in moment and force resultants (see Zannon et al, 2015) to formulate the conical shell equations for TSDTZ. Therefore, the strain-displacement equations and middle surface strains are obtained (Qatu et *al.*, 2010; Dung et *al.*, 2014; Jam & Kiani, 2015; Akbari et *al.*, 2015)

$$\varepsilon_{\alpha} = \varepsilon_{0\alpha} + z\,\kappa_{\alpha}^{(1)} + z^{2}\kappa_{\alpha}^{(2)},$$

$$\varepsilon_{\theta} = \frac{1}{(1+ z/\alpha\tan(\varphi))}(\varepsilon_{0\theta} + z\,\kappa_{\theta}^{(1)} + z^{2}\kappa_{\theta}^{(2)}),$$

$$\varepsilon_{z} = \psi_{z}(\alpha,\beta) \neq 0,$$

$$\varepsilon_{\alpha\theta} = \varepsilon_{0\alpha\theta} + z\,\kappa_{\alpha\theta}^{(1)} + z^{2}\kappa_{\alpha\theta}^{(2)},$$

$$\varepsilon_{\theta\alpha} = \frac{1}{(1+ z/\alpha\tan(\varphi))}(\varepsilon_{0\theta\alpha} + z\,\kappa_{\theta\alpha}^{(1)} + z^{2}\kappa_{\theta\alpha}^{(2)}),$$

$$\gamma_{\alpha z} = \gamma_{0\alpha z} + zG^{(1)} + z^{2}G^{(2)},$$

$$\gamma_{\theta z} = \frac{1}{(1+ z/\alpha\tan(\varphi))}(\gamma_{0\theta z} + z\,E^{(1)} + z^{2}E^{(2)}),$$

$$\varepsilon_{0\alpha} = \frac{\partial u_{0}}{\partial \alpha}$$

$$\varepsilon_{0\theta} = \frac{1}{\alpha\sin(\varphi)}\frac{\partial v_{0}}{\partial \theta} + \frac{u_{0}}{\alpha} + \frac{w_{0}}{\alpha\tan(\varphi)},$$

$$\varepsilon_{0\alpha\theta} = \frac{\partial v_{0}}{\partial \alpha},$$

$$\varepsilon_{0\theta\alpha} = \frac{1}{\alpha\sin(\varphi)}\frac{\partial u_{0}}{\partial \theta} - \frac{v_{0}}{\alpha},$$

$$\gamma_{0\alpha z} = \frac{\partial w_{0}}{\partial \alpha} + \psi_{\alpha},$$

$$\gamma_{0\theta z} = \frac{1}{\alpha\sin(\varphi)}\frac{\partial w_{0}}{\partial \theta} - \frac{v_{0}}{\alpha\tan(\varphi)} + \psi_{\beta}.$$

The curvature and twist of the conical thick shells are considered

$$\kappa_\alpha^{(1)} = \frac{\partial \psi_\alpha}{\partial \alpha},$$
$$\kappa_\theta^{(1)} = \frac{\partial \psi_\beta}{\partial \theta} + \frac{\psi_\alpha}{\alpha} + \frac{\psi_z}{\alpha \tan(\varphi)},$$
$$\kappa_{\alpha\theta}^{(1)} = \frac{\partial \psi_\beta}{\partial \alpha},$$
$$\kappa_{\theta\alpha}^{(1)} = \frac{1}{\alpha \sin(\varphi)} \frac{\partial \psi_\alpha}{\partial \theta} - \frac{\psi_\beta}{\alpha}.$$

$$\kappa_\alpha^{(2)} = \frac{\partial \phi_\alpha}{\partial \alpha},$$
$$\kappa_\theta^{(2)} = \frac{\partial \phi_\beta}{\partial \theta} + \frac{\phi_\alpha}{\alpha},$$
$$\kappa_{\alpha\theta}^{(2)} = \frac{\partial \phi_\beta}{\partial \alpha},$$
$$\kappa_{\theta\alpha}^{(2)} = \frac{1}{\alpha \sin(\varphi)} \frac{\partial \phi_\alpha}{\partial \theta} - \frac{\phi_\beta}{\alpha}.$$

Therefore, the equations of motion are (Qatu et *al.*, 2013; Zannon & Qatu, 2014b; Qatu et *al.*, 2010; Asadi & Qatu, 2012; Jam & Kiani, 2015; Duc & Cong, 2015; Akbari et *al.*, 2015)

$$\frac{\partial}{\partial \alpha}(\alpha \sin(\varphi) N_\alpha) - \sin(\varphi) N_\theta + \frac{\partial}{\partial \theta}(N_{\theta\alpha}) + \alpha \sin(\varphi) q_\alpha = \alpha \sin(\varphi)(\bar{I}_1 \ddot{u}_0 + \bar{I}_2 \ddot{\psi}_\alpha),$$

$$\frac{\partial}{\partial \alpha}(\alpha \sin(\varphi) N_{\alpha\theta}) + \sin(\varphi) N_{\theta\alpha} + \frac{\partial}{\partial \theta}(N_\theta) + \frac{\sin(\varphi)}{\tan(\varphi)} Q_\theta + \alpha \sin(\varphi) q_\theta = \alpha \sin(\varphi)(\bar{I}_1 \ddot{v}_0 + \bar{I}_2 \ddot{\psi}_\theta),$$

$$\frac{\partial}{\partial \alpha}(\alpha \sin(\varphi) Q_\alpha) + \frac{\partial}{\partial \theta} Q_\theta - \alpha \sin(\varphi)\left(\frac{N_\theta}{\alpha \tan(\varphi)}\right) + \alpha \sin(\varphi) q_n = \alpha \sin(\varphi)(\bar{I}_1 \ddot{w}_0),$$

$$\frac{\partial}{\partial \alpha}(\alpha \sin(\varphi) M_\alpha^{(1)}) - \sin(\varphi) M_\theta^{(1)} + \frac{\partial}{\partial \theta}(M_{\theta\alpha}^{(1)}) - \alpha \sin(\varphi) Q_\alpha + \alpha \sin(\varphi) m_\alpha^{(1)} = \alpha \sin(\varphi)(\bar{I}_2 \ddot{u}_0 + \bar{I}_3 \ddot{\psi}_\alpha),$$

$$\frac{\partial}{\partial \theta}(M_\theta^{(1)}) + \frac{\partial}{\partial \alpha}(\alpha \sin(\varphi) M_{\alpha\theta}^{(1)}) + \sin(\varphi) M_{\theta\alpha}^{(1)} - \alpha \sin(\varphi) Q_\theta + \alpha \sin(\varphi) m_\theta^{(1)} = \alpha \sin(\varphi)(\bar{I}_2 \ddot{v}_0 + \bar{I}_3 \ddot{\psi}_\theta),$$

$$\frac{\partial}{\partial \alpha}(\alpha \sin(\varphi) P_\alpha^{(1)}) + \frac{\partial}{\partial \theta}(P_\theta^{(1)}) - \alpha \sin(\varphi)\left(N_z + \frac{M_\theta^{(1)}}{\alpha \tan(\varphi)}\right) + \alpha \sin(\varphi) m_z = \alpha \sin(\varphi)(\bar{I}_3 \ddot{\psi}_z),$$

$$\frac{\partial}{\partial \alpha}(\alpha \sin(\varphi) M_\alpha^{(2)}) - \sin(\varphi) M_\theta^{(2)} + \frac{\partial}{\partial \theta}(M_{\theta\alpha}^{(2)}) - 2\alpha \sin(\varphi) P_\alpha^{(1)} + \alpha \sin(\varphi) m_\alpha^{(2)} = \alpha \sin(\varphi)(\bar{I}_3 \ddot{u}_0 + \bar{I}_4 \ddot{\varphi}_\alpha),$$

$$\frac{\partial}{\partial \theta}(M_\theta^{(2)}) + \frac{\partial}{\partial \alpha}(\alpha \sin(\varphi) M_{\alpha\theta}^{(2)}) - \left(\frac{\sin(\varphi)}{\tan(\varphi)} P_\theta^{(2)} + 2\alpha \sin(\varphi) P_\theta^{(1)}\right) + \alpha \sin(\varphi) m_\theta^{(2)} = \alpha \sin(\varphi)(\bar{I}_3 \ddot{v}_0 + \bar{I}_4 \ddot{\varphi}_\theta).$$

The boundary conditions are given (see Zannon et *al.*, 2015; Qatu et *al.*, 2013)

4. Mathematical Analysis

One cannot find an exact solution for a general lamination structure shell with general boundary conditions and/or lamination having series of sequence and layers (Qatu et *al.*, 2013; Zannon et *al.*, 2015; Qatu, 1994). Many researchers talked about the vibration of shells as in Leissa & Qatu, (2011), she considered a thin plate in her paper "vibration of shells". One can be permitted to obtain a fundamental frequency with good accuracy as in Qatu et *al.*, (2013) by using the classical thin plate (CPT), now using the shear deformation plate theories (SDPTs) can largely eliminate the inaccuracies. Later Qatu et *al.*, (2010) and Reddy (1994) developed this subject, Leissa & Qatu (2011) studied the exact solutions "solutions which satisfy both the equations of motion, and boundary conditions" for simply supported cross-ply thick shell.

The partial differential equations of motions can be found from Qatu et *al.* (2013) and their solution forms can be found in many sources (Qatu et *al.*, 2013; Zannon & Qatu, 2014b; Duc & Cong, 2015; Akbari et *al.*, 2015). Substituting the solution forms in (Zannon & Qatu, 2014a; Qatu et *al.*, 2010) we give a system of equations, rewrite the coefficient as an eigenvalue problem (Qatu et *al.*, 2013; Qatu et *al.*, 2010; Reddy, 1994), hence we get the form $([Z]-\lambda\,[N])\{\Delta\}=\{F(t)\}$ where $\lambda=\omega^2$, ω is the natural frequency and $\{\Delta\}$ is the displacement vector (Qatu et *al.*, 2013).The structural stiffness parameters $\{Z_{ij}\}$ of the thick conical shell are following:

$$Z_{11}=-\bar{A}_{11}\cdot A^{*^2}-\hat{A}_{66}\cdot B^{*^2},\ Z_{12}=-A_{12}\cdot A^*\cdot B^*-A_{66}\cdot A^*\cdot B^*,$$

$$Z_{13}=\frac{A_{12}\cdot A^*}{\alpha\tan(\varphi)},\quad Z_{14}=-\bar{B}_{11}\cdot A^{*^2}-\hat{B}_{66}\cdot B^{*^2},$$

$$Z_{15}=-B_{12}\cdot A^*\cdot B^*-B_{66}\cdot A^*\cdot B^*,\ Z_{16}=A_{13}\cdot A^*+\frac{B_{12}\cdot A^*}{\alpha\tan(\varphi)},$$

$$Z_{17}=-\bar{D}_{11}\cdot A^{*^2}-\hat{D}_{66}\cdot B^{*^2},Z_{18}=-D_{12}\cdot A^*\cdot B^*-D_{66}\cdot A^*\cdot B^*,$$

$$Z_{21}=-A_{12}\cdot A^*\cdot B^*-A_{66}\cdot A^*\cdot B^*,$$

$$Z_{22}=-\frac{\hat{A}_{22}\cdot B^{*2}}{\alpha\tan(\varphi)}-\bar{A}_{66}\cdot A^{*^2}+\frac{\hat{A}_{44}}{(\alpha\tan(\varphi))^2},\quad Z_{23}=\hat{A}_{22}\cdot B^*,$$

$$Z_{24}=-B_{12}\cdot A^*\cdot B^*-B_{66}\cdot A^*\cdot B^*,$$

$$Z_{25}=-\hat{B}_{22}\cdot B^{*^2}-\bar{B}_{66}\cdot A^{*^2}+\frac{\hat{A}_{44}}{\alpha\tan(\varphi)},$$

$$Z_{26}=A_{23}\cdot B^*+\frac{\hat{B}_{22}\cdot B^*}{\alpha\tan(\varphi)}+\frac{\hat{B}_{44}\cdot B^*}{\alpha\tan(\varphi)},Z_{27}=-D_{12}\cdot A^*\cdot B^*-D_{66}\cdot A^*\cdot B^*,$$

$$Z_{28}=-\hat{D}_{22}\cdot B^{*^2}-\bar{D}_{66}\cdot A^{*^2}+\frac{2\cdot\hat{B}_{44}}{\alpha\tan(\varphi)}+\frac{\hat{D}_{44}}{(\alpha\tan(\varphi))^2},$$

$$Z_{31}=\frac{A_{12}\cdot A^*}{\alpha\tan(\varphi)},\ Z_{32}=\frac{\hat{A}_{44}\cdot B^*}{\alpha\tan(\varphi)},Z_{33}=-\bar{A}_{55}\cdot A^{*^2}-\hat{A}_{44}\cdot B^{*^2},$$

$$Z_{34}=-\bar{A}_{55}\cdot A^*+\frac{B_{12}\cdot A^*}{\alpha\tan(\varphi)},\ Z_{35}=-\hat{A}_{44}\cdot B^*+\frac{\hat{B}_{22}\cdot B^*}{\alpha\tan(\varphi)},$$

$$Z_{36}=-A^{*^2}\cdot\bar{B}_{55}-B^{*^2}\hat{B}_{44}-\frac{\hat{A}_{22}}{\alpha\tan(\varphi)}+\frac{\hat{B}_{22}}{(\alpha\tan(\varphi))^2}$$

$$Z_{37} = 2 \cdot A^* \cdot \overline{B}_{55} + \frac{D_{12} \cdot A^*}{\alpha \tan(\varphi)},\ Z_{38} = -2 \cdot B^* \cdot \hat{B}_{44} + \frac{\hat{D}_{22} \cdot B^*}{\alpha \tan(\varphi)},$$

$$Z_{41} = Z_{14},\ Z_{42} = Z_{24},\ Z_{43} = Z_{34}$$

$$Z_{44} = -\overline{D}_{11} \cdot A^{*2} - \hat{D}_{66} \cdot B^{*2} - \overline{A}_{55},$$

$$Z_{45} = -D_{12} \cdot A^* \cdot B^* - D_{66} \cdot A^* \cdot B^*$$

$$Z_{46} = B_{13} \cdot A^* + \frac{D_{12} \cdot A^*}{\alpha \tan(\varphi)} - \overline{B}_{55} \cdot A^*,$$

$$Z_{47} = -\overline{E}_{11} \cdot A^* - \hat{E}_{66} \cdot B^{*2} - 2 \cdot \overline{B}_{55},$$

$$Z_{48} = -E_{12} \cdot A^* \cdot B^* - E_{66} \cdot A^* \cdot B^*, Z_{51} = Z_{15},$$

$$Z_{52} = Z_{25},\ Z_{53} = Z_{35},\ Z_{54} = Z_{45},$$

$$Z_{55} = -\hat{D}_{22} \cdot B^* - D_{66} \cdot A^{*2} - \hat{A}_{44},$$

$$Z_{56} = -\hat{B}_{44} \cdot B^*, Z_{57} = -E_{12} \cdot A^* \cdot B^* - E_{66} \cdot A^* \cdot B^*,$$

$$Z_{58} = -\hat{E}_{22} \cdot B^{*2} - \overline{E}_{66} \cdot A^{*2} - 2 \cdot \hat{B}_{44} - \frac{\hat{D}_{44}}{\alpha \tan(\varphi)},$$

$$Z_{61} = Z_{16},\ Z_{62} = Z_{26},\ Z_{63} = Z_{36},\ Z_{64} = Z_{46},\ Z_{65} = Z_{56}$$

$$Z_{66} = -\overline{D}_{55} \cdot A^{*2} - \hat{D}_{44} \cdot B^{*2} - A_{33} - \frac{B_{23}}{\alpha \tan(\varphi)} - \frac{\hat{D}_{22}}{(\alpha \tan(\varphi))^2},$$

$$Z_{67} = -2 \cdot \overline{D}_{55} \cdot A^* + \frac{E_{12} \cdot A^*}{\alpha \tan(\varphi)},$$

$$Z_{68} = -2 \cdot \hat{D}_{55} \cdot B^* + \frac{\hat{E}_{44} \cdot B^*}{R_\beta} + \frac{E_{12} \cdot B^*}{R_\alpha} + \frac{\hat{E}_{22} \cdot B^*}{R_\beta},$$

$$Z_{71} = Z_{17},\ Z_{72} = Z_{27},\ Z_{73} = Z_{37},\ Z_{74} = Z_{47},\ Z_{75} = Z_{57},$$

$$Z_{76} = Z_{67},\ Z_{81} = Z_{18}, Z_{84} = Z_{48},\ Z_{85} = Z_{58},$$

$$Z_{77} = -\overline{F}_{11} \cdot A^{*2} - \hat{F}_{66} \cdot B^{*2} - 4 \cdot \overline{D}_{55},$$

$$Z_{78} = -F_{12} \cdot A^* \cdot B^* - F_{66} \cdot A^* \cdot B^*,$$

$$Z_{82} = Z_{28},\ Z_{83} = Z_{38}, Z_{86} = Z_{68}, Z_{87} = Z_{78},$$

$$Z_{88} = -\hat{F}_{22} \cdot B^{*2} - \overline{F}_{66} \cdot A^{*2} - \frac{\hat{E}_{44}}{\alpha \tan(\varphi)} - \frac{\hat{F}_{44}}{(\alpha \tan(\varphi))^2} - 4 \cdot \hat{D}_{44}.$$

The structural mass parameters $\{N_{ij}\}$ and the external applied load vector, as a function of time $\{F_{ij}\}$ are given (see Qatu et *al.*, 2013; Zannon et *al.*, 2015).

5. Conclusions

The mathematical analysis of the third order shear deformation theory (see Zannon et *al.*, 2015; Qatu et *al.*,2013) are presented for simply supported with circular cross section of a thick conical shell. This solution will be used in further investigations to assess the results for free vibration analysis of the circular cross section thick shells.

References

Akbari, M., Kiani, Y., & Eslami, M. R. (2015). Thermal buckling of temperature-dependent FGM conical shells with arbitrary edge supports. *Acta Mechanica, 226*(3), 897-915.

ASADI, E. & QATU, M. (2012). Free vibration of thick laminated cylindrical shells with different boundary conditions using general differential quadrature. *Journal of Vibration and Control, 10*, 1177/1077546311432000.

Chih-Ping Wu, & Wei-Lun Liu. (2014). 3D buckling analysis of FGM sandwich plates under bi-axial compressive loads. *Smart Structures and Systems, 13,* 111-135.

Dao, V. D., Le, K. H., & Nguyen, T. N. (2014). On the stability of functionally graded truncated conical shells reinforced by functionally graded stiffeners and surrounded by an elastic medium. *Composite Structures, 108*, 77-90.

Duc, N. D., & Cong, P. H. (2015). Nonlinear thermal stability of eccentrically stiffened functionally graded truncated conical shells surrounded on elastic foundations. *European Journal of Mechanics-A/Solids*, *50*, 120-131.

Jam, J. E., & Kiani, Y. (2015). Buckling of pressurized functionally graded carbon nanotube reinforced conical shells. *Composite Structures*, *125*, 586-595.

Korjakin, A., Rikards, R., Chate, A., & Altenbach, H. (1998). Analysis of free damped vibrations of laminated composite conical shells. *Composite structures*, *41*(1), 39-47.

Leissa, W. I., & Qatu, M. S. (2011). Vibrations of Continuous Systems. McGraw Hill, Cataloging-in-Publication Data is on file with the Library of Congress. P (103, 367, 375).

Mukhopadhyay, T., Naskar, S., Dey, S., & Adhikari, S. (2016). On quantifying the effect of noise in surrogate based stochastic free vibration analysis of laminated composite shallow shells. *Composite Structures*. http://dx.doi.org/10.1016/j.compstruct.2015.12.037

QATU, MS. (1994). On the validity of nonlinear shear deformation theories for laminated composite plates and shells. *Composite Structure, 27,* 395–401.

Qatu, M., Zannon, M., & Mainuddin, G. (2013). Application of Laminated Composite Materials in Vehicle Design: Theories and Analyses of Composite Shells. *SAE Int. J. Passeng. Cars - Mech. Syst., 6*(2), 1347-1353. http://dx.doi.org/10.4271/2013-01-1989

Qatu, Ms., Sullivan, Rw., & Wang, W. (2010). Recent Research Advances in the Dynamic Behavior of Composite Shells: 2000-2009. *Composite Structures, 93,* 14-31.

Reddy, Jn. (1984). Exact solutions of moderately thick laminated shells. *J. Engg. Mech, 110,* 794–809.

Roh, J. H., Woo, J. H., & Lee, I. (2008). Thermal post-buckling and vibration analysis of composite conical shell structures using layerwise theory. *Journal of Thermal Stresses*, *32*(1-2), 41-64.

Viola, E., Rossetti, L., Fantuzzi, N., & Tornabene, F. (2016). Generalized Stress-strain Recovery Formulation Applied to Functionally Graded Spherical Shells and Panels Under Static Loading. *Composite Structures*. http://dx.doi.org/10.1016/j.compstruct.2015.12.060

Zannon, M., Al-Shutnawi, B., Alrabaiah, H. (2015). Theories and Analyses Thick Hyperbolic Paraboloidal Composite Shells. *American Journal of Computational Mathematics, 5,* 80-85.

Zannon, M, & Qatu, M. (2014a). Mathematical Modeling of Transverse Shear Deformation Thick Shell Theory. *International Journal of Engineering Research and Management (IJERM), 1*(7).

Zannon, M, & Qatu, M. (2014b). Free Vibration Analysis of Thick Cylindrical Composite Shells Using Higher Order Shear Deformation Theory. *International Journal of Engineering Research and Management (IJERM), 1*(7).

6

The Global Formulation of the Cauchy Problem

Mohammad Ali Bashir[1,2] & Tarig Abdelazeem Abdelhaleem[3]

[1]Department of Mathematics, Faculty of Sciences, University of Alnillin, Khartoum, Sudan.

[2]Academy of Engineering Sciences, Khartoum, Sudan.

[3]Department of Mathematics, collage of Applied and Industrial Science, University of Bahri, Khartoum, Sudan

Correspondence: Marina Victorovna Yashina, E-mail: yash-marina@yandex.ru

Abstract

A Geometrical model for the global Cauchy problem, generalizing the traditional Cauchy problem is considered .The complete correspondence between the known analytical formulation and the geometrical interpretation is described, we have utilized the generalized Green's function and the open mapping theorem appropriate to the problem.

Keywords: Cauchy problem, Green's function, Globally Hyperbolic space time, Open mapping theorem, semi-Riemannian metric.

1. Introduction

In this paper we discuss the global formulation of the Cauchy problem(Bar, Ginoux, & Pfaffe, 2007; Minguzzi & Sánchez, 2008), and its solution for globally hyperbolic space time(Beem, Ehrlich, & Easley, 1996; O'neill, 1983). Also we discuss the role of open mapping theorem(Bär & Ginoux, 2012; Kreyszig, 1989), in our solution, because of its various properties. The open mapping theorem seems to be a good tool for investigating that for general maps between topological spaces. For the formulation of the global Cauchy problem we need to know two kinds of structure, the first is a time orientation which separates future from past(Bär & Fredenhagen, 2009), the second ingredient is that of a hyper surface Σ in which we can specify the initial values. In order to approach the global existence of solutions we assume that M is globally hyperbolic with a smooth spacelike Cauchy hyper surface Σ. For every $p \in M$ we have a unique time t *with* $p \in \Sigma_t$, on each Σ_t (Mühlhoff, 2011), we also have a Riemannian metric g_t such that $g = \beta\, dt^2 - g_t$,

2. Preliminaries

2.1 Cauchy Problem in the (n-1)-dimensional Subspace E_{n-1} (Stakgold & Holst, 2011),

Let G be a domain in the (n-1)-dimensional subspace E_{n-1} of the variable, $x_1, x_2, \ldots., x_{n-1}$. then the following is a Cauchy problem :

$$\sum_{i=1}^{n-1} \frac{\partial^2 u}{\partial x_i^2} - \frac{\partial^2 u}{\partial x_n^2} = 0 \tag{1}$$

Satisfying the conditions

$$u(x_1, x_2, \ldots., x_{n-1}, 0) = f(\bar{x}) \tag{2}$$

$$\left[\frac{\partial u(x_1, \ldots, x_{n-1}, x_n)}{\partial x_n}\right]_{x_n=0} = g(\bar{x}) \tag{3}$$

For $\bar{x} = (x_1, x_2, \ldots., x_{n-1}) \in G$ and f , g are sufficiently smooth functions defined in G. Conditions (2),(3) are called Cauchy conditions or initial conditions f , g are called Cauchy data and the system (1), (2) and (3) is called a Cauchy problem, G is called the initial manifold. In the IVP G is the hypersurface obtained by the intersection of the n-dimensional region T and the hyperplane $x_n = 0$. An initial domain may not be a proper subset of the boundary, for example in E_2 consisting of point (x ,t) , the initial domain may be $t = 0$ or a subset of it. In general elliptic equations are associated with boundary conditions and hyperbolic and parabolic equations with initial conditions.

2.2 (Example) : take the PDE

$$\frac{\partial^2 u}{\partial x^2} - \frac{\partial^2 u}{\partial t^2} = 0 \text{ with ICs : } u(x,0) = f(x), \frac{\partial u}{\partial t} = g(x)$$

The D' Alembert's solution to the Cauchy problem is

$$u(x,t) = \frac{1}{2} f(x+t) + \frac{1}{2} f(x-t) + \frac{1}{2}\int_{x-t}^{x+t} g(s)ds, t > 0 \tag{4}$$

The solution exists, is unique and depends continuously on the data $f(x)$ and $g(x)$. Hence the Cauchy problem for the wave equation is well-posed.

2.3 Semi-Riemannian Metric

A section $g \in \Gamma^\infty(S^2T^*M)$ is called semi-Riemannian metric if the bilinear form $g_p \in S^2T_p^*M$ *on* T_pM is non-degenerate for all $p \in M$. If in addition g_p is positive definite for all $p \in M$ then g is called Riemannian metric. If g_p has signature $(+,-,...,-)$ then g is called Lorentz metric.

2.4 Causal Subsets

Let $U \subseteq M$ be an open subset. Then U is called causal if there is a geodesically convex open subset $U' \subseteq M$ such that $U^{cl} \subseteq U'$ and for any two points $p,q \in U^{cl}$,the diamond $J_{U'}(p,q)$ is compact and contained in U^{cl} .

2.5 A Causal and Achronal Subsets

Let $A \subseteq M$ be a subset of a time-oriented Lorentz manifold. Then A is called

i.) a chronal if every timelike curve intersects A in at most one point.

ii.) a causal if every causal curve intersects A in at most one point

2.6 (Theorem) A Chronal Hyper Surfaces

Let (M,g) be a time-oriented Lorentz manifold and $A \subseteq M$ a chronal. Then A is a topological hyper surface in M if and only if A does not contain any of its edge points.

2.7 Cauchy Hyper Surface

Let (M,g) be a time-oriented Lorentz manifold. A subset $\Sigma \subseteq M$ is called a Cauchy hyper surface if every inextensible timelike curve meets Σ in exactly one point.

2.8. Cauchy development

Let $A \subseteq M$ be a subset. The future Cauchy development $D_M^+(A) \subseteq M$ of A is the set of all those points $p \in M$ for which every past-inextensible causal curve through p also meets A Alogously, one defines the past Cauchy development $D_M^-(A)$ and we call

$$D_M(A) = D_M^+(A) \cap D_M^-(A) \tag{5}$$

the Cauchy development of A.

2.9 Globally Hyperbolic Spacetime

A time-oriented Lorentz manifold (M,g) is called globally hyperbolic if

i.) (M,g) is causal,
ii.) all diamonds $J_M(p,q)$ are compact for $p,q \in M$.

2.10. Time Function(Baer & Strohmaier, 2015)

Let (M,g) be a time-oriented Lorentz manifold and $t: M \to R$ a continuous function. Then t is called a

i.) time function if t is strictly increasing along all future directed causal curves.
ii.) temporal function if t is smooth and grad t is future directed and timelike.
i.) Cauchy time function if t is a time function whose level sets are Cauchy hypersurfaces.
v.) Cauchy temporal function if t is a temporal function such that all level sets are Cauchy hyper surfaces.

2.11. Theorem(Baer & Strohmaier, 2015),

Let (M,g) be a connected time-oriented Lorentz manifold. Then the following statements are equivalent:

i.) (M,g) is globally hyperbolic.

ii.) There exists a topological Cauchy hypersurface.

iii.) There exists a smooth spacelike Cauchy hypersurface.

In this case there even exists a Cauchy temporal function t and (M,g) is isometrically diffeomorphic to the product manifold

$$R \times \Sigma \, with \; metric \, g = \beta dt^2 - gt \quad , \tag{6}$$

where $\beta \in \ell^{\infty}(R \times \Sigma)$ is positive and $g_t \in \Gamma^{\infty}(S^2 T^* \Sigma)$ is a Riemannian metric on Σ depending smoothly on t. Moreover, each level set

$$\Sigma_t = \{(t,\sigma) \in R \times \Sigma\} \subseteq M \tag{7}$$

of the temporal function t is a smooth spacelike Cauchy hypersurface.

3. Existence of Global Solutions to the Cauchy Problem

3.1 Proposition

Let (M,g) be a time-oriented Lorentz manifold with a smooth spacelike hyper surface $\iota:\Sigma \mapsto M$ with future directed normal vector field n. Moreover, let $U \subseteq U^{cl} \subseteq U'$ be a sufficiently small causal open subset of M such that $\Sigma \cap U \mapsto U$ is a Cauchy hyper surface for U . Then there exists a unique solution $u \in \Gamma^{\infty}(E|_U)$ for given initial values $u_0, \dot{u}_0 \in \Gamma_0^{\infty}(\iota^{\#} E|_U)$ and given inhomogeneity $\upsilon \in \Gamma_0^{\infty}(E|_U)$ of the inhomogeneous wave equation

$$Du = \upsilon \tag{8}$$

with $\iota^{\#} u = u_0$, *and* $\iota^{\#} \nabla_n^E u = \dot{u}_0$. in addition we have

$$\operatorname{supp} u \subseteq J_M(\operatorname{supp} u_0 \cup \operatorname{supp} \dot{u}_0 \cup \operatorname{supp} v) \tag{9}$$

3.2 Theorem

Let (M,g) be a globally hyperbolic and let $\iota:\Sigma \mapsto M$ be a smooth spacelike Cauchy hypersurface with future directed normal vector field $n \in \Gamma^{\infty}(\iota^{\#} E|_U)$. Assume that is a solution to the wave equation $Du = 0$ with initial conditions

$$u_0 = 0 = \dot{u}_0 \tag{10}$$

then

$$u = 0 \tag{11}$$

Moreover to develop our constructing we assume that M is globally hyperbolic with Σ is smooth spacelike. For every $p \in M$ we have a unique time t with $p \in \Sigma_t$.we have a Riemannian metric g_t such that $g = \beta dt^2 - g_t$ on each Σ_t .and open Ball $B_r(p)$, such $B_r(p) \subseteq \Sigma_t$ is open n Σ_t but not in M .Then

$$d_{gt}(p,q) = \inf\left\{\int_a^b gt(\dot{\gamma}(\tau), \dot{\gamma}(\tau)) d\tau \mid \gamma(a) = p, \gamma(b) = q, \gamma(\tau) \in \Sigma_t\right\} \tag{12}$$

where γ is an at least piecewise ℓ^1 curve joining $p,q \in \Sigma_t$ inside Σ_t .consider its Cauchy development $D_M(B_r(p)) = D_M^+((p)) \cup D_M^-(B_r(p))$ in M according to Definition.Now we want to find r small enough that $D_M(B_r(p))$ is a nice open neighbourhood of p allowing a local fundamental solution[4].

3.3. Lemma,(Waldmann, 2012)

The function $\rho: M \to (0, +\infty]$ defined by

$$\rho(p) = \sup\{r > 0 | D(B_r(p)) isRCCSV\} \tag{13}$$

Is well defined and lower semi-continuous.

3.4. Lemma

for every point $p \in M$ and $r > 0$ the exists a $t > 0$ such that

$$J_m(B_{\frac{r}{2}}(p)^{cl}) \cap ([t-\tau, t+\tau] \times \Sigma) \subseteq D_M(B_r(p)) \tag{14}$$

Where $t \in R$ is the uniqe time with $p \in \Sigma_t$.

3.5. Lemma

The function $\theta_r : \mathrm{M} \to (0,\infty]$ is well-defined and lower semi-continuous, where

$$\theta_r = \sup\left\{\tau > 0 \left| J_m(B_{\frac{r}{2}}(p)^{cl}) \cap ([t-\tau, t+\tau] \times \Sigma) \subseteq D_M(B_r(p))\right.\right\}$$

3.6. Lemma (Waldmann, 2012)

let $K \subseteq \mathrm{M}$ be compact then there is a $\delta > 0$ such that for all times $t \in R$ and all $u_t, \dot{u}_t \in \Gamma^\infty(\iota_t^{\#}E)$ *on* Σ_t with support $Supp\ u_t, Supp\ \dot{u}_t \subseteq K$, We have smooth solution u of the homogeneous Wave Equation $Du = 0$, on the time slice $(t-\delta, t+\delta) \times \Sigma$, with the initial conditions $u\big|_{\Sigma_t} = u_t$ *and* $\nabla_n^E u\big|_{\Sigma_t} = \dot{u}_t$ Moreover for the support we have

$$Supp\ u \subseteq J_M(Supp\ u_t \cup Supp\ \dot{u}_t) \tag{15}$$

Proof:

Since ρ is lower semi-continous according to lemma 2-1 and positive, it admits a minimum on the compact subset K. Thus we find $r > 0$ with $\rho(p) > 2r_0$ for all $p \in k$ so for this radius the function θ_{2r0} is lower semi-continuous according lemma 3.3 and positive. And we can find $\delta > 0\ with\ \theta_{2r0} > \delta\ on\ K$, given $t \in R$, $\Sigma_t \cap \mathrm{K}$ is compact. We can cover it with open balls $B_{r0}(p_1), \ldots\ldots, B_{r0}(p_N)$ of radius r_0. Also we can find $\chi_1, \ldots, \chi_n$ subordinate to $B_{r0}(p_1) \cup, \ldots\ldots, \cup B_{r0}(p_N)$

Then we have $\chi_1, \ldots, \chi_N = 1$, and $Supp\ \chi_\alpha \subseteq B_{r0}(p_\alpha)$ for all $\alpha = 1, \ldots. N$.

let u_t *and* $\dot{u}_t$ *in* $B_{r0}(p_\alpha)$ by considering $\chi_\alpha u_t, \chi_\alpha \dot{u}_t$ respectively ,with $\chi_\alpha u_t, \chi_\alpha \dot{u}_t \in \Gamma_0^\infty(\iota_t^{\#} E)$, and $\chi_1 u_t + \ldots + \chi_N u_t = u_t$ *and* $\chi_1 \dot{u}_t + \ldots + \chi_N \dot{u}_t = \dot{u}_t$,Then $D_M(B_{2r0}(p_\alpha))$, is still RCCSV.

Then we can use the proposition 3.1 to obtain smooth solution $, u_\alpha \in \Gamma^\infty\left(E\big|_{DM(B_{2r_0}(p_\alpha))}\right)$ of the homogeneous wave equation $Du_\alpha = 0$,on $D_M(B_{2r0}(p_\alpha))$, For the initial conditions, $u_{\alpha|\Sigma_t} = \chi_\alpha\ u_t$ *and* $\nabla_n^E u_\alpha\big|_{\Sigma_t} = \chi_\alpha\ \dot{u}_t$. and Then

$$Supp\ u_\alpha \subseteq J_M(Supp\ \chi_\alpha\ u_t \cup Supp\ \chi_\alpha\ \dot{u}_t) \tag{16}$$

By the Definition of the function θ_{2ro} and the choice of δ we see that

$$J_M(B_{r0}(p_\alpha)^{cl} \cap ([t-\delta, t+\delta] \times \Sigma) \subseteq D_M(B_{2r0}(p_\alpha)),$$

Since

u_α is difined on $J_M(B_{r0}(p_\alpha)^{cl} \cap ([t-\delta, t+\delta] \times \Sigma)$, and $Supp \chi_\alpha\ u_t, Supp\ \chi_\alpha\ \dot{u}_t \subseteq (B_{r0}(p_\alpha)$ from (16) we find

$$Supp\ \ u_\alpha \subseteq J_M\ (B_{r0}(p_\alpha)^{cl}),$$

since u_α *is smooth on* $D_M\left(B_{2r0}(p_\alpha)\right)$, we can extend u_α to $([t-\delta, t+\delta] \times \Sigma)$

Then $u_\alpha \in \Gamma^\infty\left(E\big|_{([t-\delta, t+\delta] \times \Sigma}\right)$ satisfying $Suppu_\alpha \subseteq J_M(B_{r0}(p_\alpha)^{cl} \cap ([t-\delta, t+\delta] \times \Sigma)$, and $Du_\alpha = 0$ as well as $u_{\alpha|\Sigma_t} = \chi_\alpha u_t$ *and* $\nabla_n^E u_\alpha\big|_{\Sigma_t} = \chi_\alpha\ \dot{u}_t$.

since χ_α is partition of unity , finally

$$\begin{aligned} Supp\ \ u &\subseteq Supp\ \ u_1 \cup \ldots\ldots \cup Supp\ \ u_N, \\ &\subseteq J_M\left(Supp\ \chi_1\ u_t \cup Supp\ \chi_1\ \dot{u}_t\right) \cup \ldots\ldots \cup J_M\left(Supp\ \chi_N\ u_t \cup Supp\ \chi_N\ \dot{u}_t\right), \\ &\subseteq J_M\left(Supp\ \chi_1\ u_t \cup Supp\ \chi_1\ \dot{u}_t \cup Supp\ \chi_N\ u_t \cup Supp\ \chi_N\ \dot{u}_t\right), \\ &\subseteq J_M\left(Supp\ u_t \cup Supp\ \dot{u}_t\right), \end{aligned} \tag{17}$$

Since $J_M(A) \cup J_M(B) \subseteq J_M(A \cup B)$ and $Supp\ \chi_\alpha\ u_t \subseteq Supp\ u_t$ *and* $Supp\ \chi_\alpha\ \dot{u}_t \subseteq Supp\ \dot{u}_t$ for all α. this completes the proof.

3.7. Theorem (Waldmann, 2012),

Let (M, g) be a globally hyperbolic spacetime with smooth spacelike Cauchy hyper-surface $\iota : \Sigma \to M$.

i.) *for* $u_0, \dot{u}_0 \in \Gamma_0^\infty(i^\# E)$ *and* $v \in \Gamma_0^\infty(E)$, there exists a unique global solution $u \in \Gamma^\infty(E)$ of the inhomogeneous wave equation $Du = v$ with initial conditions, $\iota^\# u = u_0$ *and* $\iota^\# \nabla_n^E u = u_0$, We have

$$\operatorname{supp} u \subseteq J_M(\operatorname{supp} u_0 \cup \operatorname{supp} \dot{u}_0 \cup \operatorname{supp} v) \tag{18}$$

ii.) For $k \geq 2$ and $u_0 \in \Gamma_0^{2(k+n+1)+2}(i^\# E)$, $u_0 \in \Gamma_0^{2(k+n+1)+1}(i^\# E)$ *and* $v \in \Gamma_0^{2(k+n+1)}(E)$ there exists a unique global solution $u \in \Gamma^k(E)$ of the inhomogeneous wave equation $Du = v$ with initial conditions $\iota^\# u = u_0$ *and* $\iota^\# \nabla_n^E u = u_0$. It also satisfies (18).

proof

let $Supp\, u_0, Supp\, \dot{u}_0$, *and* $Supp\, v \subset U \subseteq U^{cl} \subseteq U'$ *(RCCSV)* and *set* $Supp\, u_0 \cup Supp\, \dot{u}_0 \cup$ *and* $Supp\, v \subseteq U$ which is compact then f we have $k \subseteq (-\in, \in) \times \Sigma$ *and* $J_M(k) \cap ((-\in, \in) \times \Sigma) \subseteq U$, for $\in > 0$ let $u \in \Gamma^\infty(E|_U)$ be the solution according to Proposition 3.1.

Since $Supp \subseteq J_M(K)$ we can extend u to the whole time slice $(-\in, \in) \times \Sigma\, by\, 0$, and we have to argue that we can extend this solution to large time slices $(-T, T) \times \Sigma$.we solve $Dw = 0$ for the initial conditions $w|_{\Sigma_t} = u|_{\Sigma_t}$ *and* $\nabla_n^E w|_{\Sigma_t} = \nabla_n^E u|_{\Sigma_t}$ by using Lemma 2.4 .

Then on $(t-\eta, t+\eta) \times \Sigma$, v is vanishes by $Supp\, \upsilon \subseteq k$ since $k \subseteq (-\in, t) \times \Sigma$.
and w, υ both solve the $Dw = 0$ with same initial condition on Σ_t,
Then $w = u$ on $(-\in, t) \times \Sigma$ by the uniqueness theorem .and shows that w extend u to the slice $(-\in, t+\infty) \times \Sigma$ in smooth way. and $w \subseteq J_M(k) \cap \Sigma_t$
For the future of t means that *Supp w is still contained in* $J_M(k)$
For the past of t we already know that $w = u$ whence in total $SUPP\ w \subseteq J_M(k)$.

4. Global Green Functions and Cauchy Problem

in this part we, show the Well-posedness of the Cauchy problem with respect to the usual locally convex topologies of smooth or ℓ^k – sec*tions*,

4.1. Open mapping Theorem

Let $\varepsilon, \widetilde{\varepsilon}$ be Fréchet spaces and let $\phi: \varepsilon \to \widetilde{\varepsilon}$ be a continuous linear map. If ϕ is surjective then ϕ is an open map. As usual, a map ϕ is called open if the images of open subsets are again open

4.2. Corollary

Let $\phi: \varepsilon \to \widetilde{\varepsilon}$ be a continuous linear bijection between Fréchet spaces. Then ϕ^{-1} is continuous as well. Indeed, let $U \subseteq \varepsilon$ be open. Then the set-theoretic $(\phi^{-1})^{-1}(U)$, i.e. the pre-image of U under ϕ^{-1}, coincides simply with $\phi(U)$ which is open by the theorem. Thus ϕ^{-1} is continuous. Take the result of theorem(2.5) .

$$\Gamma_0^\infty(i^\# E) \oplus \Gamma_0^\infty(i^\# E) \oplus \Gamma_0^\infty(E) \to \Gamma^\infty(E), \tag{19}$$

Sending $(u_0, \dot{u}_0, \upsilon)$ to the unique solution u of the Wave Equation $Du = \upsilon$ with initial conditions u_0 *and* $\dot{u}_0$.

4.3. (Theorem) Well-posed Cauchy Problem

Let (M, g) be a globally hyperbolic spacetime with smooth spacelike Cauchy hyper surface $\iota: \Sigma \to M$. Then the linear map (19) sending the initial conditions and the inhomogeneity to the corresponding solution of the Cauchy problem is continuous.

4.4. (Theorem) Well-posed Cauchy problem II

Let (M, g) be a globally hyperbolic spacetime with smooth spacelike Cauchy hyper surface $\iota: \Sigma \to M$ and let $k \geq 2$. Then the linear map

$$\Gamma_0^{2(k+n+1)+2}(\iota^\# E) \otimes \Gamma_0^{2(k+n+1)+1}(\iota^\# E) \otimes \Gamma_0^{2(k+n+1)}(E) \to \Gamma^k(E) \tag{20}$$

sending (u_0, u_0, v) to the unique solution u of the inhomogeneous wave equation $Du = v$ with initial $\iota^\# u = u_0$ *and* $\iota^\# \nabla_n^E u = u_0$ continuous.

4.5. (Theorem)

Let (M, g) be a globally hyperbolic spacetime and $D \in Diffop^2(E)$ a normally hyperbolic differential operator. For every point $P \in M$ there is a unique advanced and retarded fundamental solution $F_M^\pm(P)$ *of D at p*. Moreover, for every test section $\varphi \in \Gamma_0^\infty(E^*)$ the section.

$$\mathrm{M} \ni P \mapsto F_M^{\pm}(P)\varphi \in E_p^{\pm} \tag{21}$$

is a smooth section of E^* which satisfies the equation

$$D^T F_M^{\pm}(.)\varphi = \varphi. \tag{22}$$

Finally, the linear map

$$F_M^{\pm} : \Gamma_0^{\infty}(E^*) \ni \varphi \mapsto F_M^{\pm}(.)\varphi \in \Gamma^{\infty}(E^*) \tag{23}$$

is continuous.

4.6. (Theorem)

Let (M,g) be a globally hyperbolic spacetime and $D \in \mathit{Diffop}^2(E)$ a normally hyperbolic differential operator. Then the unique advanced and retarded Green functions $F_{\mathrm{M}}^{\pm}(p)$ of D at p are of global order

$$ord\, F_{\mathrm{M}}^{\pm}(p) \leq 2n+6. \tag{24}$$

More precisely, the linear map (23) extends to a continuous linear map

$$F_{\mathrm{M}}^{\pm} : \Gamma_0^{2(k++1)}(E^*) \ni \varphi \mapsto F_{\mathrm{M}}^{\pm}(.)\, \varphi \in \Gamma^k(E^*) \tag{25}$$

for all $k \geq 2$ such that we still have

$$D^T\ F_{\mathrm{M}}^{\pm}(.)\ \varphi = \varphi \tag{26}$$

4.7. Green Operator

Let (M,g) be a time-oriented Lorentz manifold and $D \in \mathit{Diffop}^2(E)$ a normally hyperbolic differential operator. Then a continuous linear map

$$G_{\mathrm{U}}^{\pm} : \Gamma_0^{\infty}(E) \to \Gamma^{\infty}(E) \tag{27}$$

with

i.) $DG_{\mathrm{M}}^{\pm} = id\,\Gamma_0^{\infty}(E)$,

ii.) $G_{\mathrm{M}}^{\pm} D\big|_{\Gamma_0^{\infty}} = id\,\Gamma_0^{\infty}(E)$,

iii) $\mathit{Supp}\left(G_{\mathrm{M}}^{\pm}u\right) \subseteq J_{\mathrm{M}}^{\pm}\left(\mathit{Supp}\,u\right)^{c1}$ *for all* $u \in \Gamma_0^{\infty}(E)$.

is called an advanced and retarded Green operator for D respectively

4.8. (Proposition) Green Operators and Fundamental Solutions

Let (M,g) be a time-oriented Lorentz manifold and $D \in \mathit{Diffop}^2(E)$ a normally hyperbolic differential operator.

i.) Assume $\{G_{\mathrm{M}}^{\pm}(p)\}$ is a family of global advanced or retarded fundamental solutions of D^T at every point $p \in \mathrm{M}$ with the following property: for every test section $u \in \Gamma_0^{\infty}(E)$ the section $p \mapsto G_{\mathrm{M}}^{\pm}(p)u$ is a smooth section of E depending continuously on u and satisfying $DG_{\mathrm{M}}^{\pm}(.)u = u$. Then

$$(G_{\mathrm{M}}^{\pm}u)(p) = G_{\mathrm{M}}^{\mp}(p)u \tag{28}$$

yield advanced or retarded Green operator for D, respectively.

ii.) Assume $G_{\mathrm{M}}^{\pm}$ are advanced or retarded Green operator for D, respectively. Then $G_{\mathrm{M}}^{\pm}(p) : \Gamma_0^{\infty}(E) \to C$ defined by

$$(G_{\mathrm{M}}^{\pm})(p)u = (G_{\mathrm{M}}^{\mp}u)(p) \tag{29}$$

defines a family of advanced and retarded fundamental solutions of D^T at every point $P \in \mathrm{M}$ with the properties described in i.), respectively.

4.9. (Proposition)

Let (M,g) be globally hyperbolic and let $D \in \mathit{Diffop}^2(E)$ be a normally hyperbolic differential operator with advanced and retarded Green operators $G_{\mathrm{M}}^{\pm} : \Gamma_0^{\infty}(E) \to \Gamma^{\infty}(E)$.

i.) The dual map $(G_{\mathrm{M}}^{\pm})' : \Gamma_0^{-\infty}(E^*) \to \Gamma^{-\infty}(E^*)$ is weak**continuous* and satisfies

$$D^T\,(G_{\mathrm{M}}^{\pm})'(\varphi) = \varphi = (G_{\mathrm{M}}^{\pm})' D^T\, \varphi \tag{30}$$

for all generalized sections $\varphi \in \Gamma_0^{-\infty}(E^*)$ with compact support .

ii.) for generalized section $\varphi \in \Gamma_0^{-\infty}(E^*)$ with compact support we have

$$\mathit{Supp}(G_{\mathrm{M}}^{\pm})'(\varphi) \subseteq J_{\mathrm{M}}^{\pm}(\mathit{Supp}\,\varphi). \tag{31}$$

4.10. (Lemma)

Let (M,g) be globally hyperbolic and let $D \in \mathit{Diffop}^2(E)$ be a normally hyperbolic differential operator with advanced and retarded Green operators $G_{\mathrm{M}}^{\pm}$, Moreover, denote the corresponding Green operator of

$D^T \in Diffop^2(E^*)$
by $F_M^{\pm}$ .Then we have for $\varphi \in \Gamma_0^{\infty}(E^*)$ and $u \in \Gamma_0^{\infty}(E)$

$$\int_M (F_M^{\pm}\varphi).u\,\mu_g = \int_M \varphi(G_M^{\pm}u)\,\mu_g . \quad (32)$$

4.11(Theorem)
Let (M,g) be a globally hyperbolic and $D \in Diffop^2(E)$ be normally hyperbolic differential operator. Denote the global advanced and retarded Green operator of D by $G_M^{\pm}$ and those of D^T by $F_M^{\pm}$ respectively .
i.) For the dual operators we have

$$\left(G_M^{\pm}\right)'\Big|_{\Gamma_0^{\infty}(E^*)} = F_M^{\pm} \quad (33)$$

$$\left(F_M^{\pm}\right)'\Big|_{\Gamma_0^{\infty}(E^*)} = G_M^{\pm} \quad (34)$$

ii.) The duals of the Green operators restrict to maps,

$$\left(G_M^{\pm}\right)';\Gamma_0^{\infty}(E^*) \rightarrow \Gamma^{\infty}(E^*) \quad (35)$$

$$\left(F_M^{\pm}\right)';\Gamma_0^{\infty}(E) \rightarrow \Gamma^{\infty}(E) \quad (36)$$

which are continuous with respect to the $\ell_0^{\infty}-and\ \ell^{\infty}-topology, respectively.$
iii.) The Green operators have unique $weak^*$ continuous extensions to operators

$$G_M^{\pm};\Gamma_0^{-\infty}(E) \rightarrow \Gamma^{-\infty}(E) \quad (37)$$

$$F_M^{\pm};\Gamma_0^{-\infty}(E^*) \rightarrow \Gamma^{-\infty}(E^*) \quad (38)$$

satisfying

$$Supp\left(G_M^{\pm}u\right) \subseteq J_M^{\pm}(Supp\,u) \quad (39)$$

$$Supp\left(F_M^{\pm}\varphi\right) \subseteq J_M^{\pm}(Supp\,\varphi) \quad (40)$$

respectively. for these extensions one has

$$G_M^{\pm} = \left(F_M^{\pm}\Big|_{\Gamma_0^{\infty}(E^*)}\right)' \quad (41)$$

$$F_M^{\pm} = \left(G_M^{\pm}\Big|_{\Gamma_0^{\infty}(E^*)}\right)' \quad (42)$$

4.12. (Theorem)
Let (M,g) be a globally hyperbolic spacetime and $D \in DiFFOp^2(E)$ normally hyper-bolic with advanced and retarded Green operators $G_M^{\pm}$.
i.) The Green operators $G_M^{\pm};\Gamma_0^{-\infty}(E) \rightarrow \Gamma^{-\infty}(E)$, satisfy

$$DG_M^{\pm} = id_{\Gamma_0^{-\infty}(E)} = G_M^{\pm}D\Big|_{\Gamma_0^{-\infty}(E)} \quad (43)$$

ii.) For every $v \in \Gamma_0^{-\infty}(E)$, every smooth spacelike Cauchy hypersurface $\iota:\Sigma \mapsto M$ with

$$Supp\,v \subseteq I_M^{+}(\Sigma) \quad (44)$$

and all $u_0, u_0 \in \Gamma_0^{-\infty}(\iota^{\#}E)$, there exists a unique generalized section $u \in \Gamma^{-\infty}(E)$, with

$$Du_+ = v \quad (45)$$

$$Supp\,u_+ \subseteq J_M\left(Supp\ u_0 \cup Supp\ u_0 \cup J_M^{+}(Supp\ v)\right) \quad (46)$$

$$Sing\,Supp\,u_+ \subseteq J_M^{+}(Supp\,v) \quad (47)$$

$$\iota^{\#}u_+ = u_0 \ and\ \iota^{\#}\nabla_n^E u = u_0 . \quad (48)$$

The section u_+ *depends weak* continuously on v and continuously on* u_0, u_0.
iii.) An analogous statement holds for the case $Supp\,v \subseteq I_M^{-}(\Sigma)$.

5. Conclusion

The formulation of the Cauchy problem in Euclidean space with specified boundary condition is well known. In that formulation the traditional Green's function is involved in the construction of the solution. However one need a generalization of the Cauchy problem to spaces that are not Euclidean, such as Lorentzian manifolds, with pseudo-Riemannian metric .The consideration of this problem in such a geometrical Lorentzian manifold has very important impact on wave propagation with applications cosmic wave, Thus we have treated the formulation of Cauchy problem in Lorentzian manifolds. Here we also needed a generalizing form of Green's function . In order to find the inverse of Cauchy hyperbolic differential operator on a fiber bundle we also utilized the open mapping theorem appropriate to the problem. The solution appeared as a cross section of a fiber bundle ,that may be pulled down to base Lorentzian manifold to give the traditional local solution.

References

Baer, C., & Strohmaier, A. (2015). An index theorem for Lorentzian manifolds with compact spacelike Cauchy boundary. *arXiv preprint arXiv:1506.00959.*

Bär, C., & Fredenhagen, K. (2009). *Quantum field theory on curved spacetimes: Concepts and mathematical foundations* (Vol. 786): Springer.

Bär, C., & Ginoux, N. (2012). Classical and quantum fields on Lorentzian manifolds. *Global differential geometry* (pp. 359-400): Springer.

Bar, C., Ginoux, N., & Pfaffe, F. (2007). *Wave equations on lorentzain manifold and quantization.ESL lectures in mathematics an physics* Zurich: European Mathematic Society

Beem, J. K., Ehrlich, P., & Easley, K. (1996). *Global lorentzian geometry* (Vol. 202): CRC Press.

Kreyszig, E. (1989). *Introductory functional analysis with applications* (Vol. 81): wiley New York.

Minguzzi, E., & Sánchez, M. (2008). The causal hierarchy of spacetimes. *Recent developments in pseudo-Riemannian geometry, ESI Lect. Math. Phys,* 299-358.

Mühlhoff, R. (2011). Cauchy problem and Green's functions for first order differential operators and algebraic quantization. *Journal of Mathematical Physics, 52*(2), 022303.

O'neill, B. (1983). *Semi-Riemannian Geometry With Applications to Relativity, 103* (Vol. 103): Academic press.

Stakgold, I., & Holst, M. J. (2011). *Green's functions and boundary value problems* (Vol. 99): John Wiley & Sons.

Waldmann, S. (2012). Geometric wave equations. *arXiv preprint arXiv:1208.4706.*

7

Description of the Dependence Strength of Two Variogram Models of a Spatial Structure Using Archimedean Copulas

Moumouni Diallo[1], Diakarya Barro[2]

[1] FSEG, Université des SSG. BP: 2575 Bamako, République du Mali

[2] Université Ouaga II. BP: 417 Ouagadougou 12, Burkina Faso

Correspondence: Moumouni Diallo, FSEG, Université des SSG. BP: 2575 Bamako, République du Mali. E-mail: moudiallo1@gmail.com

Abstract

Variogram is a geostatistical tool which describes how the spatial continuity changes with a given separating distance between pairs of stations. In this paper, we study the dependence structure within a same class of bivariate spatialized archimedean copulas. Specifically, we point out properties of the gaussian variogram and the exponential one. A new measure of similarity of two copulas is computed particularly between the spatial independent copula and full dependence one.

Keywords: variogram, gaussian distribution, Archimedean copulas, similarity, exponential variogram, tail dependent coefficient

2010 MSC: 62H20, 62H11, 60G15

1. Introduction

The main objective of geostatistical analysis is the characterization of spatial phenomenos that are incompletely known. Different definitions of geostatistic have been proposed by spatial statistics researchers. Thereby, geostatistics can be defined as a branch of statistics focusing on spatial or spatio-temporal datasets. While G. Matheron (Matheron, G., 1969)found in geostatistics the application of probabilistic methods to regionalized variables, Issaks E. H. et all (Helena, F., 2012) rather defined them as a way of describing phenomenons and provides adaptation of classical regression technics to take advantages of this continuity. Geostatistics covers mainly three subdomains of statistical studies: analysis of variogram, krieging and stochastic simulation. All of these subdomains use variogram models, so that variogram lies at the earth of every geostatistical activity.

The variogram function describes the degree of spatial dependence of a given spatial random field or stochastic process $\{Z(x), x \in D\}$. This tool plays with the madogram a key role in spatial modelling, estimations and inference properties given data constraints. The simple form for madogram is for all $h \in \mathbb{R}^d$

$$M_F(h) = \frac{1}{2}E(|F[Z(x+h)] - F[Z(x)]|), \quad x \in \mathbb{R}^d \tag{1}$$

where h is the average value of the separating distance between the two points (Shepard, R. N., 1987).

Moreover, the concept of F-madogram has been introduced by Cooley et al. (Shepard, R. N., 1987) to generalized is the λ-madogram associated to the distribution underlying the stochastic process $\{Z(x)\}$.

$$\gamma_F(h) = \tfrac{1}{2}E\left\{\left|[F(Y(x))]^{\lambda} - [F(Y(x+h))]^{1-\lambda}\right|\right\}; \lambda \in]0,1[\,. \tag{2}$$

In the same way, combining results of spatial statistics and multivariate dependence tools, Barro D. (Diakarya, B., 2012) used spatial extreme values copulas to characterize the λ–madogram of process distribution under a distortional assumption. Suppose H is a bivariate distribution satisfying the key assumption. If its associated multivariate EV distribution marginal are unit-Fréchet distributed, then, the λ-madogram is given by

$$\gamma_\lambda(h) = \frac{1}{D_h(\lambda, 1-\lambda) + \lambda} - c(\lambda) \text{ where } c(\lambda) = \frac{2\lambda(1-\lambda)+1}{2(\lambda+1)(2-\lambda)}; \tag{3}$$

where D_h is a conditional spatial measure convex defined on the unit simplex of $\mathbb{R}^2$. In spatial prevision in particular, the current problem is that one needs to replace empirical model by an theorical admissible variogram model. These models are characterized by a nugget, sill and range. In practice, for a same data, different geostatisticians can obtains different paramaters because of the their appreciations.

The main contribution of this study is to study the dependence structure between two models of variograms within a same class. A new measure of similarity of two copulas are computed in particular between the spatial independent copula and full dependence one.

2. Prelimaries

In this section we collect important definitions and properties on bivariate copulas, variogram tail dependence coefficient. These tools turn out to be necessary for our approach. We refer the readers to standard references for copulas analysis as Joe (1997) which provide detailed and readable introductions to copulas.

2.1 Elements of Copulas Analysis

Introduced in spatial analysis by authors as Blanchet(Shepard, R. N., 1987)and Kazianka (Kazianka, H. 2009)copulas functions constitutes the fundamental tool in dependence modeling in statistics.

Definition 1 (Diakarya, B., 2012) Let $X = (X_1, ..., X_n)$ be a random vector with multivariate continuous distribution function (c.d.f.) H and c.d.f marginal $H_1, ..., H_n$. The copula of X (of the c.d.f. H respectively) is the multivariate c.d.f. C of the random vector $U = [H_1(X_1), ..., H_n(X_n)]$. Due to the continuity of $\{H_i, 1 \leq i \leq n\}$, each component of U is standard uniformly distributed, i.e., $U_i \backsim U(0, 1)$ for $i = 1, ..., n$.

Particularly, every n-copula must satisfy the n-increasing property, that means that, for any rectangle $B = [a, b]^n \subseteq \mathbb{R}^n$, the B-volume C_B of C is positive, i.e

$$C_B = \int_B dC(u) = \sum_{i_1=1}^{2} ... \sum_{i_n=1}^{2} (-1)^{i_1+...+i_n} C(u_{1i_1}, ..., u_{1i_n}) \geq 0. \tag{4}$$

In a bivariate study, the relation (2.1) is equivalent to the rectangular inequality, that is, for all (u_1, v_1); $(u_2, v_2) \in [0, 1]^2$ with $u_1 \leq v_1$ and $u_2 \leq v_2$,

$$C(v_1, v_2) - C(u_1, v_2) - C(v_1, u_2) + C(u_1, u_2) \geq 0.$$

Two particular copulas are used in multivariate dependence modelling: the independent or product copula Π, defined for all by:

$$\Pi(u_1, ..., u_n) = u_1 \times ... \times u_n \text{ for } (u_1, ..., u_n) \in [0, 1]^n$$

and the complete dependence one, given for a parameter θ and in bivariye case by C_θ

$$C_\theta(u_1, ..., u_n) = \begin{cases} M(u, v) - \theta & \textit{if } (u, v) \in [\theta, 1-\theta]^2 \\ W(u, v) & \textit{otherwhere} \end{cases}$$

where M and W are the so called bounds of Fréchet.

Particularly, an n-dimensional copula C is an Archimedean copula, if there exists a continuous and strictly decreasing convex function ϕ, the generator of C, in the class of completely monotone functions

$$\left\{\phi : [0, +\infty] \longrightarrow [0, 1]; \phi(0) = 1; \phi(\infty) = 0; (-1)^k \frac{\partial^k \phi^{-1}(t)}{\partial t^k} \geq 0; k \in \mathbb{N}\right\}, \tag{5}$$

with generalized inverse $\phi^{-1}(y) = \inf\{t \in [0, 1], \phi(t) \leq y\}$ such that,

$$C(u_1, ..., u_n) = \phi^{-1}[\phi(u_1) + ... + \phi(u_n)]; \text{ for all } (u_1, ..., u_n) \in [0, 1]^n. \tag{6}$$

2.2 Setting of Variogram

The variogram is a measure of dissimilarity. Let $S = \left\{s_i \in \mathbb{R}^d, 1 \leq i \leq n, d = 1, 2, 3\right\}$ be the set of n spatial sites. In classic statistic the computation of the parameters is relatively possible because the repetition of independent data. However in spatial statistics one can study the most of the time a single observation in each site. Therefore, in isotropical erea one groups these sites according to their separed distance h and we compute the parameters of interest. In the practice, the paramaters depend on chosen step (espace unity).

Let $Y(s)$ be the regionalized variable, then the empirical version of variogram is computed as

$$\tilde{\gamma}(h) = \frac{1}{2\sharp N(h)} \sum_{N(h)} [Y(s+h) - Y(s)]^2$$

where $N(h)$ is the set of pair sites $\{s+h, s\}$ separated by a magnitude lag h and $\sharp N(h)$ its cardinal.

Our key assumption may be assumed and strenghtened by the intrinsec hypothesis

$$\begin{cases} \mathbb{E}[Y(s) - Y(s+h)] & = 0 \\ 0.5\mathbb{E}\left\{[Y(s) - Y(s+h)]^2\right\} & = \gamma(h) \end{cases} \tag{7}$$

and the continuous in quadratic which mean that

$$\lim_{h \to 0} \mathbb{E}\left([Y(s+h) - Y(s)]^2\right) = 0. \tag{8}$$

Next, we denote $\mathcal{H}$ the set of spatial fields that respect these previous conditions.

3. Main Results

In this paper the main objective is to describe the strenght of the dependence between two variogram functions in the same erea.

3.1 Concept of Class of Variograms

In this subsection, we study the dependence structure in two main classes which are exponential class and gaussian one. The spatial structure has about the intrinsic hypothesis. Then, any admissible variogram functions γ is supposed continuous at the origin, has a sill ϖ (asymptotically or not) and concave function.

Definition 2 Two variogram functions γ_1 and γ_2 are said to be of the same type if and only if, for all $h \in [0, +\infty]$ there exists two real quantities $\beta > 0, \alpha \in \mathbb{R}$ such that

$$\gamma_1(h) = \beta\gamma_2(\alpha h). \tag{9}$$

Moreover, the set of variograms of the same type is called a class of variograms.

Endeed, it is easy to prove that the relation(9) is an equivalent relation. Each of these classes is represented by its normalized version provided respectively by equation (10) for the exponential class

$$\gamma_E(h) = 1 - \exp(-h/a), h \geq 0, a > 0 \tag{10}$$

and by (11) for gaussian one.

$$\gamma_G(h) = \begin{cases} 1 - \exp\left(-h^2/a\right), & 0 \leq h \leq a \\ 1, & h \geq a \end{cases}. \tag{11}$$

The main objective consists in describing the strenght dependence between two variogram functions in the same region via archimedean copulas, so that, we need to find the generator of each class.

Proposition 3 Let γ be an normalized exponential or gaussian variogram, then function F defined one $\mathbb{R}$ by

$$F(x) = \begin{cases} \gamma(x), & 0 \leq x \\ 0 & otherwise \end{cases} \tag{12}$$

is a distribution function.

proof. Let γ be an exponential or gaussian variogram normalized. Then, on $[0, +\infty[$, F is continuous and increasing function because γ satisfies these properties. Morever,

$$\lim_{x \to 0} F(x) = \gamma(0) = F(0) = 0, also \lim_{x \to +\infty} F(x) = \lim_{x \to +\infty} \gamma(x) = 1$$

and $\lim_{x \to -\infty} F(x) = 0$. So, it follows that F is a distribution function. ∎

For a simplicity reason, let's use the standardized version of each class of variogram in the main assumption. There is no loss of generality in this case, this standardization does not modifie the strength spatial dependence structure. It allows to reduce all scales to one without affect the spatial dependence structure. Another advantage of this standardization is the follolling proposition.

Proposition 4 Let γ be the standardized variogram with the main asumption, their exist $\delta \in \mathbb{R}_+ \cup \{+\infty\}$ such that the function $\check{F}_s$ defined from $\mathbb{R}$ to $\mathbb{R}$ by

$$F_s(x) = \begin{cases} 0, & x \leq 0 \\ \gamma_s(x), & 0 \leq x < \delta \\ 1, & x \geq \delta \end{cases} \tag{13}$$

is a distribution function.

Proof. In the main asumption, if the spatial structure has a range then δ is 1 else $+\infty$. Suppose that the spatial structure has a range, then F becomes

$$F(x) = \begin{cases} 0, & x \leq 0 \\ \gamma(x), & 0 \leq x < 1 \\ 1, & x \geq 1 \end{cases}.$$

Since γ is a continuous and increasing function and

$$\lim_{x \to 0} F(x) = \gamma(0) = F(0) = 0 \text{ and} \lim_{x \to 1} F(x) = \gamma(1) = F(1) = 1$$

then F is continuous and increasing function. Morever, it follows that

$$\lim_{x \to -\infty} F(x) = 0 \textit{ and } \lim_{x \to +\infty} F(x) = 1.$$

Therefore F is a distribution function as asserted. Similarly, the case $\delta = +\infty$ has been proved. ■

3.2 Spatial Archimedean Copulas of a Class of Variograms

Let $X_s = \{(X_{s,1}; ...; X_{s,n}), s \in S\}$ be a continuous stochastic random vector observed at a finite number of locations $S_m = \{s_1, ..., s_m\}; s_i \in \mathbb{R}^2$. Barro et all (2016) have introdued the defintio of spatialized generator ϕ_s of Archimedean copulas as

$$C_{\phi_s}(u_1^{\check{s}}; ...; u_n^{\check{s}}) = \phi_s^{-1}\left[\phi_s(u_1^{\check{s}}) + ... + \phi_s(u_n^{\check{s}})\right]. \tag{14}$$

for a given $s \in S$ and for all $\left(u_1^{\check{s}}; ...; u_n^{\check{s}}\right) \in [0,1]^n$.

Spatial copulas are used for generating joint distributions with a variety of spatial dependence structure. This section deals with how to find a spatial copula knowing the variogram function of field. Let ϕ be a continuous and strictly decreasing function from $[0, 1]$ to $[0, +\infty]$ such that $\phi(1) = 0$. The pseudo-inverse of ϕ is the function $\phi^{[-1]}$ is given as

$$\phi(h) = \begin{cases} \phi^{-1}(h), & 0 \leq h \leq \phi(0) \\ 0, & \phi(0) \leq h \end{cases}. \tag{15}$$

Proposition 5 Let γ be an admissible standardized variogram function and F_γ its associated distribution function. The bivariate distribution function C_γ defined from $[0,1]^2$ to $[0, 1]$ as

$$C_\gamma(u, v) = \max\left[\varphi_\gamma^{[-1]}\left(\varphi_\gamma(u) + \varphi_\gamma(v)\right), 0\right] \tag{16}$$

is a spatial Archimedean copula, where φ_γ is the Laplace transform of F_γ defined as

$$\varphi_\gamma(t) = \begin{cases} \int_0^1 \exp(wt)\, dF_\gamma(w), & \textit{if } \delta < +\infty \\ \int_0^{+\infty} \exp(wt)\, dF_\gamma(w), & \textit{else} \end{cases}.$$

Proof. Previously we prove that F_γ is a distribution function. And Nelson proved that, Laplace transform of any distribution function is a Archimedean copula generator. Consequently φ_γ is a Archimedean generator. ■

3.2.1 Application

Applying the previous proposition, we compute the generators of Archimedean copulas associated to gaussian variogram [see Appendix 1] and associated to gaussian variogram [see Appendix 2]. The results are summarized in the following table.

This table put out that the dependence structure in gaussian class and exponential class are respectively descibed by the inferior bound of Fréchet-Hoeffding and Clayton copula with parameter one.

Table 1. Spatial Archimedean Copulas with Generators in exponential and gaussian class.

Classes of Variograms	$F_\gamma(h), a > 0$	$\varphi_\gamma(w)$	$C_\gamma(u,v)$
Gaussian	$\begin{cases} 1-\exp\left(-h^2/a\right), & h \geq 0 \\ 0 & h < 0 \end{cases}$	$1-\sqrt{\pi w}$	$\max(u+v-1,0)$
Exponential	$\begin{cases} 1-\exp(-h/a), & h \geq 0 \\ 0 & h < 0 \end{cases}$	$(aw+1)^{-1}$	$\left[u^{-1}+v^{-1}-1\right]^{-1}$

3.2.2 Interpretation

Let C_I and C_D denote respectively the dependence and the independent copulas. On one hand, if the two variograms in a class (gaussian or exponential class) are totally independent, then the resulting copula will be independent (i.e $C_I(u,v) = uv$). That means also that using one variogram rather than another could cause an important deviation of information. On other hand, if the two variograms of gaussian (or exponential) class was totally dependent, the resulting copula is rather dependent one (i.e $C_D(u,v) = \min(u,v)$). That situation would mean that using one variogram rather than another could not affect deeply the informations. The dependence structure of our interest classes are neither that of C_I nor C_D. Therefore, we need to know the proximity degree between our particular classes from each of C_I nor C_D.

Visually, by the bands of levels lines forms, the figures 2.4 and 2.5 in appendix 3, show that the associated copula of gaussian variogram class is more similar with the independent copula than the dependente copula. But this visual diagnotic does not allow to see the similarity of associated copula of exponential variogram with neither. We need to add another method for measuring the similarity between the resulting copulas and the reference copulas. That one is discussed in the following subsection.

3.3 Similarity between the Resulting Copulas and Reference Copulas

In this subsection we study the similarity between two copulas by their matrix. Many measures of similarity can be found in literature. Following Shepard in (Issaks, E. H., 1989), we defined the similarity between two copulas. In the following, let denote Ñ and M̃ respectively the set of the n and the m first naturals.

Proposition 6 Let $A = \{(a_{ij}); i \in \tilde{N}, j \in \tilde{M}\}$ and $B = \{(b_{ij}); i \in \tilde{N}, j \in \tilde{M}\}$ be two matrix of the set of matrix with n rows et m colons $\mathcal{M}_{n,m}(\mathbb{R})$ defined on $\mathbb{R}$. The function s defined from $[\mathcal{M}_{n,m}(\mathbb{R})]^2$ to $[0,1]$ as

$$s(A,B) = \exp\left(-\max_{i \in \tilde{N}, j \in \tilde{M}} \left|a_{ij} - b_{ij}\right|\right).$$

is similarity measure.

Proof. Since $A = \{(a_{ij}); i \in \tilde{N}, j \in \tilde{M}\}$ and $B = \{(b_{ij}); i \in \tilde{N}, j \in \tilde{M}\}$ are two elements of $\mathcal{M}_{n,m}(\mathbb{R})$, valid similarity measure obeys to four axioms [see 11].

First

$$s(A,B) = 1$$

which is equivalent to writte that

$$\exp\left(-\max_{i \in N,\, i \in M} \left|a_{ij} - b_{ij}\right|\right) = 1$$

or, in the other words

$$a_{ij} = b_{ij}, \text{with } i \in N \text{ and } i \in M.$$

In particular that proves that s obeys to the equal of self-similarity.

Moreover, since $|a_{ij} - b_{ij}| = |b_{ij} - a_{ij}|$, then $s(A,B) = s(B,A)$, that is, the measure s is symmetric.

And then $A \neq B$ means that, there exists an pair $(i, j) \in \tilde{N} \times \tilde{M}$ such as $a_{ij} \neq b_{ij}$.

Then,

$$\exp\left(-\max_{i\in\tilde{N},j\in\tilde{M}} |a_{ij} - b_{ij}|\right) < 1.$$

In other word, for all $A \neq B$ we are $s(A,B) < s(A,A)$.

Then s respect the minimality condition. The last criterion is that of triangle inequality.

Let $C = (c_{ij})\ i \in \tilde{N}, j \in \tilde{M}$ be a third of $\mathcal{M}_{n,m}$. For all pair $(i, j) \in i \in \tilde{N} \times \tilde{M}$,

$$|a_{ij} - b_{ij}| \leq |a_{ij} - c_{ij}| + |c_{ij} - b_{ij}| \iff s(B,A) \geq s(A,C).s(C,B)$$

Therefore s such is a measure of similarity. Next we use it to study the similarity between the matrix engender by two copulas. For simplicity, we will say that two matrix A et B in $\mathcal{M}_{n,m}$ are simular when $s(A,B) = 1$.

Definition 7 Let C_1 and C_2 be two copulas defined from $[0,1]^2$ to $[0,1]$. These copulas will say simular if and only if for all pair $(n,m) \in \mathbb{N}^* \times \mathbb{N}^*$ their respective engendered matrix $M_1 = (m^1_{ij})$ with $(i, j) \in \tilde{N} \times \tilde{M}$ and $M_2 = (m^2_{ij})$ with $(i, j) \in \tilde{N} \times \tilde{M}$ such as $m^1_{ij} = C_1(u_i, v_j)$ and $m^2_{ij} = C_2(u_i, v_j)$ for all $0 \leq u_i, v_j \leq 1$, are simular.

Let remark that the couple (u_i, v_j) with $0 \leq u_i, v_j \leq 1$ are the positions of the grid points discretization of $[0,1]^2$. Denote C_{γ_E} and C_{γ_G} the copulas associated to exponential variogram and gaussian variogram respectively. Likely, C_I and C_D are respectively the independent copula and dependent copula.

Proposition 8 Two copulas C_1 and C_2 are simular if and only if for all $0 \leq u, v \leq 1$, $C_1(u,v)$ and $C_2(u,v)$ are simular.

Proof. Deduce from the simularity of two copula definition.

As far as concerned the simularity between C_I, C_D, C_{γ_E} and C_{γ_G}, we are the following proposition.

Proposition 9 For all $0 \leq u, v \leq 1$, the associated copula satisfies to the following properties.

i) $s\left(C_{\gamma_G}(u,v), C_D(u,v)\right) \leq s\left(C_{\gamma_E}(u,v), C_D(u,v)\right)$.

ii) $s\left(C_{\gamma_G}(u,v), C_I(u,v)\right) \leq s\left(C_{\gamma_E}(u,v), C_I(u,v)\right)$.

Proof. To prove proposition 9 consists in proving that, for all $0 \leq u, v \leq 1$,

$$C_{\gamma_G}(u,v) \leq C_{\gamma_E}(u,v) \leq C_I(u,v) \leq C_D(u,v).$$

In fact, C_{γ_G} et C_D are Fréchet-Hoeffding bound, they are considered as universal bounds for copulas. Then for any copula, particulary C_{γ_E} and C_I, we are

$$C_{\gamma_G}(u,v) \leq C_{\gamma_E}(u,v) \leq C_D(u,v)$$

and

$$C_{\gamma_G}(u,v) \leq C_I(u,v) \leq C_D(u,v), 0 \leq u, v \leq 1.$$

Therefore, it is enough to compare $C_{\gamma_E}(u,v)$ and $C_I(u,v)$, for all $0 \leq u, v \leq 1$.

When at less one of u and v is zero, we are

$$C_{\gamma_G}(0,v) = 0 = C_{\gamma_E}(0,v) \quad \text{and } C_{\gamma_G}(u,0) = 0 = C_{\gamma_E}(u,v).$$

For $0 < u, v \leq 1$, we obtain

$$C_{\gamma_E}(u,v) = uv/(v + u - uv).$$

Since we have

$$v + u - uv - 1 = (1 - v)(1 - u),$$

then, $v + u - uv \geq 1$. Therefore

$$C_{\gamma_E}(u,v) \leq C_I(u,v)$$

That result shows that the dependence structure in the both classes is more similar with that of described by the independent copula.

3.4 Simulation

The theorical results are illustrated by a similation. With statistic sofware R (Shepard, R. N. 1987), we plot the similarities

$$s\left(C_{\gamma_H}(u,v), C_K(u,v)\right) \text{ with } 0 \le u, v \le 1 \text{ and } H \in \{G, E\}$$

and $K \in \{I, D\}$. Indeed, we discretized regularly [0, 1] in n, $1 \le n \le 1000$ nodes.

For each value of n, we ploted

$$s\left(C_{\gamma_H}(u,v), C_K(u,v) \mid n\right), 0 \le u, v \le 1, H \in \{G, E\} \text{ and } K \in \{I, D\}.$$

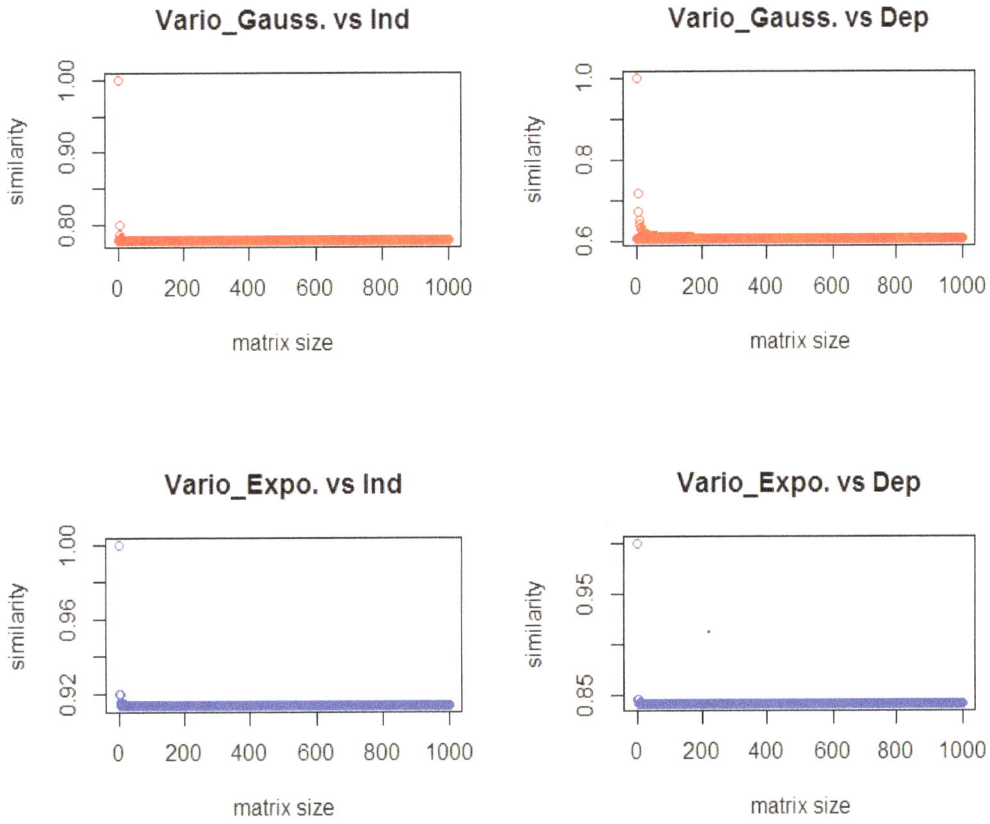

Table 2. Similarities (S. Mean) : $s\left(C_{\gamma_G}, C_I\right)$ (G.I), $s\left(C_{\gamma_G}, C_D\right)$ (G.D), $s\left(C_{\gamma_E}, C_I\right)$ (E.I), $s\left(C_{\gamma_E}, C_I\right)$ (E.I) and their standard deviations (St. D.) respectively.

	G.I.	G.D.	E.I.	E.D.
S. Mean	0.7792889	0.6082728	0.9139695	0.8426776
St.D.	0.009914057	0.018192645	0.003864092	0.007051282

3.5 Tail Dependence Coefficient of in a Class of Variograms

Like many dependence concepts, the tail dependence coefficient describes how large (or small) values of random variables appear with large (or small) value of an other. Here, this coefficient measures the strength of link between normalized variograms in the upper-right quadrant and in the lower-left quadrant of $[0, 1]^2$. Let C_γ be the Archimedean copula associated to variogram function γ which is either γ_G or γ_E.

Corollary 10 Let C_γ be an Archimedean copula with generator $\varphi_\gamma^{[-1]}$ associated to γ. Then

$$\lambda_U^\gamma = 2 - \lim_{u \to 1^-} \frac{1 - \varphi_\gamma\left(2\varphi_\gamma^{[-1]}(u)\right)}{1-u} = 2 - \lim_{h \to 0^+} \frac{1 - \varphi_\gamma(2h)}{1 - \varphi_\gamma(h)} \tag{17}$$

and

$$\lambda_L^\gamma = \lim_{u \to 0^+} \frac{\varphi_\gamma\left(2\varphi_\gamma^{[-1]}(u)\right)}{u} = \lim_{h \to +\infty} \frac{\varphi_\gamma(2h)}{\varphi_\gamma(h)} \tag{18}$$

Proof. Nelson established the link between the upper tail dependence coefficient λ_U and lower tail dependence coefficient with the copula C_γ, as

$$\lambda_U = 2 - \lim_{u \to 1^-} \frac{1 - C_\gamma(u,u)}{1-u} \quad \text{and} \quad \lambda_L = \lim_{u \to 0^+} \frac{C_\gamma(u,u)}{u}.$$

The first equalities of equation (17) and (18) are shown by Nelson (Nelson, R. B., 1999). Concerning that last one change the variables as when $u \to 1^-$ then $h = \varphi_\gamma^{[-1]}(u) \to 0^+$(remaind that $\varphi_\gamma(0) = 1$)

The results of this section confirm partially those of the previous. In fact, the upper tails dependence coefficients are zero in both of cases which means that there is an asymptotical independence.

4. Conclusion and Discussion

Our study shows that the dependence structure in the class of gaussian variogram and exponential variogram class is more near with that of independent copula than dependent copula. Therefore, as in the theorical gaussian variogram class, two theorical exponential variograms with a too little difference can engender too important disparity in the interpretation of the spatial prediction. A new measure of similarity of two copulas are computed in particular between the spatial independent copula and full dependence one.

References

Blanchet, J. & Davison, A. C. (2011). *Spatial modelling of extreme snow depth.* Annals of Applied Statistics, vol.5, N°3, 1699-1725. http://dx.xoi.org/10.1214/11-AOAS464

Carlo, G. (2008). *Modélisation et statistique spatial*, Springer.

Cooley, D. S., Poncet, P., & Naveau, P. (2006), *Variograms for max-stable random fields. In Dependence in Probability and Statistics.* Lecture Notes in Statistics 187 373–390. Springer, New York.

Diakarya, B. (2012). Analysis of stochastic spatial processes via copulas and measures of extremal dependence. *Archives des Sciences, 65*(12), 665-673.

Diakarya, (2016). Spatial Tail Dependence and Survival Stability in a Class of Archimedean Copulas. *International Journal of Mathematics and Mathematical Sciences Volume (2016)*,
Article ID 8927248, 8 pages http://dx.doi.org/10.1155/2016/8927248

Embrechts, P. *Modelling Dependence with Copula and Applications to Risk Management,* Department of Mathematics, ETHZ, CH-8092 Zürich, Switzerl.

Genest, C. (1986). Copules archimédiennes et familles des lois bidimensionnelles dont les marges sont données. *The canadian Journal of Statistics, 14*(2), 145-159.

Helena, F. (2012) *Generalized madogram and pair wise dependence of maxima over two disjoint region of random field,* Arxive prepint, arXiv:1104.2637.

Issaks, E. H. (1989), *An introduction to Applied Geostatistics,* Oxford university Press.

Kazianka, H. (2009). Spatial modeling & interpolation using copulas. PhD thesis, University of Klagenfurt. Ribatet (2011)- Statistical Modelling of Spatial Extremes A. C. Davison, S. A. Padoan and M. Ribatet October 3, 2011

Matheron, G. (1969). *Cours de géostatistique*, Ecole des Mines de Paris.

Nelson, R. B. (1999). *An Introduction to Copulas, vol.139 of Springer Series in Statistics*, springer, New York, NY,USA, 2nd edition.

Pravin, K. T. *Copula Modeling: An Introduction for Practitioners*, Published, sold and distributed by: now Publishers Inc.

Schmid, R. (2010). *Copula-based measure of multivariate association, in Copula theory and Its applications: Proceedings of the Workshop Held in Warsaw*, 25-26 September 2009, vol. 198 of lecture Notes in Statistics, 209-236, Springer, Berlin, Germany.

Shepard, R. N. (1987). *Toward a universal low of generalisation for psychological science:* Science,237,1317-1323.

Appendice

Appendice 1: Generator of Spatial Archimedean Copula of Gaussian Variogram

The normalized Gaussian variogram function, is written as

$$\gamma(h) = 1 - \exp\left(-h^2/a\right),\ h \in \mathbb{R}_+, a > 0.$$

Its associated distribution function is defined as

$$F(h) = \begin{cases} 1 - \exp\left(-h^2/a\right), & h \geq 0 \\ 0, & otherwise \end{cases}.$$

Therefore, the coresponding spatial Archimedean copula has a generator φ defined as

$$\begin{aligned} \varphi(w) &= \int_0^{+\infty} \exp(-wt)\,dF(t), \\ &= \frac{\exp\left(w^2/4\right)}{a} \int_0^{+\infty} 2t\exp\left[-(t/a + w/2)^2\right]dt. \end{aligned}$$

After variable changing $x = t/a + w/2$, we get

$$\varphi(w) = \exp\left(w^2/4\right)\left[\exp\left(-w^2/4\right) - w\int_{w/2}^{+\infty} \exp\left(-x^2\right)dx\right]. \quad (19)$$

Let I be

$$\begin{aligned} I(w) &= \int_{w/2}^{+\infty} \exp\left(-x^2\right)dx \Longrightarrow I^2(w) = \left[\int_{w/2}^{+\infty} \exp\left(-y^2\right)dy\right]\left[\int_{w/2}^{+\infty} \exp\left(-x^2\right)dx\right], \\ I^2(w) &= \int\int_{w/2}^{+\infty} \exp\left(-x^2 - y^2\right)dxdy. \end{aligned}$$

After a polar changing, we obtain

$$I(w) = \sqrt{\pi}\exp\left(-w^2/4\right). \quad (20)$$

Putting result (20) in the equation (19), we find the generator $\varphi(w) = 1 - \sqrt{\pi}w$.

Appendice 2: Generator of Spatial Archimedean Copula of Exponential Variogram

The normalized Exponential variogram function, is written as

$$\gamma(h) = 1 - \exp(-h/a),\ h \in \mathbb{R}_+, a > 0$$

Then its associated distribution function is as

$$G(h) = \begin{cases} 1 - \exp(-h/a), & h \geq 0 \\ 0, & otherwise \end{cases}.$$

And then, the coresponding spatial archimedean copula has a generator ψ defined as

$$\begin{aligned}\psi(w) &= \frac{1}{a}\int_0^{+\infty} \exp\left[-(w + 1/a)\,t\right] dt \\ &= (aw + 1)^{-1}\end{aligned}$$

Appendice 3: The Figures

In the Figure 2.4 and Figure 2.5, the levels lines are described colored bands. The colors goes from blue (mean level zero) two black (mean level one).

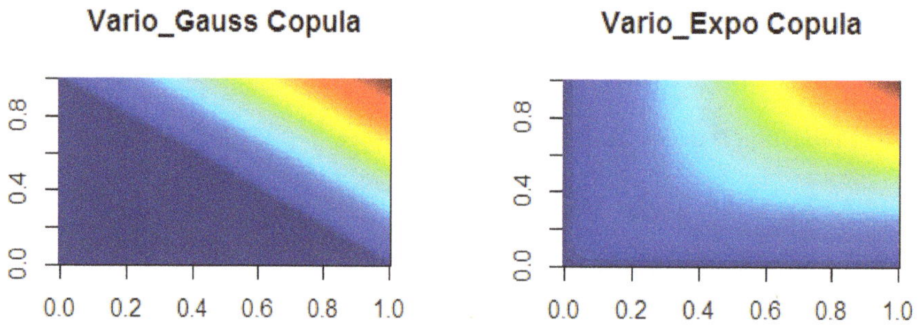

Figure 1. Images contour of: the Gaussian variogram class associated copula and Exponential variogram class associated copula

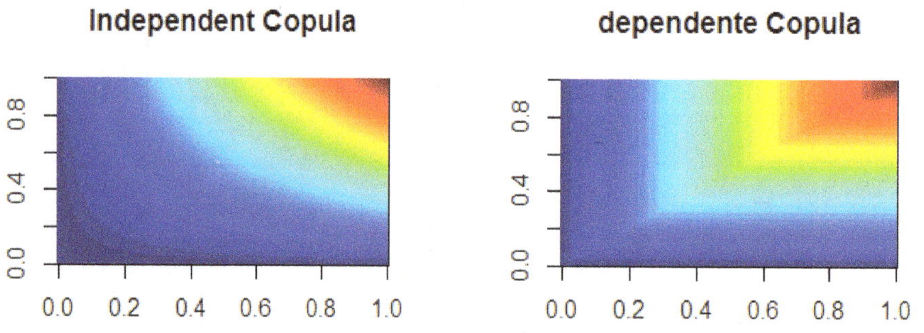

Figure 2. Images contour of Independent copula (left) and Dependent copula (rigth).

8

Finite Element Approximation and Numerical Analysis of Three-dimensional Electrical Impedance Tomography

Yirang Yuan[1], Jiuping Li[1], Changfeng Li [1,2] & Tongjun Sun[1]

[1] Institute of Mathematics, Shandong University, Jinan, P. R. China, 250100

[2] School of Economics, Shandong University, Jinan, P. R. China, 250100

Correspondence: Yirang Yuan, Institute of Mathematics, Shandong University, Jinan, P. R. China, 250100. E-mail: yryuan@sdu.edu.cn

Abstract

Electrical impedance tomography is solved by solving an inverse problem of elliptic equation, and a new numerical method or a new technique is argued to consider finite element (such as normal element and mixed element) in this paper on three dimensional region. Introducing different perturbations to boundary restrictions and using different spacial steps, the authors obtain numerical solutions and give comparison with exact solutions. Numerical data show that numerical solution can approximate exact solution well as spacial step taken small and the approximation of Neumann boundary condition is more stable than that of Dirichlet case. For Newton iterations on finite element method, a large-scaled system of massive linear equations is solved in each iteration, thus the computation is quite expensive. So two techniques are argued in the first half of this paper. Firstly, the invariance property of quasi-element stiffness matrix is used in the iterations and a type of special current model is introduced. Then the minimum number of direct problems solved is considered. Later a local conservative numerical approximation, low order mixed element (block-centered method) is presented in the latter part and the positive semi-definiteness and the existence of its solution are proved. Computational formula of error functional Jacobi matrix is derived and the least direct problems in each iteration are solved by using the symmetry of algorithm and a special current basis. This method has been applied successfully in actual numerical simulation of three-dimensional electrical impedance tomography.

Keywords: three-dimensional electrical impedance tomography, numerical simulation, normal finite element, low order mixed element, stability

MSC(2010) 35R30, 35J25, 92C55, 78A30

1. Introduction

Electrical Impedance Tomography (EIT) is a new biomedicine imagining technique. Firstly some electrodes are attached on the skin and safe current is injected. Secondly, the values of potentials are measured and the distribution is obtained. Thirdly, the data are input in a computer and the distribution of electrical resistivity inside the body is illustrated by the monochrome graph after calculations and transformations. Since electrical resistivity is assigned by different values at different places, so the graph shows where and how the inner organs are. This technique is safe and economic, so EIT can be generalized in numerical simulation of more actual problems such as geophysical exploration, hydrogeology detection, dam body detection and underwater target detection. Therefore, this method is important and arouses more attention on theoretical research and its applications (Kirsch, 1996; Cheney, 1999; Webster, 1990; Tamburrino, 2002; Breckon, 1987; Du, 1997).

Electrical impedance tomography is essentially described by an inverse problem of elliptic partial differential equation,

$$\nabla \cdot (\rho^{-1}\nabla u) = 0, \quad x \in \Omega, \tag{1a}$$

$$\rho^{-1}\frac{\partial u}{\partial \nu} = \psi, \quad x \in \partial\Omega, \tag{1b}$$

$$u = \varphi, \quad x \in \partial\Omega, \tag{1c}$$

where $u(x)$ denotes the potential, $\rho > 0$ is the electrical resistivity, $\psi(x)$ is the boundary electric density and $\varphi(x)$ denotes the boundary potential.

Given the electric resistivity $\rho(x)$ and the boundary density $\psi(x)$, then the potential distribution $u(x)$ is obtained by (1a) and (1b). This is called Neumann boundary problem. Dirichlet boundary problem is interpreted by the fact that $u(x)$ is computed by (1a) and (1c) as $\rho(x)$ and $\varphi(x)$ are given. The above two types are direct problems while EIT is an inverse problem. The functions $\rho(x)$ and $u(x)$ are unknown and ρ is an objective function as only two boundary value conditions $\psi(x)$ and $\varphi(x)$ are given. If $\rho(x)$ is known, the potential distribution $u(x)$ is determined by a proper boundary current density $\psi(x)$ only different by a constant, then the boundary potential $u(x)|_{\partial\Omega} = \varphi(x)$ is obtained. Therefore, the electrical resistivity distribution $\rho(x)$ can define a mapping on the boundary current density set Ψ into the boundary potential set $\Phi(x)$:

$$F_\rho : \psi(x) \to \varphi(x).$$

EIT shows how the electric resistivity $\rho(x)$ is obtained under the mapping $F_\rho = F_0$ satisfying

$$F_\rho(\psi) = F_0(\psi), \ \psi \in \Psi(x). \tag{2}$$

It is hard to get the exact relation of $\rho(x)$ and F_ρ (direct problem), and the boundary current density $\psi(x)$ is measured hardly. So this problem is solved by numerical approximations.

At present many numerical methods are discussed in solving this problem, and Newton iteration of normal finite method is a powerful tool. While Newton iteration gives rise to large-scaled computation especially for three-dimensional case. The method of finite element has high order of accuracy and has strong suitability in computational geometric regions. But the computation of three numerical integrals in generating coefficient matrices is generally quite expensive, where coefficient matrix is reformulated in each iteration to compute the gradient vector and Hesse matrix of objective function. Thus, in this paper we introduce the quasi-element stiffness matrix combined with normal finite element, because quasi-element has little-amounted and small-sized matrix fixed at each iteration. The element matrix has simple relation with coefficient matrix and the numerical integral computation cost is shortened. Another difficult of quasi-element stiffness matrix is to compute the gradient vector and Hesse matrix and to solve large-scaled linear equations (direct problem), so we introduce a special current forcing model. This uses sufficiently the potentials at all the nodes (solutions of direct problem), derives simply a Jacobi matrix of error vectors and finishes the computations of gradient vector and Hesse matrix. In the latter part the authors present a local conservative, low order mixed element scheme (block-centered finite volume element) (Russell, 1983; Weiser, 1988; Mishev, 1998; Lazarov, 1996; Larsson, 2003), and prove the scheme's positive semi-definiteness and the solution's existence. Computational formula is derived for Jacobi matrix of error functional and the number of direct problems is minimized at each iteration by the symmetry and special current basis vectors. Two different EIT models of continuous model and electrode model are considered. The model's correctness and the algorithm's reliability and feasibility are testified by exact solution simulation for continuous model, then actual algorithm is concluded for electrode model.

Numerical simulation is discussed in this paper for the inverse problem. On normal finite element method some discussion of iteration algorithm and numerical simulation is given in §2, and the algorithm's feasibility and the model's correctness are testified by comparing exact solutions on three dimensional domain $(0, 1)^3$ with numerical simulation data of different perturbations. In §3, considering actual simulation of EIT problem, the authors present quasi-element stiffness matrix technique to decrease numerical integral computation and discuss a handling method for Neumann boundary condition. The projection F_ρ is disretized and transformed into a mapping of currents and potentials at different electrodes, and a new current model and computational algorithm of Jacobi matrix are given. EIT problem on a cylinder is simulated numerically. In §4, a numerical method of low order mixed element (or named block-centered finite volume element) and its approximations are argued. This algorithm has the nature of mass conservation. In §5, image reconstruction and simulation experiments of mixed element method are considered.

2. Normal Finite Element Numerical Method and Simulation of Inverse Problem

An inverse problem is solved usually by the method of iterations. Firstly Neumann problem is solved by using normal finite element method. Then an ordinary least squares problem is given by combing its numerical solution and Dirichlet boundary condition.

2.1 Normal Finite Element Method of Neumann Boundary Problem

Neumann boundary problem of (1a) and (1b) is changed into a Galerkin variation as follows: to find $u \in H^1(\Omega)$ such that

$$D(u, v) = G(v), \quad \forall v \in H^1(\Omega), \tag{3}$$

where

$$D(u, v) = \int_\Omega \rho^{-1} \nabla u \cdot \nabla v dx, \ G(v) = \int_{\partial\Omega} v\rho^{-1} \frac{\partial u}{\partial \nu} dS.$$

Ω is partitioned by a series of elements $\{K_m\}_{m=1}^M$ whose vertexes make up a grid $\bar{\Omega} = \{x_i\}_{i=1}^n$, $x_i = (x_{i,1}, x_{i,2}, x_{i,3})$. A piecewise defined constant function $\rho_h(x)$ is used to approximate $\rho(x)$, $\rho_h(x)|_{K_m} = \rho_m$, and the discretized linear equations are formulated by

$$A(\rho_h)u = b, \tag{4}$$

where $A = (a_{ij})_{n\times n}$ and $b = (b_1, b_2, \cdots, b_n)^T$. Let $N_i(x)$ be the node basis function on $i = 1, 2, \cdots, n$, and let the number of nodes of each element be denoted by p,

$$\bar{a}_{ij}^{(m)} = \int_{K_m} \nabla N_i \cdot \nabla N_j dx. \tag{5}$$

The quasi-element stiffness matrix is defined as follows.

Definition 1 $\bar{A}^{(m)} = (\bar{a}_{ij}^{(m)})_{p\times p}$ is made up of $\bar{a}_{ij}^{(m)}$ ordered by the node labels of K_m, then $\bar{A}^{(m)}$ is called the quasi-element stiffness matrix of the element K_m.

The element stiffness matrix is defined by $A^{(m)} = (a_{ij}^{(m)})_{p\times p}$, where

$$a_{ij}^{(m)} = \int_{K_m} \rho^{-1}\nabla N_i \cdot \nabla N_j dx \approx \rho_m^{-1} \int_{K_m} \nabla N_i \cdot \nabla N_j dx = \rho_m^{-1}\bar{a}_{ij}^{(m)}, \tag{6}$$

and it is easy to find the relation of $A^{(m)}$ and $\bar{A}^{(m)}$

$$A^{(m)} = \rho_m^{-1}\bar{A}^{(m)}. \tag{7}$$

Noting that

$$a_{ij}^{(m)} = D(N_i, N_j) = \sum_{m=1}^{M} \int_{K_m} \rho^{-1}\nabla N_i \cdot \nabla N_j dx \approx \sum_{m=1}^{M} \rho_m^{-1} \int_{K_m} \nabla N_i \cdot \nabla N_j dx = \sum_{m=1}^{M} \rho_m^{-1}\bar{a}_{ij}^{(m)}, \tag{8}$$

$$b_i = G(N_i) = \int_{\partial\Omega} N_i \psi dS, \tag{9}$$

and letting $[A^{(m)}]$ and $[\bar{A}^{(m)}]$ of order n denote the expanded formulations of the matrices $A^{(m)}$ and $\bar{A}^{(m)}$ of order p, then we have the following relation

$$A = \sum_{m=1}^{M} [A^{(m)}] = \sum_{m=1}^{M} [\rho_m^{-1}\bar{A}^{(m)}] = \sum_{m=1}^{M} \rho_m^{-1}[\bar{A}^{(m)}]. \tag{10}$$

Similar to Neumann problem, the solutions of finite element equation (4) differ by constants, that is to say the solution exists solely by assigning the value of a node, $u_{i_0} = \psi(x_{i_0})$. The equation (4) turns into an equivalent symmetric and positive definite system

$$A_d u_d = b_d, \tag{11}$$

where A_d is obtained by deleting the (i_0, j) entries of the i_0th row and the (i, i_0) entries of the i_0th column from the matrix A, and the vectors u_d, b_d are given by deleting the i_0th component from u, b. Given $\rho_h(x)$, the equation (11) is solved by

$$u_d = A_d^{-1} b_d. \tag{12}$$

U denotes the vector of potentials at all the boundary nodes regardless the reference node x_{i_0}, whose components are included in u_d, then

$$U = P u_d = P A_d^{-1}(\rho_h) b_d. \tag{13}$$

The (i, j) entries of matrix P are assigned either by the number 0 or by the number 1, corresponding to the reference nodes and the other nodes.

2.2 Iterations of Inverse Problem

Let U_0 denote the vector of the values of $\varphi(x)$ at the boundary nodes except x_{i_0}, and $\rho = (\rho_1, \rho_2, \cdots, \rho_M)^T$ is defined by the values of $\rho_h(x)$ at all the elements. Let $r = U - U_0$, then we solve inverse problem by solving the following nonlinear optimization problem,

$$\min_{\rho\in C} f(\rho) = \frac{1}{2} r^T r, \tag{14}$$

where $C = \{\rho : \rho_m > 0, m = 1, 2, \cdots, M\}$.

The gradient vector of objective function $f(\rho)$ is given by

$$\nabla f(\rho) = (\frac{\partial f}{\partial \rho_1}, \frac{\partial f}{\partial \rho_2}, \cdots, \frac{\partial f}{\partial \rho_M})^T = J^T r, \tag{15}$$

where

$$J = \left(\frac{\partial U}{\partial \rho_1}, \frac{\partial U}{\partial \rho_2}, \cdots, \frac{\partial U}{\partial \rho_M}\right) \tag{16}$$

denotes the Jacobi matrix of r.

By (13),

$$\frac{\partial U}{\partial \rho_m} = P\frac{\partial A_d^{-1}}{\partial \rho_m}b_d, \tag{17}$$

we find that it is adequate to only compute $\frac{\partial A_d^{-1}}{\partial \rho_m}$.

Using the technique in (Cheney, 1999), and noting that $A_d^{-1}A_d = I$, we get $\frac{\partial A_d^{-1}}{\partial \rho_m}A_d + A_d^{-1}\frac{\partial A_d}{\partial \rho_m} = 0$. Then,

$$\frac{\partial A_d^{-1}}{\partial \rho_m} = -A_d^{-1}\frac{\partial A_d}{\partial \rho_m}A_d^{-1}. \tag{18}$$

For A_d is the remaining matrix of A deleted the i_0th row and the i_0th column, we get the expression of $\frac{\partial A_d}{\partial \rho_m}$ similarly from $\frac{\partial A}{\partial \rho_m}$

$$\frac{\partial A_d}{\partial \rho_m} = [\frac{\partial A}{\partial \rho_m}]_d. \tag{19}$$

From (10),

$$\frac{\partial A}{\partial \rho_m} = -\frac{1}{\rho_m^2}[\bar{A}^{(m)}]. \tag{20}$$

Substituting (18), (19) and (20) into (17), and noting that (12), we have

$$\frac{\partial U}{\partial \rho_m} = -PA_d^{-1}\frac{\partial A_d}{\partial \rho_m}A_d^{-1}b_d = -PA_d^{-1}\left[\frac{\partial A}{\partial \rho_m}\right]u_d = \frac{1}{\rho_m^2}PA_d^{-1}\left[\bar{A}^{(m)}\right]_d u_d. \tag{21}$$

Hesse matrix of objective function $f(\rho)$ is defined by $H = \nabla^2 f(\rho) = \left(\frac{\partial^2 f(\rho)}{\partial \rho_m \partial \rho_k}\right)_{M\times M} \approx J^T J$, and Levenberg-Marquardt iteration is formulated as follows for solving (14),

$$\rho^{(k+1)} = \rho^{(k)} - (H + \mu_k I)^{-1}\nabla f(\rho), \tag{22}$$

where I is an identity matrix.

2.3 Numerical Simulation and Data

In this subsection we give several experimental tests to show the successful applications. Take $\Omega = (0,1)^3$, and let Γ_i denote boundary surfaces,

$$\begin{array}{ll}\Gamma_1 : (0,1)\times(0,1)\times\{0\}, & \Gamma_2 : (0,1)\times(0,1)\times\{1\},\\ \Gamma_3 : (0,1)\times\{0\}\times(0,1), & \Gamma_4 : (0,1)\times\{1\}\times(0,1),\\ \Gamma_5 : \{0\}\times(0,1)\times(0,1), & \Gamma_6 : \{1\}\times(0,1)\times(0,1),\end{array}$$

then we find that $\partial\Omega = \bigcup_{i=1}^{6}\bar{\Gamma}_i$. Each $(0,1)$ interval is divided into n_1 subintervals with equal step $h = 1/n_1$, and Ω is divided into n_1^3 cubes. Basis node functions $\{N_i(x)\}$ are defined by three piecewise defined linear interpolation functions. Relative error function is defined by

$$r = \frac{||\rho_h(x) - \rho(x)||_{L^2(\Omega)}}{||\rho(x)||_{L^2(\Omega)}}.$$

2.3.1 Experiment 1

Suppose that exact solution of (1a) is $\rho(x) = u(x) = e^{x_1+x_2+x_3}$, satisfying two boundary conditions (1b) and (1c), where $\psi = (-1)^i, x \in \Gamma_i$. Initial iteration values are defined by $\rho_m^{(0)} \equiv 1, m = 1, 2, \cdots, M$, and numerical data are shown in Table 1.

Table 1. Without perturbation

h	Relative error r	Time cost (second)	Iteration times
1/4	13.35%	10	5
1/8	6.77%	545	8
1/12	4.83%	5988	9
1/16	3.96%	35578	10

To show how the initial data affect numerical results, we consider $\psi(x)$ and $\varphi(x)$ perturbed in the simulation by

$$\psi_{\delta_1}(x) = \psi(x)(1 + \delta_1(\sin \omega\pi x_1 + \sin \omega\pi x_2 + \sin \omega\pi x_3)), \tag{23}$$

$$\varphi_{\delta_2}(x) = \varphi(x)(1 + \delta_2(\cos \omega\pi x_1 + \cos \omega\pi x_2 + \cos \omega\pi x_3)), \tag{24}$$

and give the perturbation effects in Table 2.

Table 2. Perturbation effects ($\omega = 10$)

h	δ_1	δ_2	Relative error r	Time cost(second)	Iteration times
1/8	5%	0	7.34%	550	8
1/8	10%	0	8.39%	555	8
1/12	5%	0	5.52%	6125	9
1/12	10%	0	7.53%	6835	10
1/8	0	0.5%	10.80%	550	8
1/8	0	1%	17.58%	578	8
1/12	0	0.5%	11.06%	6335	9
1/12	0	1%	24.02%	17986	10

2.3.2 Experiment 2

Exact solutions of (1a) are $\rho(x) = (2.5 + x_1^2 - 2x_2^2 + x_3^2)^{-1}$ and $u(x) = \ln(2.5 + x_1^2 - 2x_2^2 + x_3^2)$, where $\psi(x) = ((-1)^i + 1)(\frac{3}{4}(i - 1)^2 - 2)$, $x \in \Gamma_i$. Initial approximations are taken by $\rho_m^{(0)} \equiv 1, m = 1, 2, \cdots, M$, and numerical data are illustrated in Table 3. The boundary conditions $\psi(x)$ and $\varphi(x)$ are perturbed by (23) and (24) (see Table 4).

Table 3. Without perturbation

h	Relative error r	Time cost (second)	Iteration times
1/4	22.36%	17	8
1/8	15.05%	945	12
1/12	14.05%	15368	14
1/16	14.69%	125930	16

Table 4. Perturbation effects ($\omega = 10$)

h	δ_1	δ_2	Relative error r	Time cost(second)	Iteration times
1/8	1%	0	15.13%	959	12
1/8	5%	0	15.96%	947	12
1/12	1%	0	14.12%	15211	14
1/12	5%	0	15.26%	14043	13
1/8	0	0.5%	15.88%	965	12
1/8	0	1%	17.32%	884	11
1/12	0	0.5%	15.41%	14618	13
1/12	0	1%	18.31%	14245	12

2.3.3 Experiment 3

The functions $\rho(x), u^{(1)}(x), \psi^{(1)}(x)$ are taken as $\rho(x), u(x), \psi(x)$ of Experiment 2. $u^{(2)}(x) = x_1x_2x_3$ is exact solution of (1a) corresponding to $\rho(x)$, and Neumann boundary condition is defined by

$$\psi^{(2)}(x) = \begin{cases} (-1)^i x_1x_2(i + 1.5 + x_1^2 - 2x_2^2), & x \in \Gamma_i, \quad i = 1,2, \\ (-1)^i x_1x_3(-2i + 8.5 + x_1^2 + x_3^2), & x \in \Gamma_i, \quad i = 3,4, \\ (-1)^i x_2x_3(i - 2.5 - 2x_2^2 + x_3^2), & x \in \Gamma_i, \quad i = 5,6. \end{cases} \tag{25}$$

The objective function (14) is modified by $f(\rho) = \frac{1}{2}\sum_{l=1}^{2}\left[r^{(l)}\right]^T r^{(l)}$. Let $r^{(l)} = U^{(l)} - U_0^{(l)}$, where $U^{(l)}$ is defined by (13) and $U_0^{(l)}$ is defined by the values at boundary nodes of $u^{(l)}(x)$. Gradient operator and Hesse matrix are formulated by

$$\nabla f(\rho) = \sum_{l=1}^{2}\left[J^{(l)}\right]^T r^{(l)}, \; H = \nabla^2 f(\rho) \approx \sum_{l=1}^{2}\left[J^{(l)}\right]^T J^{(l)},$$

where $J^{(l)}$ denotes Jacobi matrix of $r^{(l)}$. Numerical data are shown in Table 5 and Table 6 after repeated simulations of Experiment 1 and Experiment 2.

Table 5. Without perturbation

h	Relative error r	Time cost (second)	Iteration times
1/4	23.54%	16	7
1/8	13.55%	926	12
1/12	9.87%	14300	14
1/16	8.55%	105410	16

Table 6. Perturbation effects ($\omega = 1$)

h	δ_1	δ_2	Relative error r	Time cost(second)	Iteration times
1/8	1%	0	14.06%	928	12
1/8	5%	0	16.65%	916	12
1/12	1%	0	10.46%	13643	13
1/12	5%	0	14.20%	12783	12
1/8	0	0.5%	17.60%	775	9
1/8	0	1%	21.36%	752	8
1/12	0	0.5%	16.94%	13717	10
1/12	0	1%	20.64%	14055	9

Notes. Since the problem has distinct semi-convergence, so numerical data in Table 5 and Table 6 denote optimal values of iteration computations.

2.3.4 Conclusions and Discussions

From numerical experiments, we conclude that numerical solution of inverse problem can approximate exact solution well as h approaches zero. However, the convergence rate is slow and the computational cost increases rapidly for three-dimensional problems. The problem is ill-conditioned, while numerical solution of Neumann boundary condition is more stable than that of Dirichlet case. Comparing Experiment 2 with Experiment 3, we can develop the accuracy by considering boundary condition but don't increase supplemental work, because the computation cost on modifying the L-M parameter μ_k decreases. The model (2) is testified correctly. For numerical simulation of inverse problems, at present there only has the primary regularization theory of Tikhonov and Lavnentiv (Borcea, 2003,Newman2000; Quarteroni, 2000; Stoer, 1993). Therefore, it is necessary to give a careful discussion on the convergence and stability analysis, and this is a major problem in theoretical research on inverse problem (Wang, 2007; Liu, 2005; Xiao, 2003).

3. Application of Normal Finite Element in Electrical Impedance Tomography

In actual imagining problem, some computation and measures should be solved as follows.

3.1 Quasi-element Stiffness Matrix Strategy

In the previous section, the iteration algorithm is formulated on finite element method, where massive three-dimensional numerical integrals are considered in expressing coefficient matrix, and this is most expensive. Furthermore, coefficient matrices are different in the iterations and give a large amount of additional computation. Thus we introduce quasi-element stiffness matrix and use constant function to approximate $\rho(x)$. Quasi-element stiffness matrix only depends on the values of the node and basis function (5), and doesn't vary during the iteration. The square matrix of order p is described by (10), where p denotes the number of nodes in an element, and all the small-scaled matrices derived from a numerical integral calculation are stored and expanded into the coefficient matrix with $\rho_h(x)$. As shown in the previous numerical experiments, the domain is divided uniformly and all the quasi-element stiffness matrices have a same definition (5). So only an 8×8 square matrix is computed and stored, and this computation cost can be ignored.

Actual imaging region Ω is usually irregular, and it is partitioned into a series of rectangular subdomains and an irregular subdomain. Rectangular domain is divided equally into lots of small rectangles, and the element size determines a quasi-element stiffness matrix only stored one time in the computation. Irregular domain is partitioned by isoparametric element method and all the quasi-element stiffness matrices are stored. For regular elements are more greatly than irregular elements, so the size and quality of stored quasi-element stiffness matrices are very small and their storage and computation are simple.

The quasi-element stiffness matrix of irregular element is determined as follows. In (Borcea, 2003), let $\hat{K}$ denote reference unit, and let node basis function $N_i(\xi)$ be defined by a three piecewise linear interpolation. Define

$$\nabla u = \begin{bmatrix} \frac{\partial u}{\partial x_1} \\ \frac{\partial u}{\partial x_2} \\ \frac{\partial u}{\partial x_3} \end{bmatrix}, \quad \hat{\nabla} u = \begin{bmatrix} \frac{\partial u}{\partial \xi_1} \\ \frac{\partial u}{\partial \xi_2} \\ \frac{\partial u}{\partial \xi_3} \end{bmatrix}, \quad B_m = \begin{bmatrix} \frac{\partial x_1}{\partial \xi_1} & \frac{\partial x_2}{\partial \xi_1} & \frac{\partial x_3}{\partial \xi_1} \\ \frac{\partial x_1}{\partial \xi_2} & \frac{\partial x_2}{\partial \xi_2} & \frac{\partial x_3}{\partial \xi_2} \\ \frac{\partial x_1}{\partial \xi_3} & \frac{\partial x_2}{\partial \xi_3} & \frac{\partial x_3}{\partial \xi_3} \end{bmatrix},$$

then $\nabla u = B_m^{-1} \hat{\nabla} u$. B_m is Jacobi matrix of isoparametric transformation of reference unit $\hat{K}$ to grid unit K_m, and by (5), the quasi-element stiffness matrix $\bar{A}^{(m)}$ is defined by

$$\bar{a}_{ij}^{(m)} = \int_{K_m} (B_m^{-1} \hat{\nabla} N_i) \cdot (B_m^{-1} \hat{\nabla} N_j) |\det(B_m)| d\xi. \tag{26}$$

3.2 Neumann Boundary Condition Argument

Boundary current density $\psi(x)$ (Neumann boundary condition) is immeasurable, so it is necessary to be properly handled. Boundary potential $\varphi(x)$ (Dirichlet boundary condition) can be measured by U_0 only at the electrodes (partial boundary nodes).

Electrodes, whose total quantity is n_0, are supposed to locate at boundary nodes (see Fig. 1), and each electrode only covers a node. The curved face covered by the kth electrode is denoted by γ_k, where the node is x_{j_k} and the measurable current intensity is $I_k = \int_{\gamma_k} \psi dS$. Let I represent the current vector of I_k.

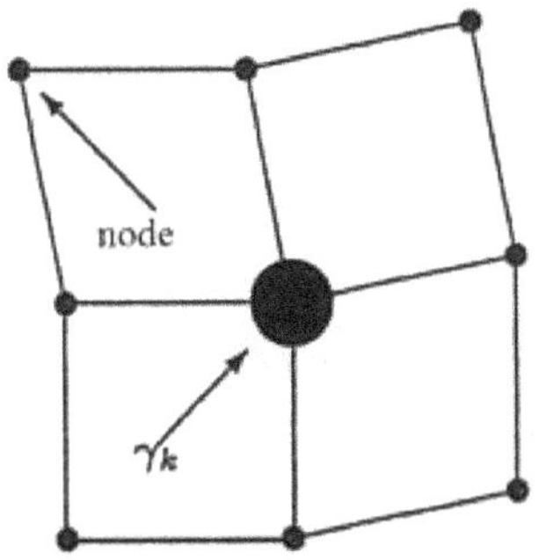

Figure 1. Electrode and grids

Because the current in-out takes place only at electrodes, $\psi(x) = 0, x \in \partial\Omega \setminus \bigcup_{k=1}^{n_0} \gamma_k$, we make integrals on both sides of

(1a) on Ω and use Gauss formula to get

$$\int_{\Omega}\nabla\cdot(\rho^{-1}\nabla u)dx = \int_{\partial\Omega}\rho^{-1}\frac{\partial u}{\partial n}dS = \int_{\partial\Omega}\psi dS = \sum_{k=1}^{n_0}\int_{\gamma_k}\psi dS = \sum_{k=1}^{n_0} I_k = 0. \quad (27)$$

It is obviously seen that the input equals the output, and permissible current vector can be expanded to an n_0-1 dimensional space. The right-hand side term of (4) is treated by

$$b_i = G(N_i) = \int_{\partial\Omega} N_i\psi dS = \sum_{k=1}^{n_0}\int_{\gamma_k} N_i\psi dS \approx \sum_{k=1}^{n_0} N_i(x_{j_k})\int_{\gamma_k}\psi dS = q_{ik}I_{k_i}, \quad (28)$$

where k_i is the electrode number corresponding to the node x_i,

$$q_{ik} = \begin{cases} 1, & j_k = i, \text{ i.e., } x_i \in \gamma_k, \\ 0, & \text{otherwise.} \end{cases} \quad (29)$$

That is to say that

$$b_i = \begin{cases} I_k, & x_i \in \gamma_k, \\ 0, & \text{otherwise.} \end{cases} \quad (30)$$

Let $b = [I]$, where b is an n dimensional vector expanded from the n_0 dimensional vector I according the node numbers. Therefore, (4) is changed into

$$Au = [I]. \quad (31)$$

The right expression is given by the measured current directly, that is to say that the function $\psi(x)$ and numerical integrals are not involved.

Denote $Q = (q_{ij})_{n\times n_0}$, then $[I] = QI$. (31) turns into

$$Au = QI. \quad (32)$$

Similar analysis as in §2 is considered,

$$u_d = A_d^{-1}Q_dI, \quad (33)$$

$$U = Pu_d = PA_d^{-1}Q_dI. \quad (34)$$

Q_d is the remaining matrix of Q deleted the i_0th row. P is an $(n_0-1)\times(n_0-1)$ matrix made up of the numbers 0 and 1, and it can extract the potentials at the electrodes except the k_{i_0} electrode of u_d. U denotes an n_0-1 dimensional vector consisting of the electrode potentials except the reference node.

3.3 Current Model and Simplified Calculation of Jacobi Matrix and Hesse Matrix

(34) shows that the relation of current vector and boundary potential vector (a discretization of F_ρ) is linear. Different boundary currents $I^{(l)}, l = 1, 2, \cdots, L$ are imposed on Ω, where $\{I^{(l)}\}$ at least includes a basis. Measuring the potential at the boundary electrodes $U_0^{(l)}$, then we give an objective function

$$f(\rho) = \frac{1}{2}\sum_{l=1}^{L}[r^{(l)}]^T r^{(l)}, \quad (35)$$

where $r^{(l)} = U^{(l)} - U_0^{(l)}$ and $U^{(l)}$ is defined by (34), $L \geq n_0 - 1$. Similarly,

$$\nabla f(\rho) = \sum_{l=1}^{L}[J^{(l)}]^T r^{(l)}, \quad H = \nabla^2 f(\rho) \approx \sum_{l=1}^{L}[J^{(l)}]^T J^{(l)},$$

where $J^{(l)}$ denotes the Jacobi matrix of $r^{(l)}$. Adopting the expression of (31), then we reformulate (33) and (34) as follows

$$u_d = A_d^{-1}[I]_d, \quad (36)$$

$$U = Pu_d = PA_d^{-1}[I]_d, \quad (37)$$

where $[I]_d$ is an $n-1$ dimensional vector by deleting the i_0th element from $[I]$.

Jacobi matrix, gradient vector and Hesse matrix are computed by (21), and the following equation system is solved

$$A_d y = [\bar{A}^{(m)}]_d u_d. \tag{38}$$

The quantity is determined by the number of partitioned elements and the measurement times, and there at least $n\times(n_0-1)$ systems are solved at each iteration for finite element problem. So its computation is greatly complicated. A technique of (Du, 1997) is introduced and improved. Noting that A_d and $(A_d^{-1})^T = (A_d^T)^{-1} = A_d^{-1}$ are symmetric and positive definite, we give another expression of (21),

$$\frac{\partial U}{\partial \rho_m} = \frac{1}{\rho_m^2} P A_d^{-1} \left[\bar{A}^{(m)}\right]_d u_d = \frac{1}{\rho_m^2} \left(A_d^{-1} P^T\right)^T \left[\bar{A}^{(m)}\right]_d u_d. \tag{39}$$

P^T is an $(n-1)\times(n_0-1)$ matrix, thus the calculation of $A_d^{-1}P^T$ turns into solving (n_0-1) systems of equations. Moreover, the current model is arranged as follows. A unit current is input at the reference electrode (the i_0th node) and is output at the others. The reference electrode is supposed to be labeled by the node n, and the current model is equivalent to the following current basis,

$$I^{(1)} = (1,0,0,\cdots,0,-1)^T, I^{(2)} = (0,1,0,\cdots,0,-1)^T, \cdots, I^{(n_0-1)} = (0,0,0,\cdots,1,-1)^T. \tag{40}$$

Collecting the vectors as $W = ([I^{(1)}],[I^{(2)}],\cdots,[I^{(n_0-1)}])$, and deleting the i_0th row, we have

$$W_d = ([I^{(1)}]_d,[I^{(2)}]_d,\cdots,[I^{(n_0-1)}]_d). \tag{41}$$

This shows the entries -1 are deleted from W, and $W_d = P^T$. Then,

$$A_d^{-1}P^T = A_d^{-1}W_d = (A_d^{-1}[I^{(1)}]_d, A_d^{-1}[I^{(2)}]_d,\cdots,A_d^{-1}[I^{(n_0-1)}]_d) = (u_d^{(1)},u_d^{(2)},\cdots,u_d^{(n_0-1)}) = V. \tag{42}$$

$u_d^{(l)}$ computed previously is the solution of (32), and it is the potential vector of the lth time imposing current $I_d^{(l)}$ except the reference node i_0. Then (39) is equivalent to

$$\frac{\partial U}{\partial \rho_m} = \frac{1}{\rho_m^2} V^T \left[\bar{A}^{(m)}\right]_d u_d. \tag{43}$$

Therefore, it is adequate to solve $n_0 - 1$ finite element problems (systems of equations) at each iteration, and the computations of Jacobi matrix, gradient vector and Hesse matrix end out of the complicated formula (38).

A numerical experiment follows on a cylinder domain $\Omega = \{x : x_1^2 + x_2^2 \le 1, 0 \le x_3 \le 1\}$ with $\rho(x) = 2 - x_1^2 - x_2^2 + x_3^2$. The function $\psi(x)$ has no analytical expression, and $I^{(l)}$ is known. Using this method above we can get numerical solutions of $u(x)$ and boundary potential $U^{(l)}$ to replace measurement solution $U_0^{(l)}$. Initial approximation is taken by $\rho_m \equiv 1$, and the electrical resistivity ρ is solved by L-M method. It aims to testify how the approximation $\rho_h(x)$ approaches the preestablished function $\rho(x)$. The domain Ω is divided into 10 element strips from the top to the bottom, 300 units in each strip and the total number is 3000. The domain is separated by 11 node layers, 331 nodes in each layer, and the total number is 3641. In the simulation 261 electrodes are arranged and numerical data are shown in Table 7.

Table 7. Simulation results on a cylinder

Figure No.	Boundary current perturbation	Boundary potential perturbation	Relative error r	Time cost (second)	Iteration times
3	0	0	2.26%	18906	20
4	0.1%	0	3.81%	19117	20
5	0	0.1%	3.92%	19177	20

For convenience, we take the section $x_3 = 0.55$ as an illustration, and the comparisons of true data and numerical simulations are shown in Fig. 2–Fig. 5. All the numerical data are obtained by Matlab software in our personal computer.

4. Low Order Mixed Element Numerical Method of Inverse Problem and Numerical Simulation

4.1 Low Order Mixed Element Method

A local conservative low order mixed element method, or named block-centered finite volume method, is discussed in this section (Russell, 1983,Weiser1988; Mishev, 1998; Lazarov, 1996; Larsson, 2003). The domain Ω is enclosed by several

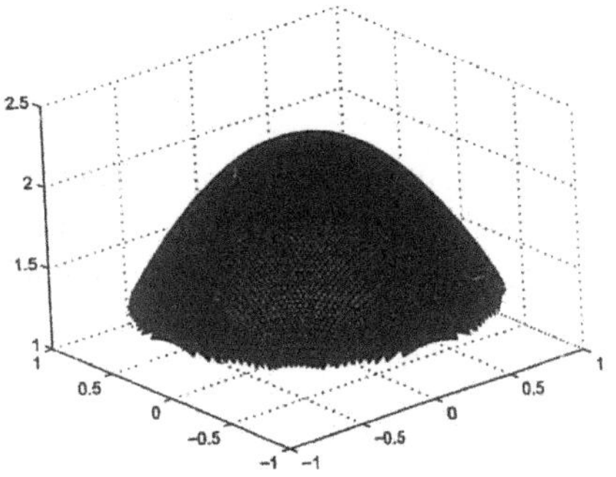

Figure 2. True distribution

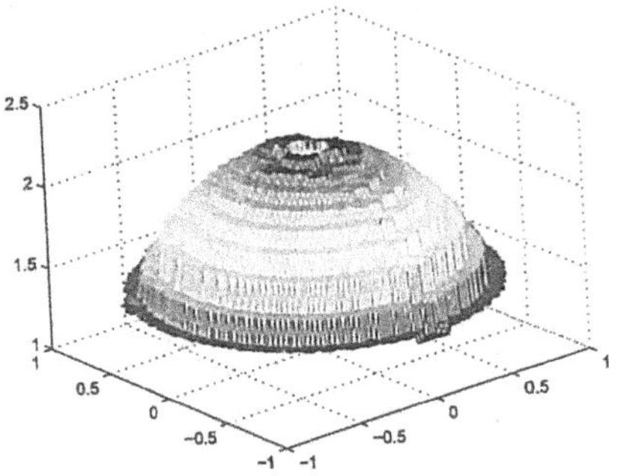

Figure 3. EIT simulation without perturbation

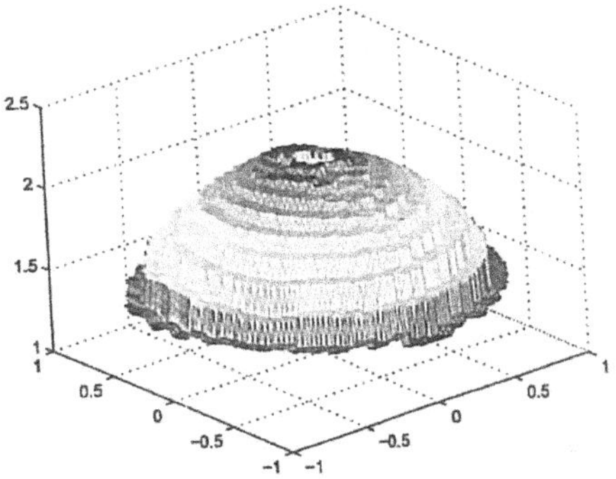

Figure 4. EIT simulation with current perturbation

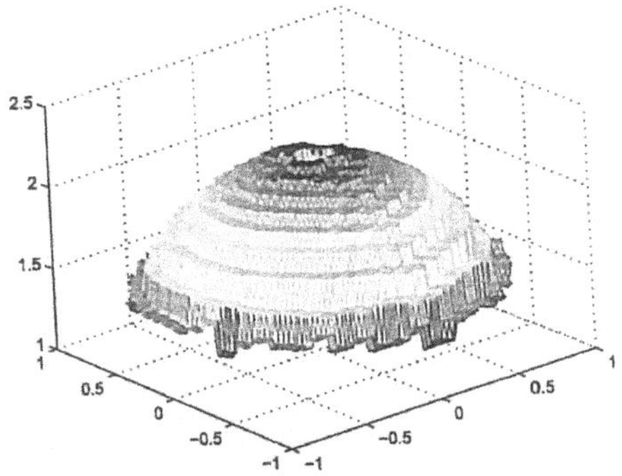

Figure 5. EIT simulation with potential perturbation

faces parallel to the coordinate planes. The space R^3 is partitioned by lots of rectangular solids with the length, the weight and the height equal to (h_1, h_2, h_3), whose centers $x_i = (x_{i,1}, x_{i,2}, x_{i,3})$ form a grid $\bar{\omega} = \{x_i\}_{i=1}^n$, and the center of boundary block is supposed to lie in the boundary $\partial\Omega$ (see Fig. 6 and Fig. 7). The subset of cuboid and Ω is called control volume, in symbol V_i. It holds obviously

$$\bar{\Omega} = \bigcup_{i=1}^{n} \bar{V}_i,\ V_i \cap V_j = \emptyset,\ \bar{V}_i \cap \bar{V}_j = \gamma_{ij}, i \neq j,$$

where γ_{ij} is a rectangle and its area is denoted by $m(\gamma_{ij})$. Let $\omega = \bar{\omega} \cap \Omega$, $\partial\omega = \bar{\omega} \cap \partial\Omega$. Introduce the notations,

$$d(x_i, x_j) = [\sum_{k=1}^{3}(x_{i,k} - x_{j,k})^2]^{1/2},\ \Sigma(i) = \{j :\ j \neq i, m(\gamma_{ij}) \neq 0\}.$$

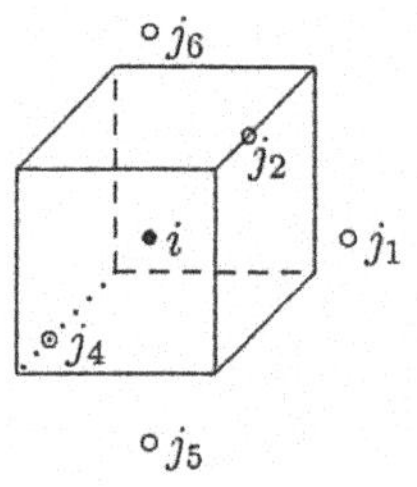

Figure 6. Inner nodes, control volume V_i and adjacent points $\Sigma(i)$

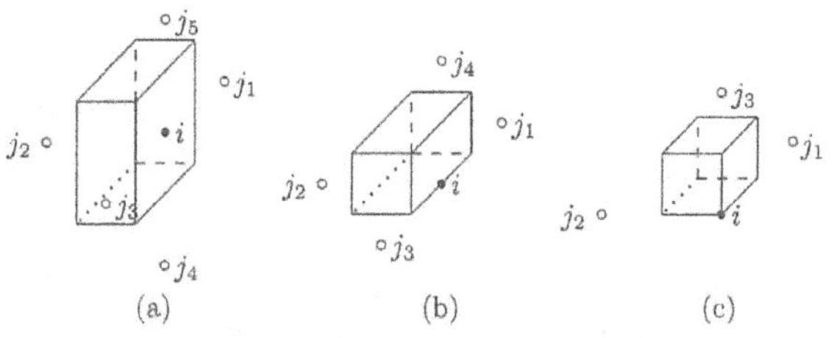

Figure 7. Boundary points, control volume V_i and adjacent points $\Sigma(i)$

Making integral on both sides of (1a) on any subset $V \subset \Omega$, and using Gauss formula, we get the electric quantity balance equation,

$$\int_{\partial V} \rho^{-1} \frac{\partial u}{\partial \nu} dS = 0. \tag{44}$$

Taking $V = V_i$, and letting

$$d_{ij} = d(x_i, x_j),\ \rho_{ij} = \frac{\rho(x_i) + \rho(x_j)}{2}, \tag{45}$$

then, we have

$$\begin{aligned} -\int_{\partial V_i} \rho^{-1}\frac{\partial u}{\partial \nu}dS &= -\int_{\partial V_i|_\Omega} \rho^{-1}\frac{\partial u}{\partial \nu}dS - -\int_{\partial V_i \cap \partial\Omega} \rho^{-1}\frac{\partial u}{\partial \nu}dS \\ &= -\sum_{j\in\Sigma(i)}\int_{\gamma_{ij}} \rho^{-1}\frac{\partial u}{\partial \nu}dS - \int_{\partial V_i\cap\partial\Omega}\psi(x)dS \\ &\approx -\sum_{j\in\Sigma(i)}\frac{1}{\rho_{ij}}\frac{u_{h,j}-u_{h,i}}{d_{ij}}m(\gamma_{ij}) - \int_{\partial V_i\cap\partial\Omega}\psi(x)dS. \end{aligned} \tag{46}$$

As $x_i \in \bar{\omega}$, it is true that $\partial V_i \cap \partial\Omega = \emptyset$, $\int_{\partial V_i\cap\partial\Omega}\psi(x)dS = 0$.

Therefore, we get the algorithm of finite volume element,

$$\mathcal{L}_h u_h = \psi_h, \tag{47}$$

where

$$\mathcal{L}_h u_{h,i} \equiv -\sum_{j\in\Sigma(i)}\frac{1}{\rho_{ij}}\frac{u_{h,j}-u_{h,i}}{d_{ij}}m(\gamma_{ij}),\ i = 1, 2, \cdots, n, \tag{48}$$

$$\psi_{h,i} = \int_{\partial V_i\cap\partial\Omega}\psi(x)dS. \tag{49}$$

Let

$$a_{ii} = \sum_{j\in\Sigma(i)}\frac{1}{\rho_{ij}}\frac{m(\gamma_{ij})}{d_{ij}},\ a_{ij} = -\frac{1}{\rho_{ij}}\frac{m(\gamma_{ij})}{d_{ij}},\ j \in \Sigma(i), \tag{50}$$

then, we have the following statement

Theorem 1 Suppose that the grid $\bar{\omega}$ is connected and the coefficient matrix of (47) is denoted by $A = (a_{ij})_{n\times n}$, then A is symmetric and positive semi-definite, the rank $r(A) = n - 1$, and any equation of (47) can be derived from the other $n - 1$ equations.

Proof: From (50) it follows clearly that A is symmetric.

$y(x)$ and $z(x)$ are supposed to be grid functions on $\bar{\omega}$, then by the symmetry of A we have

$$\begin{aligned} (\mathcal{L}_h y, z) &= -\sum_{x_i\in\bar{\omega}} z_i \sum_{j\in\Sigma(i)}\frac{1}{\rho_{ij}}\frac{y_j - y_i}{d_{ij}}m(\gamma_{ij}) \\ &= \frac{1}{2}\sum_{x_i\in\bar{\omega}}\sum_{j\in\Sigma(i)}\frac{1}{\rho_{ij}}\frac{(y_j-y_i)z_j - (y_j-y_i)z_i}{d_{ij}}m(\gamma_{ij}) \\ &= \frac{1}{2}\sum_{x_i\in\bar{\omega}}\sum_{j\in\Sigma(i)}\frac{1}{\rho_{ij}}\frac{(y_j-y_i)(z_j-z_i)}{d_{ij}}m(\gamma_{ij}). \end{aligned} \tag{51}$$

Take $z = y$ in the above expression, we have

$$(\mathcal{L}_h y, y) = \frac{1}{2}\sum_{x_i\in\bar{\omega}}\sum_{j\in\Sigma(i)}\frac{1}{\rho_{ij}}\frac{(y_j-y_i)^2}{d_{ij}}m(\gamma_{ij}) \geq 0.$$

Then A is positive semi-definite.

Considering $\mathcal{L}_h y = 0$ and the above expression together, we have $y_j = y_i$, $j \in \Sigma(i)$, $x_i \in \bar{\omega}$, then by the connected property of $\bar{\omega}$ we have $y_i \equiv C$. Then the kernel space of A is one-dimensional, therefore $r(A) = n - 1$.

Making summation on $i = 1, 2, \cdots, n$ of (47),

$$\sum_{i=1}^{n}\mathcal{L}_h u_{h,i} = -\sum_{i=1}^{n}\sum_{j\in\Sigma(i)}\frac{1}{\rho_{ij}}\frac{u_{h,j}-u_{h,i}}{d_{ij}}m(\gamma_{ij}) = (\mathcal{L}_h u_h, 1) = 0.$$

Making integrals on both sides of (1a) on Ω, and using Gauss formula and (1b), we have

$$\int_\Omega \nabla\cdot(\rho^{-1}\nabla u)dx = \int_{\partial\Omega}\rho^{-1}\frac{\partial u}{\partial v}dS = \int_{\partial\Omega}\psi(x)dS = 0,$$

then we get

$$\sum_{i=1}^{n}\psi_{h,i} = \sum_{x_i\in\partial\omega}\int_{\partial V_i\cap\partial\Omega}\psi(x)dS = \int_{\partial\Omega}\psi(x)dS = 0.$$

That is to say that any equation of (47) can be expressed by the other $n-1$ equations. □

From the proof of Theorem 1, we can get the following corollary.
Corollary 1: Any two solutions of (47) are different by a constant.

Given $\rho(x)$, it is next to solve the equation (47). Let $u = (u_{h,1}, u_{h,2}, \cdots, u_{h,n})^T$, and express (47) in matrices

$$Au = b. \tag{52}$$

By Theorem 1 and Corollary 1, the potential at a node x_{i_0} (electrode) is taken as the reference potential $u_{h,i_0} = 0$, then the solution of (47) exists and is unique. Deleting the i_0th equation (as a redundant equation), we have

$$A_d u_d = b_d, \tag{53}$$

where the matrix A_d is obtained by deleting the entries of the i_0th row and the i_0th column and it is symmetric and positive definite. u_d and b_d are obtained from u and b similarly. Its solution is

$$u_d = A_d^{-1}b_d. \tag{54}$$

U denotes a potential vector of all the boundary nodes except the reference node, and U is made up of the partial components of u_d. Therefore,

$$U = Pu_d = PA_d^{-1}b_d, \tag{55}$$

where the entries of P are assigned by the number 0 or by the number 1, and the values correspond to the potentials of the nodes except the reference node.

4.2 Mixed Element Iteration Algorithm of Inverse Problem

Let $U_0^{(l)}$ denote the vector of the values of $\phi^{(l)}$ at the boundary nodes except x_{i_0}, $l = 1, 2, \cdots, L$. U is defined by (55), an approximation to $U_0^{(l)}$ by using finite volume element, and $\rho = (\rho_1, \rho_2, \cdots, \rho_n)^T$ is defined by the values of $\rho_h(x)$ at all the grids. Let $r^{(l)} = U^{(l)} - U_0^{(l)}$, then we solve inverse problem by solving the following nonlinear optimization problem,

$$\min_{\rho_h\in\mathscr{R}^n} f(\rho_h) = \frac{1}{2}\sum_{l=1}^{L}[r^{(l)}]^T r^{(l)}, \tag{56}$$

where $\mathscr{R}_+^n = \{z\in\mathscr{R}^n : z_i > 0, i = 1, 2, \cdots, n\}$.

Here we adopt the method of Levenberg-Marquardt, and we have to formulate the gradient vector and the Hesse matrix of objective function.

The gradient vector of $f(\rho)$, $\nabla f(\rho) = (\frac{\partial f}{\partial\rho_1}, \frac{\partial f}{\partial\rho_2}, \cdots, \frac{\partial f}{\partial\rho_n})^T$, is calculated later,

$$\frac{\partial f}{\partial\rho_m} = \sum_{l=1}^{L}(U^{(l)} - U_0^{(l)})^T\frac{\partial U^{(l)}}{\partial\rho_m}. \tag{57}$$

For convenience to discuss, we omit all the superscript (l). By the definition (55), it holds

$$\frac{\partial U}{\partial\rho_m} = P\frac{\partial A_d^{-1}}{\partial\rho_m}b_d. \tag{58}$$

and it is adequate to only compute $\frac{\partial A_d^{-1}}{\partial\rho_m}$.

By $A_d^{-1}A_d = I$, we have $\frac{\partial A_d^{-1}}{\partial \rho_m}A_d + A_d^{-1}\frac{\partial A_d}{\partial \rho_m} = 0$, then we have $\frac{\partial A_d^{-1}}{\partial \rho_m} = -A_d^{-1}\frac{\partial A_d}{\partial \rho_m}A_d^{-1}$. Then it follows from (58) and (54),

$$\frac{\partial U}{\partial \rho_m} = -PA_d^{-1}\frac{\partial A_d}{\partial \rho_m}A_d^{-1}b_d = -PA_d^{-1}\frac{\partial A_d}{\partial \rho_m}u_d. \tag{59}$$

From the definition of A_d, we get that $\frac{\partial A_d}{\partial \rho_m}$ is derived from $\frac{\partial A}{\partial \rho_m}$ by deleting the i_0th row and the i_0th column,

$$\frac{\partial A_d}{\partial \rho_m} = [\frac{\partial A}{\partial \rho_m}]_d. \tag{60}$$

Let $\frac{\partial A}{\partial \rho_m} = (\alpha_{ij}^m)_{n\times n}$, then

$$\alpha_{ij}^m = \begin{cases} -\sum\limits_{k\in\Sigma(m)} \frac{2}{(\rho_m+\rho_k)^2}\cdot\frac{m(r_{mk})}{d_{mk}}, & i=j=m,\\ -\frac{2}{(\rho_m+\rho_i)^2}\cdot\frac{m(r_{mi})}{d_{mi}}, & i=j\in\Sigma(m),\\ \frac{2}{(\rho_m+\rho_j)^2}\cdot\frac{m(r_{mj})}{d_{mj}}, & i=m, j\in\Sigma(m),\\ \frac{2}{(\rho_m+\rho_i)^2}\cdot\frac{m(r_{mi})}{d_{mi}}, & j=m, i\in\Sigma(m),\\ 0, & \text{otherwise.}\end{cases} \tag{61}$$

Hesse matrix of objective function $f(\rho)$ is defined by $\nabla^2 f(\rho) = (\frac{\partial^2 f(\rho)}{\partial\rho_m\partial\rho_k})_{n\times n}$, where

$$\begin{aligned}\frac{\partial^2 f(\rho)}{\partial\rho_m\partial\rho_k} &= \sum_{l=1}^{L}(\frac{\partial U^{(l)}}{\partial\rho_k})^T\frac{\partial U^{(l)}}{\partial\rho_m} + \sum_{l=1}^{L}(U^{(l)} - U_0^{(l)})^T\frac{\partial^2 U^{(l)}}{\partial\rho_m\partial\rho_k}\\ &\approx \sum_{l=1}^{L}(\frac{\partial U^{(l)}}{\partial\rho_k})^T\frac{\partial U^{(l)}}{\partial\rho_m}.\end{aligned} \tag{62}$$

The Levenberg-Marquardt iteration is formulated as follows to solve (56)

$$\rho^{(k+1)} = \rho^{(k)} - (H + \mu_k I)^{-1}\nabla f(\rho), \tag{63}$$

where I is an identity matrix.

4.3 Mixed Finite Element Numerical Approximation of Inverse Problem

In this subsection we give several experimental tests to show the feasibility. Take $\Omega = (0, 0.8)\times(0, 1)\times(0, 1.2)$, and define discrete norms on $\bar{\omega}$ as follows

$$|v(x)|_{0,h} = (\sum_{x_i\in\bar{\omega}} v(x_i)^2 m(V_i))^{1/2}, \tag{64}$$

where $m(V_i)$ denotes the volume of V_i, and relative error is

$$r = \frac{|\rho_h(x) - \rho(x)|_{0,h}}{|\rho(x)|_{0,h}}. \tag{65}$$

4.3.1 Experiment 1

Suppose that exact solution of inverse problem is defined by $\rho(x) = e^{x_1+x_2+x_3}$, and exact solution of (1a) satisfying two types of boundary conditions is

$$u^{(l)}(x) = e^{x_1+x_2+x_3}\hat{u}^{(l)}(x),$$

where $\hat{u}^{(l)}(x) = (a^{(l)}x_1 + b^{(l)}x_2 + c^{(l)}x_3 + d^{(l)}), a^{(l)} + b^{(l)} + c^{(l)} = 0, l = 1, 2, 3, 4$. Then,

$$\psi^{(l)}(x) = [\hat{u}^{(l)}(x)(1, 1, 1) + (a^{(l)}, b^{(l)}, c^{(l)})]\cdot\nu,$$

and and the values of $a^{(l)}, b^{(l)}, c^{(l)}, d^{(l)}$ are given in Table 8.

Table 8. The values of related coefficients

l	$a^{(l)}$	$b^{(l)}$	$c^{(l)}$	$d^{(l)}$
1	0	0	0	1
2	−1	0.5	0.5	0
3	0.5	−1	0.5	0
4	0.5	0.5	−1	0

To show how initial data affect numerical results, we give the perturbations of $\psi(x)$ and $\varphi(x)$ as follows

$$\psi^{(l)}_{\delta_1}(x) = \psi^{(l)}(x)(1 + \delta_1(\sin \omega\pi x_1 + \sin \omega\pi x_2 + \sin \omega\pi x_3)), \tag{66}$$

$$\varphi^{(l)}_{\delta_2}(x) = \varphi^{(l)}(x)(1 + \delta_2(\cos \omega\pi x_1 + \cos \omega\pi x_2 + \cos \omega\pi x_3)), \tag{67}$$

replacing $\psi(x)$, $\varphi(x)$ for numerical simulation. Initial approximations are taken by $\rho_i^{(0)} \equiv 1, i = 1, 2, \cdots, n$, and numerical data are illustrated in Table 9.

Table 9. Perturbation effects ($\omega = 10$)

(h_1, h_2, h_3)	δ_1	δ_2	r	Time cost(second)	Iteration times
(0.1600, 0.1250, 0.1500)	0	0	4.71%	365	11
(0.0800, 0.0625, 0.0750)	0	0	3.32%	23587	15
(0.1600, 0.1250, 0.1500)	5%	0	5.16%	368	11
(0.0800, 0.0625, 0.0750)	5%	0	4.73%	20660	13
(0.1600, 0.1250, 0.1500)	0	1%	11.71%	295	9
(0.0800, 0.0625, 0.0750)	0	1%	13.87%	12179	8

4.3.2 Experiment 2

Exact solution of inverse problem is defined by $\rho(x) = (2.5 + x_1^2 - 2x_2^2 + x_3^2)^{-1}$, and exact solution of (1a) and Neumann boundary condition are given as follows,

$$u^{(1)}(x) = \ln(2.5 + x_1^2 - 2x_2^2 + x_3^2), x \in \Omega,\ \psi^{(1)}(x) = (2x_1, -4x_2, 2x_3) \cdot \nu, x \in \partial\Omega,$$
$$u^{(2)}(x) = x_1x_2x_3, x \in \Omega,\ \psi^{(2)}(x) = (2.5 + x_1^2 - 2x_2^2 + x_3^2)(x_2x_3, x_1x_3, x_1x_2) \cdot \nu, x \in \partial\Omega.$$

Initial approximations are taken by $\rho_i^{(0)} \equiv 1, i = 1, 2, \cdots, n$, and numerical data are shown in Table 10 after a similar process of the above experiment.

Table 10. Perturbation effect ($\omega = 10$)

(h_1, h_2, h_3)	δ_1	δ_2	r	Time cost(second)	Iteration times
(0.1600, 0.1250, 0.1500)	0	0	26.34%	518	14
(0.0800, 0.0625, 0.0750)	0	0	21.43%	43718	20
(0.1600, 0.1250, 0.1500)	5%	0	27.92%	486	13
(0.0800, 0.0625, 0.0750)	5%	0	24.34%	37338	16
(0.1600, 0.1250, 0.1500)	0	1%	32.95%	425	11
(0.0800, 0.0625, 0.0750)	0	1%	31.68%	29301	11

Notes. Since the problem is distinctly semi-convergent, so numerical data in Table 9 and Table 10 denote optimal values of iteration computations.

4.3.3 Conclusions and Discussions

From numerical experiments, we conclude that numerical solution of inverse problem can approximate exact solution well as h approaches zero. Meanwhile, the convergence rate is slow and the computational cost increases rapidly for three-dimensional problems. The problem is ill-posed, while numerical solution of Neumann boundary condition is more stable than that of Dirichlet case. For direct problems, low order mixed element (block-centered finite volume element) is argued as a powerful tool in numerical simulation of oil reservoir (Russell, 1995; Jones, 1995; Rui, 2012; Rui, 2013; Rui, 20122; Yuan, 2016; Yuan, 2015), which originates from the principle of mass conservation in physical science and is expressed in a discrete formulation. It has many advantages such as the simplicity and the small-scaled calculation of the scheme. The coefficient matrix is symmetric and positive definite and the problem is solved easily. Its numerical solution is convergent of second order accuracy. Therefore, the convergence and stability analysis of inverse problem should be paid more attention.

5. Mixed Element Simulation Experiment of Image Reconstruction

In actual image reconstruction, Neumann boundary condition $\psi(x)$ denotes the boundary current density and is immeasurable. What we can measure is the current intensity through the electrodes, and it is just handled by finite volume element. The electrodes only cover partial boundary nodes, so the boundary potentials (boundary condition) we measure denote the potential values of partial boundary nodes. Generally, the computational region is not a cuboid. The simulation experiment of image reconstruction is discussed and the simple algorithm of Jacobi matrix is shown.

Let $\Omega_1 = (0,1)\times(0,1)\times(0,0.5)$, $\Omega_2 = (0.25,0.75)\times(0.25,0.75)\times(0.5,1)$, and take $\bar{\Omega} = \bar{\Omega}_1 \cup \bar{\Omega}_2$. N denotes the number of electrodes, and Γ_i denotes the contact surface of the ith electrode and Ω (see Fig. 8 and Fig. 9). For any interface Γ_k, it is supposed to find $x_i \in \partial\omega$ such that $x_i \in \Gamma_k \subset (\partial V_i \cap \partial\Omega)$, i.e., the electrodes are supposed to be set at the boundary nodes, then $\Gamma_i \subset \partial\Omega, i = 1,2,\cdots,N, \bar{\Gamma}_i \cap \bar{\Gamma}_j = \emptyset, i \neq j$. Let $\Gamma_0 = \partial\Omega \setminus \cup_{i=1}^N \Gamma_i$, then $\partial\Omega = \cup_{i=0}^N \Gamma_i$.

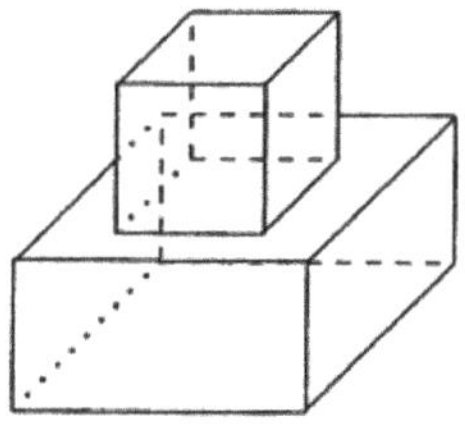

Figure 8. Stepped domain Ω

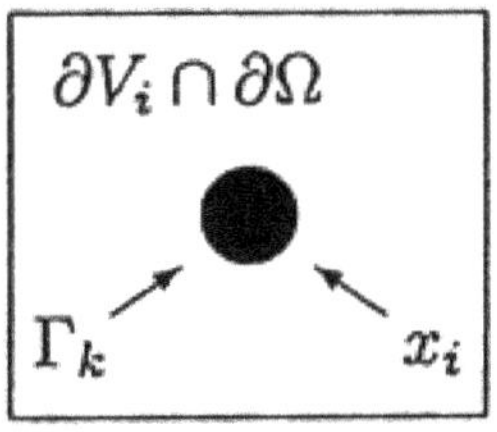

Figure 9. Electrode set

In electrical impedance tomograph problems, electric current flows in and out only through the interface, then we give the following restriction,

$$\psi(x) = 0, \ x \in \Gamma_0. \tag{68}$$

Define $J_i = \int_{\Gamma_i} \psi(x)dS, i = 1,2,\cdots,N$, and electric current vector $J = (J_1,\cdots,J_N)^T$, then we have

Lemma 1. Suppose that $u(x)$ is the solution of (1a) and (1b) satisfying Neumann boundary condition, and $\psi(x)$ satisfies (68), then we have $\sum_{i=1}^N J_i = 0$.

Proof: Making integrals on both sides of (1a) on Ω, applying Gauss formula, and using the assumption (68), we have

$$\int_\Omega \nabla\cdot(\rho^{-1}\nabla u)dx = \int_{\partial\Omega} \rho^{-1}\frac{\partial u}{\partial\nu}dS = \sum_{i=1}^N \int_{\Gamma_i} \psi(x)dS = \sum_{i=1}^N J_i = 0.$$

□

From Lemma 1 we see that the power inside Ω is conservative and allowable current vector can be expanded into an $(N-1)$ dimensional vector.

Later we discuss the relation of current vector J and the right-hand side terms of (47) or (52). By (49), we have for $x_i \in \omega$,

$$b_i = 0. \tag{69}$$

For $x_i \in \partial\omega$, there is not any electrode at x_i, $\partial V_i \cap \partial\Omega \subset \Gamma_0$, then we have

$$b_i = \int_{\partial V_i\cap\partial\Omega} \psi(x)dS = \int_{\partial V_i\cap\partial\Omega} 0dS = 0. \tag{70}$$

Considering $x_i \in \partial\omega$, an electrode k is set at x_i, $x_i \in \Gamma_k, k \neq 0$, then we have

$$b_i = \int_{\partial V_i \cap \partial\Omega} \psi(x)dS = \int_{\partial\Gamma_k} \psi(x)dS = J_k. \quad (71)$$

From (69)-(71), it follows that b is an n-dimensional vector expanded by an N-dimensional vector J in the order of nodes, denoted by $b = [J]$. Eq. (53) is rewritten by

$$A_d u_d = [J]_d. \quad (72)$$

Jacobi matrix, gradient vector and Hesse matrix are computed by (59), where a series of large-scaled equations are argued such as

$$A_d y = z. \quad (73)$$

The number of equations is determined by the number of grid points and measurement times. Using (59), we solve at least $n \times (N-1)$ equations during an iteration of direct problem. We give some proper modifications to solve the large-scaled computations. Noting that A_d is symmetric and positive definite, we see that $(A_d^{-1})^T = (A_d^T)^{-1} = A_d^{-1}$ is still symmetric and positive definite, and we change (59) into

$$\frac{\partial U}{\partial \rho_m} = -PA_d^{-1}[\frac{\partial A}{\partial \rho_m}]_d u_d = -(A_d^{-1}P^T)^T[\frac{\partial A}{\partial \rho_m}]_d u_d. \quad (74)$$

P^T is an $(n-1)\times(N-1)$ matrix, and $A_d^{-1}P^T$ is computed by solving $(N-1)$ equations. Electric current model is taken as follows. Unit current is input at reference electrode i_0 and is output at other electrodes in turns. The reference electrode is set at the node n, the current model is interpreted by a basis

$$J^{(1)} = (1,0,0,\cdots,0,0,-1)^T, J^{(2)} = (0,1,0,\cdots,0,0,-1)^T, \cdots, J^{(N-1)} = (0,0,0,\cdots,0,1,-1)^T. \quad (75)$$

Collecting all the n-dimensional current vectors together as $W = ([J^{(1)}],[J^{(2)}],\cdots,[J^{(N-1)}])$, then deleting the i_0th row, we have

$$W_d = ([J^{(1)}]_d,[J^{(2)}]_d,\cdots,[J^{(N-1)}]_d). \quad (76)$$

It means deleting the row of the number -1 of W, then we have $W_d = P^T$. Thus,

$$A_d^{-1}P^T = A_d^{-1}W_d = (A_d^{-1}[J^{(1)}]_d, A_d^{-1}[J^{(2)}]_d,\cdots,A_d^{-1}[J^{(N-1)}]_d) = (u_d^{(1)},u_d^{(2)},\cdots,u_d^{(N-1)}) = G. \quad (77)$$

In the above expressions $u_d^{(l)}$ denotes the potential at different nodes except the reference node according to the lth imposed current $J^{(l)}$, and it is the solution of finite volume element equation (52). The potential has been computed previously. Eq. (59) is changed into

$$\frac{\partial U}{\partial \rho_m} = -G^T[\frac{\partial A}{\partial \rho_m}]u_d. \quad (78)$$

Therefore, it is adequate to solve $N-1$ direct equations in every iteration of EIT and (73) is not used to solve Jacobi matrix. By (57) and (62), we compute gradient vectors and Hesse matrix similarly.

We apply Matlab software and discuss image reconstruction simulation to testify the feasibility of this type of algorithms. Spacial step is taken equally by $h_1 = h_2 = h_3 = 0.0625$, the number of total nodes is $n = 3249$, the number of boundary nodes is 1282 and the number of electrodes is $N = 610$.

Numerical simulation consists of two different modules.

One is observation simulation module. The electrical resistivity distribution is assigned by $\rho(x) = 1 + x_1^2 + x_2^2 + x_3^2$, and a current basis is defined by (75). At electrode N a unit current is input and is output from the electrodes $1,2,\cdots,n$. The boundary conditions of two types and analytical solution of (1a) are not formulated explicitly, while we can use the method of finite volume element to solve Neumann problems. By (55) we get the boundary potential vector $U^{(i)}$ to replace measured values $U_0^{(i)}, i = 1,2,\cdots,N-1$.

Another is image reconstruction module. Giving proper perturbations for electric field intensity or potential, taking initial values by $\rho_0^{(i)} \equiv 0, i = 1,2,\cdots,n$, we use L-M algorithm to get electrical resistivity distribution ρ, then we want to confirm whether the resulting distribution approximates the preestablished values well or not. The comparisons are illustrated at $x_3 = 0.25$ in the following figures (Fig. 10–Fig. 13), whose simulation data are shown in Table 11.

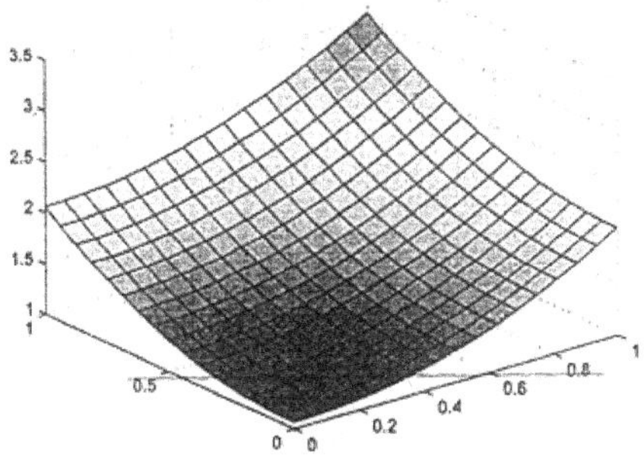

Figure 10. Actual electrical resistivity distribution

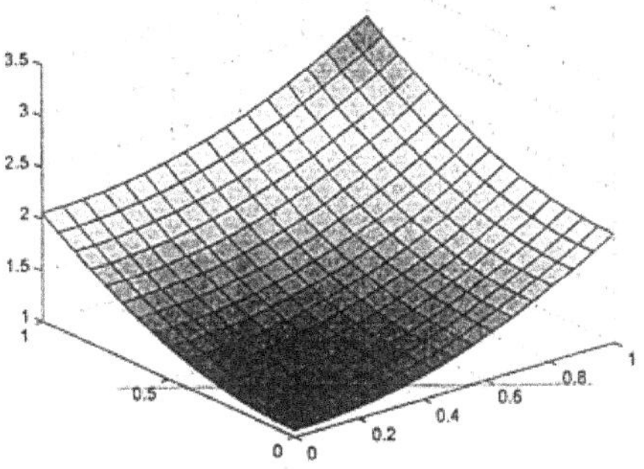

Figure 11. Numerical distribution without perturbation

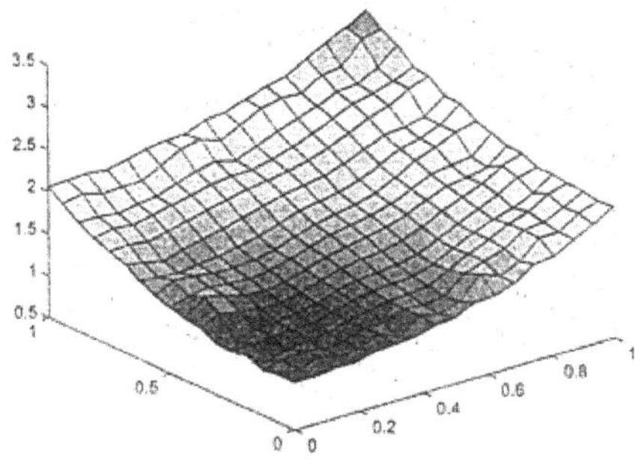

Figure 12. Numerical distribution with 0.01% current perturbation

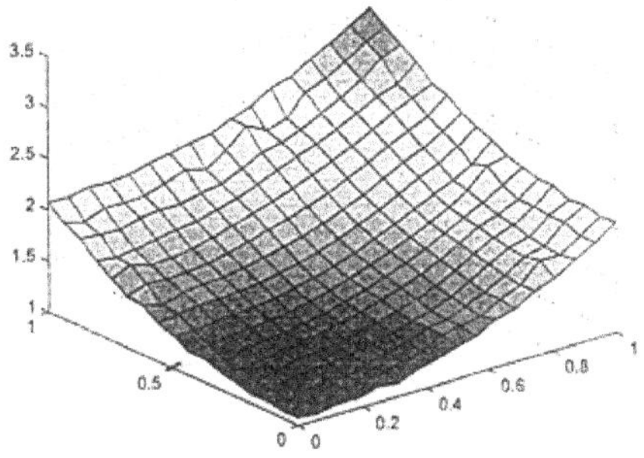

Figure 13. Numerical distribution with 0.01% potential perturbation

Table 11. Simulation data on stepped region

Figure No.	Current perturbation	Potential perturbation	Relative error r	Time cost (second)	Iteration times
11	0	0	0.03%	118650	40
12	0.01%	0	3.35%	18319	6
13	0	0.01%	1.85%	23776	8

6. Discussions

In this paper the highlights are concluded as follows.

(I) Electrical impedance tomography is a new technique in simulating biological medicine imagining technology. So this research has important theoretical and applicable values.

(II) Quasi-element stiffness matrix is applied in the iterations and has invariance property, so numerical integral computation of coefficient matrix is shortened greatly.

(III) A special current forcing model is put forward to compute gradient vector and Hesse matrix by using enough data. It is not necessary to consider large-scale linear equations.

(IV) Based on standard finite element method, a conservative low-order mixed finite element is established and its semi-positive definiteness and existence are testified.

(V) From lots numerical simulations and data, we conclude that this method is a powerful, efficient and effective tool to solve some science problems such as geophysical exploration and hydrogeology.

Acknowledgements

The authors express their deep appreciation to Prof. J. Douglas Jr, and Prof. Jiang Lishang for their helpful suggestions in this paper.

References

Borcea, L., Gray, G. A., & Zhang, Y. (2003). Variationally constrained numerical solution of electrical impedance tomography. *Inverse problems, 19*(5), 1159-1184. http://dx.doi.org/10.1088/0266-5611/19/5/309

Breckon, W. R., & Pidcock, M. K. (1987). Mathematical aspects of impedance imaging. *Clin. Phys. Physiol. Meas. 8*(Suppl), 77-84. http://dx.doi.org/10.1088/0143-0815/8/4A/010

Cheney, M., Isaacson, D., & Newell, J. C. (1999). Electrical impedance tomograph. *SIAM Review, 41*(l), 85-101. http://dx.doi.org/10.1137/S0036144598333613

Du, Y., Cheng, J. & Liu Z.(1997). Combined variable metric method for electrical impedance tomography problem. *Chin. J. Biomed. Eng. (in Chinese), 16*(2), 167-173.

Jones, J. E. (1995). *A mixed finite volume method for accurate computation of fluid velocities in porous media.* Ph. D. Thesis. University of Colorado, Denver. Co..

Kirsch, A. (1995). *An introduction to the mathematical theory of inverse problems.* Springer-Verlag, New York, 65-124. http://dx.doi.org/10.1007/978-1-4612-5338-9_3

Larsson, S., & Thomee, V. (2003). *Partial differential equations with numerical methods.* Springer-Verlag, Berlin, 43-50.

Lazarov, R. D., Mishev, I. D., & Vassilevski P S. (1995). Finite volume method for convection-diffusion problems. *SIAM J. Numer. Anal., 33*(1). 31-55. http://dx.doi.org/10.1137/0733003

Liu, J. J. (2005). *Regularization methods for ill-posed problems and its application (in Chinese).* Beijing: Science Press.

Mishev, I. D. (1998). Finite volume methods on Voronoi meshes. *Numer. Methods P. D. E., 14,* 193-212. http://dx.doi.org/10.1002/(SICI)1098-2426(199803)14:2<193::AID-NUM4>3.0.CO;2-J

Newman, G. A., & Hoversten, G. M. (2000). Strategies for two- and three-dimensional electromagnetic inverse problems. *Inverse Problems, 16*(5), 1357-1375. http://dx.doi.org/10.1088/0266-5611/16/5/314

Pan, H., & Rui, H. X. (2012). Mixed element method for two-dimensional Darcy-Forchheimer Model. *Journal of Scientific Computing, 52*(3), 563-587. http://dx.doi.org/10.1007/s10915-011-9558-3

Quarteroni, A., Sacco, R., & Saleri, F. (2000). *Numerical mathematics.* Springer-Verlag, New York, 2000, 287-316.

Rui, H. X., & Pan, H. (2012). A block-centered finite difference method for the Darcy-Forchheimer model. *SIAM J Numer. Anal., 50*(5), 2612-2631. http://dx.doi.org/10.1137/110858239

Rui, H. X., & Pan, H. (2013).Block-centered finite difference methods for parabolic equation with time-dependent coefficient. *Japan Journal of Industrial and Applied Mathematics, 30*(3), 681-699. http://dx.doi.org/10.1007/s13160-013-0114-4

Russell, T. F., & Wheeler, M. F. (1983). *Finite element and finite difference methods for continuous flows in porous media.* In: Ewing R E. The Mathematics of Reservoir Simulation. SIAM Phialdelphia. http://dx.doi.org/10.1137/1.9781611971071.ch2

Russell, T. F. (1995). *Rigorous block-centered discritization on irregular grids: Improved simulation of complex reservoir systems.* Project Report, Research Corporation. Tulsa.

Stoer, J., & Bulirsch, R. (1993). *Introduction to numerical analysis.* Springer-Verlag, New York, 260-329. http://dx.doi.org/10.1007/978-1-4757-2272-7_5

Tamburrino, A., & Rubinacci, G. (2002). A new non-iterative inversion for electrical resistance tomography. *Inverse Problems, 18*(6), 1809-1829. http://dx.doi.org/10.1088/0266-5611/18/6/323

Wang, Y. F. (2007). *Computational methods for inverse problems and their applications.* Beijing: Higher Education Press.

Webster, J. G. (1990). *Electrical Impedance Tomography.* Adam Hilger, Bristol England, 97-137.

Weiser, A., & Wheeler, M. F. (1988). On convergence of block-centered finite difference for elliptic problems. *SIAM J. Numer. Anal., 25*(2), 351-375. http://dx.doi.org/10.1137/0725025

Xiao, T. Y., Yu, S. G., & Wang, Y. F. (2003). *Numerical methods for inverse problems*. Beijing: Science Press.

Yuan, Y. R., Cheng, A. J., Yang, D. P., & Li, C. F. (2015). Applications, theoretical analysis, numerical method and mechanical model of the polymer flooding in porous media. *Special Topics & Reviews in Porous Media-An International Journal, 6*(4), 383-401. http://dx.doi.org/10.1615/SpecialTopicsRevPorousMedia.v6.i4.60

Yuan, Y. R., Liu, Y. X., Li, C. F., Sun, T. J., & Ma, L. Q. (2016). Analysis on block-centered finite differences of numerical simulation of semiconductor device detector. Applied Mathematics and Computation , 279: 1-15. http://dx.doi.org/10.1016/j.amc.2016.01.011

9

γ-Max Labelings of Graphs

Supaporn Saduakdee[1] & Varanoot Khemmani[2]

[1] Department of Mathematics, Srinakharinwirot University, Bangkok, Thailand

Correspondence: Supaporn Saduakdee, Department of Mathematics, Srinakharinwirot University, Sukhumvit 23, 10110 Bangkok, Thailand. E-mail: aa_o_rr@hotmail.com

Abstract

Let G be a graph of order n and size m. A γ-labeling of G is a one-to-one function $f : V(G) \to \{0, 1, 2, \ldots, m\}$ that induces an edge-labeling $f' : E(G) \to \{1, 2, \ldots, m\}$ on G defined by

$$f'(e) = |f(u) - f(v)|\,, \quad \text{for each edge } e = uv \text{ in } E(G)\,.$$

The value of f is defined as

$$\text{val}(f) = \sum_{e \in E(G)} f'(e)\,.$$

The maximum value of a γ-labeling of G is defined as

$$\text{val}_{\max}(G) = \max\{\text{val}(f) : f \text{ is a -labeling of } G\};$$

while the minimum value of a γ-labeling of G is

$$\text{val}_{\min}(G) = \min\{\text{val}(f) : f \text{ is a } \gamma\text{-labeling of } G\}.$$

In this paper, we give an alternative short proof by mathematical induction to achieve the formulae for $\text{val}_{\max}(K_{r,s})$ and $\text{val}_{\max}(K_n)$.

Keywords: γ-labeling, value of a γ-labeling

1. Introduction

Let G be a graph of order n and size m. A *γ-labeling* of G is defined in (Chartrand, Erwin, VanderJagt & Zhang, 2005) as a one-to-one function $f\colon V(G) \to \{0, 1, \ldots, m\}$ that induces an *edge-labeling* $f'\colon E(G) \to \{1, \ldots, m\}$ on G defined by $f'(e) = |f(u) - f(v)|$ for each edge $e = uv$ of G. The *value* of f is defined by

$$\text{val}(f) = \sum_{e \in E(G)} f'(e)\,.$$

If the edge-labeling f' of a γ-labeling f of a graph is also one-to-one, then f is a *graceful labeling*. Among all labelings of graphs, graceful labelings are probably the best known and most studied. Graceful labelings originated with a paper of Rosa (Rosa, 1966), who used the term β-valuations. A few years later, Golomb (Golomb, 1972) called these labelings "graceful" and this is the terminology that has been used since then.

Gallian (Gallian, 2009) has written an extensive survey on labelings of graphs. The subject of γ-labelings of graphs was studied in (Bullington, Eroh, & Winters, 2010; Chartrand, Erwin, VanderJagt, & Zhang, 2005; Crosse, Okamoto, Saenpholphat, & Zhang, 2007; Fonseca, Saenpholphat, & Zhang, 2013; Fonseca, Khemmani, & Zhang, 2015; Fonseca, Saenpholphat, & Zhang, 2011; Khemmani & Saduakdee, 2015, 2016).

Obviously, since γ-labeling f of a graph G of order n and size m is one-to-one, it follows that $f'(e) \geq 1$, for any edge e, and therefore, $\text{val}(f) \geq m$. Moreover, G has a γ-labeling if and only if $m \geq n - 1$ and every connected graph has a γ-labeling.

The *maximum value* and the *minimum value* of a γ-labeling of G are defined in (Chartrand, Erwin, VanderJagt, & Zhang, 2005) as

$$\text{val}_{\max}(G) = \max\{\text{val}(f)\colon\ f \text{ is a } \gamma\text{-labeling of } G\}$$

and

$$\text{val}_{\min}(G) = \min\{\text{val}(f)\colon\ f \text{ is a } \gamma\text{-labeling of } G\}\,,$$

respectively. A γ-labeling g of G is a γ-max *labeling* if $\text{val}(g) = \text{val}_{\max}(G)$ and a γ-labeling h is a γ-min *labeling* if $\text{val}(h) = \text{val}_{\min}(G)$. Figure 1 shows nine γ-labelings $f_1, f_2, \ldots, f_9$ of the path P_5 of order 5 (where the vertex labels are shown above each vertex and the induced edge labels are shown below each edge). The value of each γ-labeling is shown in Figure 1 as well.

Since $\text{val}(f_1) = 4$ and the size of $P_5 = 4$, it follows that f_1 is a γ-min labeling of P_5. It is shown in (Chartrand, Erwin, VanderJagt, & Zhang, 2005) that the γ-labeling f_9 is a γ-max labeling of P_5.

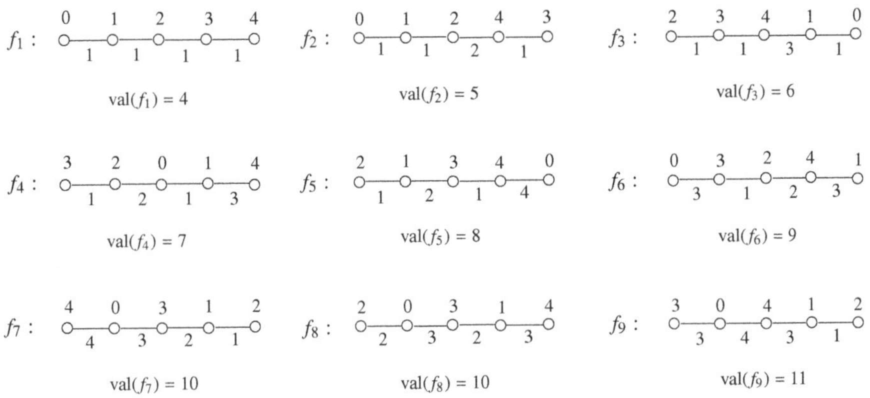

Figure 1. Some γ-labelings of P_5

By the *γ-spectrum* of a graph G, we mean the set

$$\text{spec}(G) = \{\text{val}(f) : f \text{ is a } \gamma\text{-labeling of } G\}.$$

Observe that $\text{val}_{\min}(G), \text{val}_{\max}(G) \in \text{spec}(G)$ for every graph G.

For integers a and b with $a < b$, let

$$[a, b] = \{a, a+1, \ldots, b\}$$

be a *consecutive set* of integers between a and b.

Thus for every graph G,

$$\text{spec}(G) \subseteq [\text{val}_{\min}(G),\ \text{val}_{\max}(G)].$$

The *span* of a γ-labeling f of a graph G is defined as

$$\text{span}(f) = \max\{f(v) : v \in V(G)\} - \min\{f(v) : v \in V(G)\}.$$

Consequently, if $G \cong P_5$, then $\text{spec}(G) = \{4, 5, 6, 7, 8, 9, 10, 11\}$ and for each γ-labeling f of G, $\text{span}(f) = 4 - 0 = 4$.

For a γ-labeling f of a graph G of size m, the *complementary labeling* $\bar{f}\colon V(G) \to \{0, 1, \ldots, m\}$ of f is defined by

$$\bar{f}(v) = m - f(v) \text{ for } v \in V(G).$$

Not only is $\bar{f}$ a γ-labeling of G as well but $\text{val}(\bar{f}) = \text{val}(f)$. This gives us the following.

Observation 1 (Chartrand, Erwin, VanderJagt, & Zhang, 2005) *Let f be a γ-labeling of a graph G. Then f is a* γ-max *labeling* (γ-min *labeling*) *of G if and only if $\bar{f}$ is a* γ-max *labeling* (γ-min *labeling*) *of G.*

The following result appeared in (Fonseca, Khemmani, & Zhang, 2015) is useful to us.

Theorem 1 (Fonseca, Khemmani, & Zhang, 2015) *If f is a* γ-max *labeling of a nontrivial graph G of order n and size m, then $\{0, m\} \subseteq f(V(G))$.*

In (Bullington, Eroh & Winters, 2010; Chartrand, Erwin, VanderJagt, & Zhang, 2005), the maximum and minimum values of a γ-labeling of path P_n, cycle C_n, complete graph K_n, double star $S_{p,q}$ and complete bipartite graph $K_{r,s}$ are determined.

For any positive integers n, m, Δ with $\Delta \geq m$, let G be a nontrivial graph of order n and size m, a *γ^{Δ}-labeling* of G is defined in (Fonseca, Saenpholphat, & Zhang, 2013) as a one-to-one function $f\colon V(G) \to \{0, 1, \ldots, m, m+1, \ldots, \Delta\}$ that induces an *edge-labeling* $f': E(G) \to \{1, 2, \ldots, \Delta\}$ on G defined by $f'(e) = |f(u) - f(v)|$ for each edge $e = uv$ of G. The *value* of f is defined by

$$\text{val}(f) = \sum_{e \in E(G)} f'(e).$$

The *maximum value* of a γ^{Δ}-labeling of G is

$$\text{val}^{\Delta}_{\max}(G) = \max\{\text{val}(f): \ f \text{ is a } \gamma^{\Delta}\text{-labeling of } G\}.$$

The *minimum value* of a γ^{Δ}-labeling of G is

$$\text{val}^{\Delta}_{\min}(G) = \min\{\text{val}(f): \ f \text{ is a } \gamma^{\Delta}\text{-labeling of } G\}.$$

A γ^{Δ}-labeling g of G is a γ^{Δ}-max *labeling* if $\text{val}(g) = \text{val}^{\Delta}_{\max}(G)$ and a γ^{Δ}-labeling h is a γ^{Δ}-min *labeling* if $\text{val}(h) = \text{val}^{\Delta}_{\min}(G)$.
Note that $\text{val}_{\max}(G) = \text{val}^{\Delta}_{\max}(G)$ and $\text{val}_{\min}(G) = \text{val}^{\Delta}_{\min}(G)$ when $\Delta = m$.

We first make the following observation for γ^{Δ}-max and γ^{Δ}-min labelings of graphs.

Observation 2 *Let f be a γ^{Δ}-labeling of a graph G. Then f is a γ^{Δ}-*max *labeling (γ^{Δ}-*min *labeling) of G if and only if $\bar{f}$ is a γ^{Δ}-*max *labeling (γ^{Δ}-*min *labeling) of G.*

In 2010, the explicit formula for $\text{val}_{\max}(K_{r,s})$ and the standard form for γ-max labeling of complete bipartite graph $K_{r,s}$ were determined by Bullington, Eroh and Winters (Bullington, Eroh, & Winters, 2010). Later, Fonseca, Khemmani and Zhang (Fonseca, Khemmani, & Zhang, 2015) in 2015 presented the alternative proof of the formula for $\text{val}_{\max}(K_{r,s})$ that employs γ-min labelings of complete graphs, as we state next.

Theorem 2 (Bullington, Eroh & Winters, 2010) *For any two positive integers $r \geq s$,*

$$\text{val}_{\max}(K_{r,s}) = rs\left(rs - \frac{1}{2}(r+s) + 1\right).$$

Theorem 3 (Bullington, Eroh, & Winters, 2010) *Let f be a γ-labeling of complete bipartite graph $K_{r,s}$ with partite sets V_r and V_s of cardinality r and s, respectively. Then f is a γ-*max *labeling of $K_{r,s}$ if and only if*

$$1. f(V_r) = [0, r-1] \text{ and } f(V_s) = [rs - (s-1), rs], \text{ or}$$

$$2. f(V_r) = [rs - (r-1), rs] \text{ and } f(V_s) = [0, s-1].$$

In 2005, the maximum value of γ-labeling of complete graph K_n was determined by Chartrand et al. (Chartrand, Erwin, VanderJagt, & Zhang, 2005). As well, the authors (Fonseca, Khemmani, & Zhang, 2015) characterized the γ-max labeling of complete graph K_n in 2015.

Theorem 4 (Chartrand, Erwin, VanderJagt, & Zhang, 2005) *For every positive integer n,*

$$\text{val}_{\max}(K_n) = \begin{cases} \frac{(n^2-1)(3n^2-5n+6)}{24} & \text{if } n \text{ is odd} \\ \frac{n(3n^3-5n^2+6n-4)}{24} & \text{if } n \text{ is even.} \end{cases}$$

Theorem 5 (Fonseca, Khemmani, & Zhang, 2015) *Let f be a γ-labeling of a complete graph K_n. Then f is a γ-*max *labeling of K_n if and only if*

$$f(V(K_n)) = \begin{cases} \left[0, \lfloor \frac{n}{2} \rfloor - 1\right] \cup \left[\binom{n}{2} - \lfloor \frac{n}{2} \rfloor + 1, \binom{n}{2}\right] & \text{if } n \text{ is even} \\ \left[0, \lfloor \frac{n}{2} \rfloor - 1\right] \cup \left[\binom{n}{2} - \lfloor \frac{n}{2} \rfloor + 1, \binom{n}{2}\right] \cup \{k\} & \text{if } n \text{ is odd,} \end{cases}$$

where $k \in \left[\lfloor \frac{n}{2} \rfloor, \binom{n}{2} - \lfloor \frac{n}{2} \rfloor\right]$.

The goal of this paper is to present an alternative approach to formulae for $\text{val}_{\max}(K_{r,s})$ and $\text{val}_{\max}(K_n)$ proved by mathematical induction.

The reader is referred to Chartrand and Zhang (Chartrand & Zhang, 2005) for basic definitions and terminology not mentioned here.

2. γ^{Δ}-max Labelings of Graphs

In this section, we begin our investigation for γ^{Δ}-max labeling of any nontrivial graph by presenting a useful lemma.

Lemma 1 *Let f be a γ^{Δ}-max labeling of a nontrivial graph G of order n and size m. Let $u, w \in V(G)$ with $f(u) = \min\{f(v): v \in V(G)\}$ and $f(w) = \max\{f(v): v \in V(G)\}$. Then neighborhoods of u and w are not empty.*

Proof. For any nontrivial connected graph G, it is obvious that $|N(u)|$ and $|N(w)|$ are not empty. Let G be a disconnected graph. We will show that $|N(u)| \neq 0$ and $|N(w)| \neq 0$ Assume, to the contrary, that $|N(u)| = 0$ or $|N(w)| = 0$.

Case 1. $|N(u)| = 0$.
Then u is an isolated vertex of G. Since G has a γ^{Δ}-max labeling, $\Delta \geq m \geq n-1$. Therefore, there is a component G_1 of G with $|V(G_1)| \geq 2$. Let $x \in V(G_1)$ with $f(x) = \min\{f(v): v \in V(G_1)\}$. Let g be a γ^{Δ}-labeling of G defined by

$$g(v) = \begin{cases} f(x) & \text{if } v = u \\ f(u) & \text{if } v = x \\ f(v) & \text{if } v \neq u, x. \end{cases}$$

Then

$$\begin{aligned} \text{val}(g) &= \text{val}(f) - \sum_{v \in N(x)} (f(v) - f(x)) + \sum_{v \in N(x)} (g(v) - g(x)) \\ &= \text{val}(f) - \sum_{v \in N(x)} (f(v) - f(x)) + \sum_{v \in N(x)} (f(v) - f(u)) \\ &= \text{val}(f) + |N(x)|(f(x) - f(u)) \\ &> \text{val}(f), \end{aligned}$$

which is a contradiction.

Case 2. $|N(w)| = 0$.
By a similar argument, this leads to a contradiction with the maximum value of a γ^{Δ}-labeling of G. □

We now show formula for span of γ^{Δ}-max labelings of graphs.

Proposition 1 *Let G be a nontrivial graph of order n and size m and f a γ^{Δ}-labeling of G. If f is a γ^{Δ}-max labeling of G, then* $\text{span}(f) = \Delta$.

Proof. Let f be a γ^{Δ}-max labeling of G. Let $u, w \in V(G)$ with $f(u) = \min\{f(v): v \in V(G)\}$ and $f(w) = \max\{f(v): v \in V(G)\}$. Then $f(u) \geq 0$ and $f(w) \leq \Delta$. Assume, to the contrary, that $span(f) < \Delta$. Then $f(w) - f(u) < \Delta$. Therefore, $f(u) > 0$ or $\Delta - \text{f(w)} > 0$.

Case 1. $f(u) > 0$.
Let g be a γ^{Δ}-labeling of G defined by

$$g(v) = \begin{cases} 0 & \text{if } v = u \\ f(v) & \text{if } v \neq u. \end{cases}$$

Then

$$\begin{aligned} \text{val}(g) &= \text{val}(f) - \sum_{v \in N(u)} (f(v) - f(u)) + \sum_{v \in N(u)} (g(v) - g(u)) \\ &= \text{val}(f) - \sum_{v \in N(u)} (f(v) - f(u)) + \sum_{v \in N(u)} (f(v) - 0) \\ &= \text{val}(f) + |N(u)|(f(u) - 0) \\ &> \text{val}(f) \quad \text{(by Lemma 1)}, \end{aligned}$$

which is a contradiction.

Case 2. $\Delta - \text{f(w)} > 0$.
A similar argument to the one used in Case 1 leads to a contradiction with the maximum value of a γ^{Δ}labeling of G. □

This also provides the following corollary.

Corollary 1 *Let G be a nontrivial graph of order n and size m and f a γ^{Δ}-labeling of G. If f is a γ^{Δ}-max labeling of G, then* $\{0, \Delta\} \subseteq f(V(G))$.

3. γ-max Labelings of Complete Bipartite Graphs

We define the γ^Δ-*spectrum* of a graph G by

$$\text{spec}^\Delta(G) = \{\text{val}(f)\colon\ f \text{ is a } \gamma^\Delta\text{-labeling of } G\}.$$

Consequently, $\{\text{val}^\Delta_{\min}(G), \text{val}^\Delta_{\max}(G)\} \subseteq \text{spec}^\Delta(G)$ for every graph G. As an illustration, we now establish the γ^Δ-spectrum of a star $K_{1,s}$.

Proposition 2 *For positive integers s, Δ with $\Delta \geq s$,*

$$\text{spec}^\Delta(K_{1,s}) = \left\{\binom{\Delta-k+1}{2} - \binom{\Delta-s+1}{2} + \binom{k+1}{2} : 0 \leq k \leq \Delta\right\}.$$

Proof. Let $K_{1,s}$ be a star with $V(K_{1,s}) = \{v\} \cup V_s$ where v is a central vertex and $V_s = \{v_1, v_2, \ldots, v_s\}$ and f a γ^Δ-labeling of a graph $K_{1,s}$ with $f(v) = k$ where $0 \leq k \leq \Delta$.
If $k = 0$, then we may assume that $f(v_i) = \Delta - (s - i)$ for all $1 \leq i \leq s$. Then

$$\text{val}(f) = \sum_{i=1}^{s} |f(v_i) - f(v)| = \sum_{i=1}^{s} (\Delta - (s-i)) = \binom{\Delta+1}{2} - \binom{\Delta-s+1}{2}.$$

If $k = \Delta$, then by Observation 2,

$$\text{val}(f) = \binom{\Delta+1}{2} - \binom{\Delta-s+1}{2}.$$

If $0 < k < \Delta$, then we may assume that

$$f(v_i) = \begin{cases} i-1 & \text{if } 1 \leq i \leq k \\ \Delta - (s-i) & \text{if } k+1 \leq i \leq s. \end{cases}$$

Therefore,

$$\begin{aligned} \text{val}(f) &= (k + (k-1) + \cdots + 1) + ((\Delta - (s-1)) + (\Delta - (s-2)) + \cdots + (\Delta - k)) \\ &= \tbinom{k+1}{2} + \tbinom{\Delta-k+1}{2} - \tbinom{\Delta-s+1}{2}, \end{aligned}$$

as desired. □

In Proposition 2, we considered γ^Δ-spectrum of a star $K_{1,s}$. We are now ready to compute the maximum value of a γ^Δ-labeling of $K_{1,s}$.

Corollary 2 *For positive integers s, Δ with $\Delta \geq s$,*

$$\text{val}^\Delta_{\max}(K_{1,s}) = \binom{\Delta+1}{2} - \binom{\Delta-s+1}{2}.$$

Moreover, let f be a γ^Δ-labeling of $K_{1,s}$ with

$$f(v) = 0 \quad \textit{and} \quad f(V_s) = [\Delta - (s-1), \Delta].$$

*Then f and $\bar{f}$ are only γ^Δ-*max *labelings of $K_{1,s}$.*

Next, we show an alternative and yet simple proof employing mathematical induction of Theorem 3 which is proposed by Bullington, Eroh and Winters (Bullington, Eroh & Winters, 2010) in 2010 and by Fonseca, Khemmani and Zhang (Fonseca, Khemmani & Zhang, 2015) in 2015. In order to do this, first, let $K_{r,s}$ be a complete bipartite graph with partite sets V_r and V_s of cardinalities r and s, respectively, where $V_r = \{u_1, u_2, \ldots, u_r\}$ and $V_s = \{v_1, v_2, \ldots, v_s\}$ and then we discuss γ^Δ-max labelings of $K_{r,s}$ as follows.

Theorem 6 *Let f be a γ^Δ-labeling of a complete bipartite graph $K_{r,s}$ with*

$$f(V_r) = [0, r-1] \quad \textit{and} \quad f(V_s) = [\Delta - (s-1), \Delta],$$

*where $\Delta \geq rs$. Then f and $\bar{f}$ are only two γ^Δ-*max *labelings of $K_{r,s}$.*

Proof. We proceed by induction on $r+s$. The result is certainly true for $r+s=2$. Assume that $r+s \geq 3$ and the result holds for $K_{r',s'}$ when $2 \leq r'+s' < r+s$. By Corollary 2, hence the theorem holds when $r=1$. Suppose that $r \geq 2$. Let f be a γ^{Δ}-max labeling of $K_{r,s}$ with $f(u_1) < f(u_2) < \cdots < f(u_r)$ and $f(v_1) < f(v_2) < \cdots < f(v_s)$.

Assume that $f(u_1) < f(v_1)$. By Corollary 1, $f(u_1) = 0$. Furthermore, for each $j \in \{1, 2, \ldots, s\}$ it follows that $f(v_j) \leq \Delta-(s-j)$. Let $K_{r-1,s}$ be a complete bipartite graph with vertex set $V(K_{r-1,s}) = V(K_{r,s})-\{u_1\}$ and partite sets $V_{r-1} = V_r-\{u_1\}$ and V_s. Consequently, let f_1 be a $\gamma^{\Delta-1}$-labeling of $K_{r-1,s}$ defined by

$$f_1(u) = f(u) - 1 \quad \text{for each} \quad u \in V(K_{r-1,s}).$$

Then

$$\operatorname{val}(f_1) = \sum_{\substack{2\leq i\leq r\\ 1\leq j\leq s}} f_1'(u_iv_j) \leq \operatorname{val}_{\max}^{\Delta-1}(K_{r-1,s}).$$

Let g_1 be a $\gamma^{\Delta-1}$-max labeling of $K_{r-1,s}$. Since $\Delta - 1 \geq (r-1)s$, by induction hypothesis, we have

$$g_1(V_{r-1}) = [0, r-2] \quad \text{and} \quad g_1(V_s) = [(\Delta-1)-(s-1), (\Delta-1)].$$

We can extend g_1 to a γ^{Δ}-labeling g of $K_{r,s}$ defined by

$$g(u) = \begin{cases} 0 & \text{if } u = u_1 \\ g_1(u) + 1 & \text{otherwise}. \end{cases}$$

Since

$$\begin{aligned}
\operatorname{val}_{\max}^{\Delta}(K_{r,s}) &= \operatorname{val}(f) \\
&= \sum_{\substack{2\leq i\leq r\\ 1\leq j\leq s}} f'(u_iv_j) + \sum_{j=1}^{s} |f(v_j) - f(u_1)| \\
&\leq \operatorname{val}(f_1) + \sum_{j=1}^{s} |\Delta - (s-j) - 0| \\
&\leq \operatorname{val}_{\max}^{\Delta-1}(K_{r-1,s}) + \sum_{j=1}^{s} |\Delta - (s-j) - 0| \\
&= \operatorname{val}(g_1) + \sum_{j=1}^{s} |\Delta - (s-j) - 0| \\
&= \operatorname{val}(g) \\
&\leq \operatorname{val}_{\max}^{\Delta}(K_{r,s}),
\end{aligned}$$

it follows that

$$\sum_{j=1}^{s} |f(v_j) - f(u_1)| = \sum_{j=1}^{s} |\Delta - (s-j) - 0| \tag{1}$$

and

$$\operatorname{val}(f_1) = \operatorname{val}_{\max}^{\Delta-1}(K_{r-1,s}). \tag{2}$$

From (1), we have

$$f(V_s) = [\Delta - (s-1), \Delta].$$

From (2), we have $f_1(V_{r-1}) = [0, r-2]$, hence

$$f(V_{r-1}) = [1, r-1]$$

and we have $f(u_1) = 0$. Therefore,

$$f(V_r) = [0, r-1] \quad \text{and} \quad f(V_s) = [\Delta - (s-1), \Delta].$$

On the other hand, if $f(v_1) < f(u_1)$, then a similar argument to the one used shows that

$$f(V_s) = [0, s-1] \quad \text{and} \quad f(V_r) = [\Delta - (r-1), \Delta].$$

□

The following result is the consequence of Theorem 6 when $\Delta = rs$.

Theorem 7 *Let $K_{r,s}$ be a complete bipartite graph with partite sets V_r and V_s of cardinalities r and s, respectively, let f be a γ-labeling of $K_{r,s}$ with*

$$f(V_r) = [0, r-1] \quad and \quad f(V_s) = [rs - (s-1), rs].$$

*Then f and $\bar{f}$ are only two γ-*max *labelings of $K_{r,s}$.*

4. γ-max Labelings of Complete Graphs

The γ-max labelings of complete graphs K_n were characterized in (Fonseca, Khemmani & Zhang, 2015). In this section, we present characterization of γ^{Δ}-max labelings and γ-max labeling of complete graphs K_n, by applying a similar fashion to the one used in the proof of Theorem 6.

Theorem 8 *Let f be a γ^{Δ}-labeling of a complete graph K_n with*

$$f(V(K_n)) = \begin{cases} \left[0, \left\lfloor \frac{n}{2} \right\rfloor - 1\right] \cup \left[\Delta - \left\lfloor \frac{n}{2} \right\rfloor + 1, \Delta\right] & \text{if } n \text{ is even} \\ \left[0, \left\lfloor \frac{n}{2} \right\rfloor - 1\right] \cup \left[\Delta - \left\lfloor \frac{n}{2} \right\rfloor + 1, \Delta\right] \cup \{k\} & \text{if } n \text{ is odd} \end{cases}$$

*where $\Delta \geq \binom{n}{2}$ and $k \in \left[\left\lfloor \frac{n}{2} \right\rfloor, \Delta - \left\lfloor \frac{n}{2} \right\rfloor\right]$. Then f and $\bar{f}$ are only two γ^{Δ}-*max *labelings of K_n.*

Proof. Let K_n be a complete graph with $V(K_n) = \{u_1, u_2, \ldots, u_n\}$. Assume that n is even. We use mathematical induction on n. When $n = 2$, the result is obvious. Assume that $n \geq 4$ and the result holds for $K_{n'}$ when n' is even and $2 \leq n' < n$. Let f be a γ^{Δ}-max labeling of K_n with $f(u_1) < f(u_2) < \cdots < f(u_n)$. By Corollary 1, $f(u_1) = 0$ and $f(u_n) = \Delta$. Let f_1 be a $\gamma^{\Delta-2}$-labeling of a complete graph K_{n-2} with vertex set $V(K_{n-2}) = \{u_2, u_3, \ldots, u_{n-1}\}$ defined by

$$f_1(u_i) = f(u_i) - 1 \quad \text{for each } 2 \leq i \leq n-1.$$

Let g_1 be a $\gamma^{\Delta-2}$-max labeling of K_{n-2}. Since $\Delta - 2 \geq \binom{n-2}{2}$, by induction hypothesis, we have

$$g_1(V(K_{n-2})) = \left[0, \left\lfloor \frac{n-2}{2} \right\rfloor - 1\right] \cup \left[(\Delta - 2) - \left\lfloor \frac{n-2}{2} \right\rfloor + 1, (\Delta - 2)\right].$$

We can extend g_1 to a γ^{Δ}-labeling g of K_n defined by

$$g(u) = \begin{cases} 0 & \text{if } u = u_1 \\ \Delta & \text{if } u = u_n \\ g_1(u) + 1 & \text{if } u \neq u_1, u_n. \end{cases}$$

Since

$$\begin{aligned} \text{val}^{\Delta}_{\max}(K_n) &= \text{val}(f) \\ &= \text{val}(f_1) + \sum_{i=2}^{n-1} (f(u_i) - f(u_1)) + \sum_{i=2}^{n-1} (f(u_n) - f(u_i)) + (f(u_n) - f(u_1)) \\ &\leq \text{val}^{\Delta-2}_{\max}(K_{n-2}) + \sum_{i=2}^{n-1} (\Delta - 0) + (\Delta - 0) \\ &= \text{val}(g_1) + \sum_{i=2}^{n-1} (g(u_i) - g(u_1)) + \sum_{i=2}^{n-1} (g(u_n) - g(u_i)) + (g(u_n) - g(u_1)) \\ &= \text{val}(g) \\ &\leq \text{val}^{\Delta}_{\max}(K_n), \end{aligned}$$

it follows that

$$\text{val}(f_1) = \text{val}^{\Delta-2}_{\max}(K_{n-2}).$$

Thus,

$$f_1(V(K_{n-2})) = \left[0, \left\lfloor \frac{n-2}{2} \right\rfloor - 1\right] \cup \left[(\Delta - 2) - \left\lfloor \frac{n-2}{2} \right\rfloor + 1, (\Delta - 2)\right].$$

Hence

$$f(V(K_{n-2})) = \left[1, \left\lfloor \frac{n-2}{2} \right\rfloor\right] \cup \left[(\Delta - 2) - \left\lfloor \frac{n-2}{2} \right\rfloor + 2, (\Delta - 1)\right]$$

and we have $f(u_1) = 0$, $f(u_n) = \Delta$. Therefore,

$$f(V(K_n)) = \left[0, \left\lfloor \frac{n}{2} \right\rfloor - 1\right] \cup \left[\Delta - \left\lfloor \frac{n}{2} \right\rfloor + 1, \Delta\right].$$

On the other hand, if n is odd, then by a similar argument, this shows that

$$f(V(K_n)) = \left[0, \left\lfloor \frac{n}{2} \right\rfloor - 1\right] \cup \left[\Delta - \left\lfloor \frac{n}{2} \right\rfloor + 1, \Delta\right] \cup \{k\}$$

where $k \in \left[\left\lfloor \frac{n}{2} \right\rfloor, \Delta - \left\lfloor \frac{n}{2} \right\rfloor\right]$. □

The following result is the consequence of Theorem 8 when $\Delta = \binom{n}{2}$.

Theorem 9 *Let f be a γ-labeling of a complete graph K_n with*

$$f(V(K_n)) = \begin{cases} \left[0, \left\lfloor \frac{n}{2} \right\rfloor - 1\right] \cup \left[\binom{n}{2} - \left\lfloor \frac{n}{2} \right\rfloor + 1, \binom{n}{2}\right] & \text{if } n \text{ is even} \\ \left[0, \left\lfloor \frac{n}{2} \right\rfloor - 1\right] \cup \left[\binom{n}{2} - \left\lfloor \frac{n}{2} \right\rfloor + 1, \binom{n}{2}\right] \cup \{k\} & \text{if } n \text{ is odd} \end{cases}$$

where $k \in \left[\left\lfloor \frac{n}{2} \right\rfloor, \binom{n}{2} - \left\lfloor \frac{n}{2} \right\rfloor\right]$. Then f and $\bar{f}$ are only two γ-max labelings of K_n.

5. Open Question

The characterization of γ-max labelings of $K_{r,s}$ and K_n were determined. The main open question is to characterize γ-min labelings of those graphs.

References

Bullington, G. D., Eroh, L. L., & Winters, S. J. (2010). γ-labelings of complete bipartite graphs. *Discuss. Math. Graph Theory, 30*, 45-54. https://doi.org/10.7151/dmgt.1475

Chartrand, G., Erwin, D., VanderJagt, D. W., & Zhang, P. (2005). γ-labelings of graphs. *Bull. Inst. Combin. Appl, 44*, 51-68.

Chartrand, G., Erwin, D., VanderJagt, D. W., & Zhang, P. (2005). On γ-labelings of trees. *Discuss. Math. Graph Theory, 25*(3), 363-383. https://doi.org/10.7151/dmgt.1289

Chartrand, G., & Zhang, P. (2005). *Introduction to Graph Theory*. The Walter Rudin student series in advanced mathematics, McGraw-Hill Higher Education, Boston.

Crosse, L., Okamoto, F., Saenpholphat, V., & Zhang, P. (2007). On γ-labelings of oriented graphs. *Math. Bohem, 132*, 185-203.

Fonseca, C. M., Saenpholphat, V., & Zhang, P. (2013). Extremal values for a γ-labeling of a cycle with a triangle. *Utilitas Math, 92*, 167-185.

Fonseca, C. M., Khemmani, V., & Zhang, P. (2015). On γ-labelings of graphs. *Utilitas Math, 98*, 33-42.

Fonseca, C. M., Saenpholphat, V., & Zhang, P. (2011). The γ-spectrum of a graph. *Ars Combin, 101*, 109-127.

Gallian, J. A. (2009). A dynamic survey of graph labeling. *Electron. J. Combin, 16*(6), 1-219.

Golomb, S. W. (1972). How to number a graph, In: *Graph Theory and Computing*, 23-37. Academic Press, New York.

Hegde, S. M. (2000). On (k, d)-graceful graphs. *J. Combin. Inform. System Sci, 25*, 255-265.

Khemmani, V., & Saduakdee, S. (2015). The γ-spectrum of cycle with one chord. *International Journal of Pure and Applied Mathematics, 105*(4), 835-852. https://doi.org/10.12732/ijpam.v105i4.22

Rosa, A. (1966). On certain valuations of the vertices of a graph, In: *Theory of Graphs* (Internat. Sympos., Rome, 349-355, Gordon and Breach, New York, Dunod, Paris.

Saduakdee, S., & Khemmani, V. (2016). γ-labeling of a cycle with one chord. *Discrete and Computational Geometry and Graphs, Lecture Notes in Computer Science, 9943*, 155-166. https://doi.org/10.1007/978-3-319-48532-4_14

10

Tensor Product Of Zero-divisor Graphs With Finite Free Semilattices

Kemal Toker[1]

[1] Faculty Of Arts And Sciences Department Of Mathematics, Harran University, Şanlıurfa-Turkey

Correspondence: Kemal Toker, Faculty Of Arts And Sciences Department Of Mathematics, Harran University, Şanlıurfa-Turkey.E-mail: ktoker@harran.edu.tr

Abstract

$\Gamma(SL_X)$ is defined and has been investigated in (Toker, 2016). In this paper our main aim is to extend this study over $\Gamma(SL_X)$ to the tensor product. The diameter, radius, girth, domination number, independence number, clique number, chromatic number and chromatic index of $\Gamma(SL_{X_1}) \otimes \Gamma(SL_{X_2})$ has been established. Moreover, we have determined when $\Gamma(SL_{X_1}) \otimes \Gamma(SL_{X_2})$ is a perfect graph.

Keywords: Tensor product, finite free semilattice, zero-divisor graph, clique number, domination number, perfect graph

1. Introduction

Let G be a graph then edge set of G denoted by $E(G)$ and vertex set of G denoted by $V(G)$. Let G_1 and G_2 be graphs, tensor product of G_1 and G_2 has vertex set $V(G_1) \times V(G_2)$ and has edge set $\{(u_1, v_1)(u_2, v_2) : u_1u_2 \in E(G_1) \text{ and } v_1v_2 \in E(G_2)\}$, and it is denoted by $G_1 \otimes G_2$. Let G_1 and G_2 be connected graphs then $G_1 \otimes G_2$ is connected if and only if either G_1 or G_2 contains an odd cycle (Weichsel, 1962). Also it is clear that $G_1 \otimes G_2 \simeq G_2 \otimes G_1$.

Firstly zero-divisor graph on a commutative semigroup S with 0 was studied by Demeyer and his friends (DeMeyer et all., 2002; DeMeyer et all, 2005). Let the set of zero divisor elements in S is $Z(S)$, the zero-divisor graph $\Gamma(S)$ is defined as an undirected graph with vertices $Z(S) \setminus \{0\}$ and the vertices x and y are adjacent with a single edge if and only if $xy = 0$. Always $\Gamma(S)$ is a connected graph (DeMeyer et all., 2002).

Let X be a finite non-empty set. The free semilattice on a set X is the finite powerset of X except the empty set and operation is union of sets. We show it with SL_X. Then SL_X is a commutative semigroup of idempotents with the multiplication $A \cdot B = A \cup B$ for all $A, B \in SL_X$. In zero-divisor graph of SL_X, any two distinct vertices A and B are adjacent with the rule $A \cup B = X$. In a recent study, $\Gamma(SL_X)$ has been investigated in (Toker, 2016).

We know that if $|X| \geq 3$ then $\Gamma(SL_X)$ contains an odd cycle (Toker, 2016). Let X_1 and X_2 be non-empty and finite sets and let $\Gamma(SL_{X_1})$ be zero-divisor graph associated to SL_{X_1} and $\Gamma(SL_{X_2})$ be zero-divisor graph associated to SL_{X_2}. In this paper, without loss of generality we assume that $|X_1| = n$, $|X_2| = m$ and we suppose that $X_1 = \{x_1, \ldots, x_n\}$ and $X_2 = \{y_1, \ldots, y_m\}$. So if $|X_1| \geq 2$, $|X_2| \geq 3$ then $\Gamma(SL_{X_1}) \otimes \Gamma(SL_{X_2})$ is connected graph and in this paper we have researched girth, diameter, radius, dominating number, clique number, chromatic number, chromatic index, independence number and perfectness of this graph.

For graph theory see (Gross & Yellen, 2004), and for semigroup theory (Howie, 1995).

2. Some Properties of $\Gamma(SL_{X_1}) \otimes \Gamma(SL_{X_2})$

Let G be a simple graph, the distance (length of the shortest path) between two vertices u, v in G is denoted by $d_G(u, v)$. In a connected simple graph the maximum distance (lenght of the shortest path) between v and any other vertex u in G is eccentricity of a vertex v,it is denoted by ecc(v), so that is

$$\mathrm{ecc}(v) = \max\{d_G(u, v) \mid u \in V(G)\}.$$

The diameter of G is defined by

$$\max\{\mathrm{ecc}(v) \mid v \in V(G)\}$$

and it is denioted by diam(G). Moreover radius of G is defined by

$$\min\{\mathrm{ecc}(v) \mid v \in V(G)\}$$

and it is denoted by rad(G). The girth of a graph is the length of a shortest cycle contained in the graph, and it is denoted by gr(G). If there is not any cycle in a graph, then its girth is defined to be infinity.

The degree (or valency) of a vertex of a graph is the number of edges incident to the vertex, with loops counted twice, degree of vertex $v \in V(G)$ is denoted by $\deg_G(v)$. Among all degrees, the maximum degree is denoted by $\Delta(G)$ and the minimum degree is denoted by $\delta(G)$. In a graph, the vertex of maximum degree is called delta-vertex and the set of delta-vertices of G denoted by Λ_G. In a graph, an independent set or stable set is a set of vertices in a graph, no two of which are adjacent. Independence number of G is denoted by $\alpha(G)$ and it is defined by

$$\alpha(G) = \max\{|I| \mid I \text{ is an independent set of } G\}.$$

In a graph, a dominating set for a graph G is a subset D of $V(G)$ such that every vertex not in D is adjacent to at least one member of D. The domination number of G is the number of vertices in a smallest dominating set for G, and it is denoted by $\gamma(G)$, so dominating number of G is

$$\gamma(G) = \min\{|D| \mid D \text{ is a dominating set of } G\}.$$

The open neighbourhood of a vertex $v \in V(G)$ is the set of vertices which are adjacent to v and it is denoted by $N_G(v)$, the closed neighbourhood of v is $N_G(v) \cup \{v\}$ and it is denoted by $N_G[v]$. It is clear that $|N_G[v] \cap D| \geq 1$ for each dominating set D, and for each $v \in V(G)$.

In this section we mainly deal with some graph properties of $\Gamma(SL_{X_1}) \otimes \Gamma(SL_{X_2})$ namely diameter, radius, girth, domination number and independence number.

We use the notation $\overline{A} = (X_i \setminus A)$ for all $A \subseteq X_i$ $(i = 1, 2)$, and we use the notation $d(u, v)$ instead of $d_{\Gamma(SL_{X_1}) \otimes \Gamma(SL_{X_2})}(u, v)$. Moreover for convenience we use $\Gamma_1 \otimes \Gamma_2$ instead of $\Gamma(SL_{X_1}) \otimes \Gamma(SL_{X_2})$. Notice that, for $u = (A_1, B_1), v = (A_2, B_2) \in V(\Gamma(SL_{X_1}) \otimes \Gamma(SL_{X_2}))$ there exists a single edge $u - v$ in $\Gamma(SL_{X_1}) \otimes \Gamma(SL_{X_2})$ if and only if $A_1 \supseteq \overline{A_2}$ and $B_1 \supseteq \overline{B_2}$.

Theorem 2.1

(i) If $|X_1| \geq 3$ and $|X_2| \geq 3$ then $diam(\Gamma_1 \otimes \Gamma_2) = 4$.

(ii) If $|X_1\dagger = 2$ and $|X_2| \geq 3$ then $diam(\Gamma_1 \otimes \Gamma_2) = 5$.

Proof. (i) Let $|X_1| \geq 3$, $|X_2| \geq 3$ and $u = (A_1, B_1), v = (A_2, B_2) \in V(\Gamma_1 \otimes \Gamma_2)$. If $A_1 \cup A_2 = X_1$ and $B_1 \cup B_2 = X_2$ then $d(u, v) = 1$. It is clear that in other cases $d(u, v) \geq 2$. Second case is $A_1 \cup A_2 \neq X_1$ and $B_1 \cup B_2 \neq X_2$. In second case if $A_1 \cap A_2 \neq \varnothing$ and $B_1 \cap B_2 \neq \varnothing$, we take $C_1 = \overline{A_1} \cup \overline{A_2}$, $C_2 = \overline{B_1} \cup \overline{B_2}$ thus we have a path $(A_1, B_1) - (C_1, C_2) - (A_2, B_2)$ and $d(u, v) = 2$. In second case let $A_1 \cap A_2 \neq \varnothing$ and $B_1 \cap B_2 = \varnothing$. In this case since $\overline{B_1} \cup \overline{B_2} = X_2$ thus $d(u, v) \neq 2$. If $A_1 \neq A_2$ then $A_1 \setminus A_2 \neq \varnothing$ or $A_2 \setminus A_1 \neq \varnothing$. If $A_1 \setminus A_2 \neq \varnothing$ then we have a path $(A_1, B_1) - (\overline{A_1} \cup A_2, \overline{B_1}) - (\overline{A_2}, \overline{B_2}) - (A_2, B_2)$ and if $A_2 \setminus A_1 \neq \varnothing$ then we have a path $(A_1, B_1) - (\overline{A_1}, \overline{B_1}) - (A_1 \cup \overline{A_2}, \overline{B_2}) - (A_2, B_2)$ so $d(u, v) = 3$. If $A_1 = A_2$ and $|A_1| \geq 2$ we take any 2−partition of A_1, say C and D. We have a path $(A_1, B_1) - (\overline{A_1} \cup C, \overline{B_1}) - (\overline{A_1} \cup D, \overline{B_2}) - (A_2, B_2)$ so $d(u, v) = 3$. If $A_1 = A_2$ and $|A_1| = 1$ then we have a path $(A_1, B_1) - (\overline{A_1}, \overline{B_1}) - (A_1, B_1 \cup B_2) - (\overline{A_1}, \overline{B_2}) - (A_2, B_2)$ so $d(u, v) \leq 4$. Also (A_1, B) has adjacent form of $(\overline{A_1}, C)$ where $\overline{C} \subseteq B$ so $d(u, v) \neq 3$ then $d(u, v) = 4$. In second case if $A_1 \cap A_2 = \varnothing$ and $B_1 \cap B_2 \neq \varnothing$ is similar. In second case if $A_1 \cap A_2 = \varnothing$ and $B_1 \cap B_2 = \varnothing$ then we have a path $(A_1, B_1) - (\overline{A_1}, \overline{B_1}) - (\overline{A_2}, \overline{B_2}) - (A_2, B_2)$ and since $\overline{A_1} \cup \overline{A_2} = X_1$ and $\overline{B_1} \cup \overline{B_2} = X_2$ so $d(u, v) = 3$. Third case is $A_1 \cup A_2 = X_1$ and $B_1 \cup B_2 \neq X_2$. In third case if $A_1 \cap A_2 \neq \varnothing$ and $B_1 \cap B_2 \neq \varnothing$ then we have a path $(A_1, B_1) - (\overline{A_1} \cup \overline{A_2}, \overline{B_1} \cup \overline{B_2}) - (A_2, B_2)$, so $d(u, v) = 2$. If $A_1 \cap A_2 \neq \varnothing$ and $B_1 \cap B_2 = \varnothing$ then we have a path $(A_1, B_1) - (A_2, \overline{B_1}) - (A_1, \overline{B_2}) - (A_2, B_2)$ and since $\overline{B_1} \cup \overline{B_2} = X_2$ so $d(u, v) = 3$. Let $A_1 \cap A_2 = \varnothing$ and $B_1 \cap B_2 \neq \varnothing$, in this case since $\overline{A_1} \cup \overline{A_2} = X_1$ thus $d(u, v) \geq 3$. If $B_1 \neq B_2$, so $B_1 \setminus B_2 \neq \varnothing$ or $B_2 \setminus B_1 \neq \varnothing$. If $B_1 \setminus B_2 \neq \varnothing$ then we have a path $(A_1, B_1) - (A_2, \overline{B_1} \cup B_2) - (A_1, \overline{B_2}) - (A_2, B_2)$ and if $B_2 \setminus B_1 \neq \varnothing$ then we have a path $(A_1, B_1) - (A_2, \overline{B_1}) - (A_1, B_1 \cup \overline{B_2}) - (A_2, B_2)$, so $d(u, v) = 3$. If $B_1 = B_2$ and $|B_1| \geq 2$, we take 2−partition of B_1, we say E and F, then we have a path $(A_1, B_1) - (A_2, \overline{B_1} \cup E) - (A_1, \overline{B_1} \cup F) - (A_2, B_2)$ so $d(u, v) = 3$. If $B_1 = B_2$ and $|B_1| = 1$ in this case $|A_1| \neq 1$ or $|A_2| \neq 1$ since $A_1 \cup A_2 = X_1$ and $|X_1| \geq 3$, if $|A_1| \neq 1$ then there exists $\varnothing \neq C \subsetneq A_1$, and we have path $(A_1, B_1) - (A_2 \cup C, \overline{B_1}) - (\overline{C}, B_1) - (A_1, \overline{B_1}) - (A_2, B_2)$ and adjacent of (A, B_1) is $(D, \overline{B_1})$ where $\overline{D} \subseteq A$ so $d(u, v) = 4$ and if $|A_2| \neq 1$ is similar. In third case if $A_1 \cap A_2 = \varnothing$ and $B_1 \cap B_2 = \varnothing$ then we have a path $(A_1, B_1) - (A_2, \overline{B_1}) - (A_1, \overline{B_2}) - (A_2, B_2)$ so $d(u, v) = 3$. Last case is $A_1 \cup A_2 \neq X_1$ and $B_1 \cup B_2 = X_2$ is similar with third case. Thus if $|X_1| \geq 3$ and $|X_2| \geq 3$ then $diam(\Gamma_1 \otimes \Gamma_2) = 4$.

(ii) Let $|X_1| = 2$, $|X_2| \geq 3$ and $u = (A_1, B_1), v = (A_2, B_2) \in V(\Gamma_1 \otimes \Gamma_2)$. In here different case is $A_1 \cup A_2 = X_1$ and $B_1 \cup B_2 \neq X_2$ with $B_1 = B_2$ and $|B_1| = 1$, in other cases we have same results with i). This case we take 2−partition of $\overline{B_1}$, we say M and N. We have a path $(A_1, B_1) - (A_2, \overline{B_1}) - (A_1, B_1 \cup M) - (A_2, B_1 \cup N) - (A_1, \overline{B_1}) - (A_2, B_2)$ so $d(u, v) \leq 5$. (A_1, B_1) has only one adjacent and it is $(A_2, \overline{B_1})$ and (A_2, B_1) has only one adjacent and it is $(A_1, \overline{B_1})$ and they are different vertices and they are not adjacent, moreover $d((A_2, \overline{B_1}), (A_1, \overline{B_1})) = 3$, thus $d(u, v) = 5$. So if $|X_1| = 2$ and $|X_2| \geq 3$ then $diam(\Gamma_1 \otimes \Gamma_2) = 5$.

Theorem 2.2

(*i*) If $|X_1| \geq 3$ and $|X_2| \geq 3$ then $gr(\Gamma_1 \otimes \Gamma_2) = 3$.

(*ii*) If $|X_1| = 2$, $|X_2| = 3$ then $gr(\Gamma_1 \otimes \Gamma_2) = 6$ and if $|X_1| = 2$, $|X_2| \geq 4$ then $gr(\Gamma_1 \otimes \Gamma_2) = 4$.

(*iii*) If $|X_1| \geq 3$ and $|X_2| \geq 3$ then $rad(\Gamma_1 \otimes \Gamma_2) = 3$.

(*iv*) If $|X_1| = 2$ and $|X_2| \geq 3$ then $rad(\Gamma_1 \otimes \Gamma_2) = 4$.

Proof. (i) Let $|X_1| \geq 3$, $|X_2| \geq 3$ and $(A, B) \in V(\Gamma_1 \otimes \Gamma_2)$. Assume that $|A| \geq 2$, $|B| \geq 2$ so there exists 2−partition of A, we say A_1 and A_2 and there exists 2−partition of B, we say B_1 and B_2. Thus $(A, B)-(\overline{A}\cup A_1, \overline{B}\cup B_1)-(\overline{A}\cup A_2, \overline{B}\cup B_2)-(A, B)$ is a cycle. Let $|A| = 1$, $|B| \geq 2$ then there exists $\varnothing \neq C \subsetneq \overline{A}$ so we have a cycle $(A, B)-(\overline{A}, \overline{B}\cup B_1)-(A\cup C, B)-(\overline{A}, \overline{B}\cup B_2)-(A, B)$. If $|A| \geq 2$, $|B| = 1$, we can find a cycle similar way. Moreover $\Gamma_1\otimes\Gamma_2$ is simple graph and from its definition $gr(\Gamma_1\otimes\Gamma_2) = 3$.
(ii)

η :

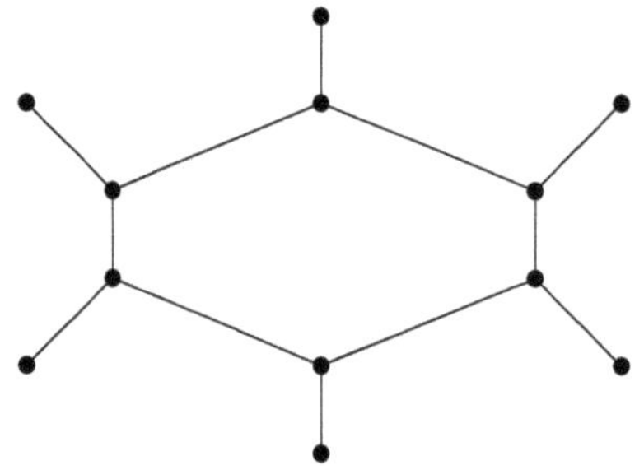

If $|X_1| = 2, |X_2| = 3$ then $\Gamma_1 \otimes \Gamma_2$ is η thus in this case $gr(\Gamma_1 \otimes \Gamma_2) = 6$. If $|X_1| = 2, |X_2| = m \geq 4$, $gr(\Gamma_1 \otimes \Gamma_2)$ can not be odd number since $|X_1| = 2$. So $gr(\Gamma_1 \otimes \Gamma_2) \geq 4$. Let $(A, B) \in V(\Gamma_1 \otimes \Gamma_2)$, if $2 \leq |B| \leq m - 2$ then there exists $k \in \overline{B}$ and 2−partition of B is E and F. So $(A, B) - (\overline{A}, \overline{B} \cup E) - (A, \{k\}) - (\overline{A}, \overline{B} \cup F) - (A, B)$ is a cycle. If $|B| = m - 1$ then without loss of generality we assume that $y_1, y_2 \in B$ then $(A, B) - (\overline{A}, \overline{B} \cup \{y_1, y_2\}) - (A, B \setminus \{y_1\}) - (\overline{A}, \overline{B} \cup \{y_1\}) - (A, B)$ is a cycle. Thus if $|X_1| = 2, |X_2| \geq 4$ then $gr(\Gamma_1 \otimes \Gamma_2) = 4$.

(iii) Let $|X_1| \geq 3$, $|X_2| \geq 3$ and $v = (A, B) \in V(\Gamma_1 \otimes \Gamma_2)$, we can determine $ecc(v)$. If $|A| \geq 2$, $|B| \geq 2$ then $ecc(v) \leq 3$ from proof of Theorem 2.1 (i) because let $u \in V(\Gamma_1 \otimes \Gamma_2)$, we found that if $|A| \geq 2$, $|B| \geq 2$ then $d(u, v) \leq 3$, so $ecc(v) \leq 3$. Moreover there exists $\varnothing \neq C \subsetneq B$, if we choose $u =(\overline{A}, C)$ so $d(u, v) = 3$ it follows that $ecc(v) = 3$. If $|A| = 1$ we choose $u = (A, \overline{B})$ and $d(u, v) = 4$. If $|B| = 1$ is similar. So $rad(\Gamma_1 \otimes \Gamma_2) = 3$.

(iv) Let $|X_1| = 2$, $|X_2| \geq 3$ and $v = (A, B) \in V(\Gamma_1 \otimes \Gamma_2)$, we can determine $ecc(v)$. If $|A| = |B| = 1$ then $ecc(v) = 5$. If $|B| \geq 2$ it is clear that $ecc(v) \leq 4$, if we choose $u = (A, \overline{B})$ then $d(u, v) = 4$, so $ecc(v) = 4$. Thus $rad(\Gamma_1 \otimes \Gamma_2) = 4$.

Theorem 2.3 *If* $|X_1| = n \geq 2$, $|X_2| = m \geq 3$ *then* $\gamma(\Gamma_1 \otimes \Gamma_2) = n.m$

Proof. Let $|X_1| = n \geq 2$, $|X_2| = m \geq 3$ and $v = (A, B) \in V(\Gamma_1 \otimes \Gamma_2)$, D be a dominating set for $\Gamma_1 \otimes \Gamma_2$. It is clear that $\deg_{(\Gamma_1\otimes\Gamma_2)}(v) = \deg_{\Gamma_1}(A).\deg_{\Gamma_2}(B)$ so if $|A| = |B| = 1$ then $\deg_{(\Gamma_1\otimes\Gamma_2)}(v) = 1$ and adjacent of v is only $u = (\overline{A}, \overline{B})$ so $N_{(\Gamma_1\otimes\Gamma_2)}[v] = \{u, v\}$. Since $|N_{(\Gamma_1\otimes\Gamma_2)}[\{v\}] \cap D| \geq 1$, either $v \in D$ or $u \in D$. Let $x_i, x_k \in X_1$ and $y_j, y_l \in X_2$. Moreover since $|X_2| \geq 3$ if $(\{x_i\}, \{y_j\}) \neq (\{x_k\}, \{y_l\})$ then

$$N_{(\Gamma_1\otimes\Gamma_2)}[(\{x_i\}, \{y_j\})] \cap N_{(\Gamma_1\otimes\Gamma_2)}[(\{x_k\}, \{y_l\})] = \varnothing.$$

Thus $\gamma(\Gamma_1 \otimes \Gamma_2) \geq n.m$. If we choose $D = \{u = (A, B); u \in V(\Gamma_1 \otimes \Gamma_2)$ and $|A| = n - 1, |B| = m - 1\}$, it is easy to see that $|D| = n.m$ and D is a dominating set for $\Gamma_1 \otimes \Gamma_2$. It follows that $\gamma(\Gamma_1 \otimes \Gamma_2) = n.m$

Theorem 2.4 *If* $|X_1| = n \geq 2$, $|X_2| = m \geq 3$ *then*

$$\alpha(\Gamma_1 \otimes \Gamma_2) = \frac{(2^n - 2)(2^m - 2)}{2}$$

Proof. Let $x_i \in X_1$ and $C = \{(A_i, B_j) : \varnothing \neq A_i \subseteq X_1 \setminus \{x_i\}, B_j \in V(\Gamma_2)\}$ and $D = \{(X_1 \setminus A_i, X_2 \setminus B_j) : (A_i, B_j) \in C\}$. It is clear that $C \cap D = \varnothing$ and $|C| = |D| = \frac{(2^n-2)(2^m-2)}{2}$ thus $C \cup D = V(\Gamma_1 \otimes \Gamma_2)$. Let I be an independence set of $\Gamma_1 \otimes \Gamma_2$, then from the pigeonhole principle $|I| \leq \frac{(2^n-2)(2^m-2)}{2}$, and C is an independence set for $\Gamma_1 \otimes \Gamma_2$, moreover $|C| = \frac{(2^n-2)(2^m-2)}{2}$.

3. Perfectness of $\Gamma(SL_{X_1}) \otimes \Gamma(SL_{X_2})$

Let G be a graph. Clique is the each of the maximal complete subgraphs of G. The number of all the vertices in any clique of G is clique number and it is denoted by $\omega(G)$. The chromatic number of a graph G is the smallest number of colors needed to color the vertices of G so that no two adjacent vertices share the same color and it is denoted by $\chi(G)$. It is well-known that

$$\chi(G) \geq \omega(G) \tag{1}$$

for any graph G (Chartrand & Zhang, 2009). Let $V' \subseteq V(G)$, then induced subgraph $G' = (V', E')$ is a subgraph of G such that E' consists of those edges whose endpoints are in V'. For each induced subgraph H of G, if $\chi(H) = \omega(H)$, then G is called a perfect graph.

The complement or inverse of a graph G is a graph on the same vertices such that two distinct vertices are adjacent if and only if they are not adjacent in G, the complement of G is denoted by G^c.

A graph G is called Berge if no induced subgraph of G is an odd cycle of length at least five or the complement of one.

The edges are called adjacent if they share a common end vertex. An edge coloring of a graph is an assignment of colors to the edges of the graph so that no two adjacent edges have the same color. The minimum required number of colours for and edge colouring of G is called the chromatic index of G and it is denoted by $\chi'(G)$. Vizing gave a fundamental theorem for that, for any graph G, we have

$$\Delta(G) \leq \chi'(G) \leq \Delta(G) + 1$$

(Vizing, 1964). Graph G is called class-1 if $\Delta(G) = \chi'(G)$ and called class-2 if $\chi'(G) = \Delta(G) + 1$.

The core of a graph G is defined to be the largest induced subgraph of G such that each edge in core is part of a cycle and it is denoted by G_Δ. Finally, let M be a subset of $E(G)$ for a graph G, if there is no two edges in M which are adjacent then M is called a matching.

Conjecture 3.1 *Let G and H be graphs then* $\chi(G \otimes H) = min\{\chi(G), \chi(H)\}$. (Hedetniemi, 1966)

Theorem 3.2 *If* $|X_1| = n \geq 2$ *and* $|X_2| = m \geq 3$ *then*

$$\omega(\Gamma_1 \otimes \Gamma_2) = \chi(\Gamma_1 \otimes \Gamma_2) = \min\{n, m\}.$$

Proof. Let $X_1 = \{x_1, \ldots, x_n\}$ and $X_2 = \{y_1, \ldots, y_m\}$ with $n \geq 2$, $m \geq 3$. We assume that K is one the maximal complete subgraph of $\Gamma_1 \otimes \Gamma_2$ and

$$V(K) = \{(A_{i_1}, B_{j_1}), (A_{i_2}, B_{j_2}), \ldots, (A_{i_k}, B_{j_k})\}.$$

It is clear that for $1 \leq p \neq q \leq k$, $A_{i_p} \neq A_{i_q}$, $B_{j_p} \neq B_{j_q}$ and graph of spanned by the vertices $V_1 = \{A_{i_1}, A_{i_2}, \ldots, A_{i_k}\}$ is complete subgraph of Γ_1 and graph of spanned by the vertices $V_2 = \{B_{i_1}, B_{i_2}, \ldots, B_{i_k}\}$ is complete subgraph of Γ_2. It follows that $k \leq \min\{n, m\}$ from (Toker, 2016). Assume that $m \geq n$ so if we choose

$$V(\Pi) = \{((x_2, x_3, \ldots, x_n), (y_2, y_3, \ldots, y_m)), ((x_1, x_3, \ldots, x_n),$$

$$(y_1, y_3, \ldots, y_m)), \ldots, ((x_1, x_2, x_3, \ldots, x_{n-1}), (y_1, y_2, \ldots, y_{n-1}, y_{n+1}, \ldots y_m))\}$$

and if $n \geq m$ is similar. Let Π be spanned graph by $V(\Pi)$. It is easy to see that Π is $\min\{n, m\}$ vertices complete subgraph of $\Gamma_1 \otimes \Gamma_2$ and so $\omega(\Gamma_1 \otimes \Gamma_2) = \min\{n, m\}$.

Thus $\chi(\Gamma_1 \otimes \Gamma_2) \geq \min\{n, m\}$ from equation (1). Assume that $m \geq n$ and $\forall 1 \leq i \leq n$; $A_i = X_1 \setminus \{i\}$. In addition let

$$\begin{aligned}
Q_1 &= \{(B, v) \mid \varnothing \neq B \subseteq A_1 \text{ and } v \in V(\Gamma_2)\}, \\
Q_2 &= \{(B, v) \mid \varnothing \neq B \subseteq A_2 \text{ and } (B, v) \notin Q_1 \text{ and } v \in V(\Gamma_2)\}, \\
&\vdots \\
Q_n &= \{(B, v) \mid \varnothing \neq B \subseteq A_n \text{ and } (B, v) \notin \bigcup_{i=1}^{n-1} Q_i \text{ and } v \in V(\Gamma_2)\}.
\end{aligned}$$

It is clear if $(u, v) \in V(\Gamma_1 \otimes \Gamma_2)$ then $(u, v) \in Q_s$ $(1 \leq s \leq n)$ for unique s, moreover $\bigcup_{i=1}^{n} Q_i = V(\Gamma_1 \otimes \Gamma_2)$ and $Q_i \cap Q_j = \varnothing$ for $1 \leq i \neq j \leq n$. For each $1 \leq k \leq n$ if we choose a different colour for each Q_k and assign the chosen colour to the all vertices in Q_k, there is no two adjacent vertices have same colour, and so $\chi(\Gamma_1 \otimes \Gamma_2) \leq n$. Thus $\chi(\Gamma_1 \otimes \Gamma_2) = \min\{n, m\}$. If $n \geq m$ is similar. So conjecture holds for $\Gamma_1 \otimes \Gamma_2$.

Lemma 3.3 *A graph is perfect if and only if it is Berge* (Chudnovsky et all., 2006).

Therefore, a graph G is perfect if and only if neither G nor G^c contains an odd cycle of length at least 5 as an induced subgraph.

Theorem 3.4 *If* $|X_1| = 2, |X_2| \geq 3$ *then* $\Gamma_1 \otimes \Gamma_2$ *is perfect graph but if* $|X_1| \geq 3, |X_2| \geq 3$ *then* $\Gamma_1 \otimes \Gamma_2$ *is not perfect graph.*

Proof. We assume that $X_1 = \{x_1, \ldots, x_n\}$ and $X_2 = \{y_1, \ldots, y_m\}$. Let $|X_1| = 2$ and $|X_2| \geq 3$. It is clear that there is not any odd cycle at least 5 as induced subgraph of $\Gamma_1 \otimes \Gamma_2$ since $|X_1| = 2$. Let $G = (\Gamma_1 \otimes \Gamma_2)^c$. For $k \geq 3$ we assume that there is an induced subgraph of G which is cycle with $2k - 1$ vertices, say

$$C_1 - C_2 - \cdots - C_{2k-1} - C_1.$$

Without loss of generality assume that $C_1 = (x_1, B_1)$. Then $C_3 = (x_2, B_3)$ and $C_{2k-2} = (x_2, B_{2k-2})$ because C_1 and C_3 adjacent vertices in $\Gamma_1 \otimes \Gamma_2$ and C_1 and C_{2k-2} adjacent vertices in $\Gamma_1 \otimes \Gamma_2$. Thus $C_2 = (x_1, B_2)$ and $C_{2k-1} = (x_1, B_{2k-1})$ because C_2 and C_{2k-2} adjacent vertices in $\Gamma_1 \otimes \Gamma_2$ and C_3 and C_{2k-1} adjacent vertices in $\Gamma_1 \otimes \Gamma_2$. But in this case C_2 and C_{2k-1} adjacent vertices in G which is a contradiction. So if $|X_1| = 2, |X_2| \geq 3$ then $\Gamma_1 \otimes \Gamma_2$ is perfect graph. Let $|X_1| \geq 3, |X_2| \geq 3$ and $Y_1 = X_1 \setminus \{x_1, x_2, x_3\}$, $Y_2 = X_2 \setminus \{y_1, y_2, y_3\}$. Let $H = (\{x_1, x_2\} \cup Y_1, \{y_1, y_2\} \cup Y_2) - (\{x_3\} \cup Y_1, \{y_2, y_3\} \cup Y_2) - (\{x_1, x_2\} \cup Y_1, \{y_1, y_3\} \cup Y_2) - (\{x_1, x_3\} \cup Y_1, \{y_2\} \cup Y_2) - (\{x_2, x_3\} \cup Y_1, \{y_1, y_3\} \cup Y_2) - (\{x_1, x_2\} \cup Y_1, \{y_1, y_2\} \cup Y_2)$. So H is cycle of length of 5 which subinduced graph of $\Gamma_1 \otimes \Gamma_2$. Thus if $|X_1| \geq 3, |X_2| \geq 3$ then $\Gamma_1 \otimes \Gamma_2$ is not perfect graph.

Lemma 3.5 *Consider the graphs* $G_1 = (V, E_1)$ *and* $G_2 = (V, E_2)$ *with the same vertex set. Suppose that* E_1 *is a matching such that no edge has both endvertices in* $N_{G_2}[\Lambda_{G_2}]$*. If the union graph* $G = G_1 \cup G_2$ *has maximum degree* $\Delta(G) = \Delta(G_2) + 1$ *then* G *is class*-1 (Machado & Figueiredo, 2010).

Theorem 3.6 *If* $|X_1| = n \geq 2$, $|X_2| = m \geq 3$ *then*

$$\chi'(\Gamma_1 \otimes \Gamma_2) = \chi'(\Gamma_1).\chi'(\Gamma_2) = (2^{n-1} - 1).(2^{m-1} - 1)$$

Proof. Let $(u, v) \in V(\Gamma_1 \otimes \Gamma_2)$, if $|u| \geq 2, |v| \geq 2$ or if $|u| = 1, |v| \geq 2$ or $|u| \geq 2, |v| = 1$ then (u, v) is in core at graph from Theorem 2.2. Let $B = \{(\{x_i\}, \{y_j\}) - (X_1 \setminus \{x_i\}, X_2 \setminus \{y_j\}) : x_i \in X_1, y_j \in X_2\}$, $G_1 = (V(\Gamma_1 \otimes \Gamma_2), B))$ and $G_2 = (V(\Gamma_1 \otimes \Gamma_2), E(\Gamma_1 \otimes \Gamma_2)_\Delta)$. So B is a matching such that no edge has both endvertices in $N_{G_2}[\Lambda_{G_2}]$. Also $(\Gamma_1 \otimes \Gamma_2) = G_1 \cup G_2$ and $\Delta(\Gamma_1 \otimes \Gamma_2) = \Delta(G_2) + 1$ so from Lemma 3.5, $\Gamma_1 \otimes \Gamma_2$ is class-1.

References

Chartrand, G., & Zhang, P. (2009). *Chromatic Graph Theory.* Charpman & Hall/CRC, London.

Chudnovsky, M., Robertson, N., Seymour, P., & Thomas, R. (2006). The strong perfect graph theorem. *Ann. Math., 164,* 51-229. http://dx.doi.org/10.4007/annals.2006.164.51

DeMeyer, F., & DeMeyer, L. (2005). Zero divisor graphs of semigroups. *J. Algebra, 283,* 190-198. http://dx.doi.org/10.1016/j.jalgebra.2004.08.028

DeMeyer, F., McKenzie, T., & Schneider., K. (2002). The zero-divisor graph of a commutative semigroup. *Semigroup Forum, 65,* 206-214. http://dx.doi.org/10.1007/s002330010128

Gross, J. L., & Yellen, J. (2004). *Handbook of Graph Theory.* Charpman & Hall/CRC, London.

Howie, J. M. (1995). *Fundamentals of Semigroup Theory.* Oxford University Press, New York.

Hedetniemi, S. (1966). *Homomorphisms of graphs and automata.* Technical Report, 03105-44-T, University of Michigan.

Machado£ R. C. S., & Figueiredo de, C. M. H. (2010). Decompositions for edge-coloring join graphs and cobipartite graphs. *Discrete Applied Mathematics, 158,* 1336-1342. http://dx.doi.org/10.1016/j.dam.2009.01.009

Toker, K. (2016). On the zero-divisor graphs of finite free semilattices. *Turkish Journal of Mathematics, 40*(4), 824-831. http://dx.doi.org/10.3906/mat-1508-38

Vizing, V. G. (1964). On an estimate of the chromatic class of a p-graph. *Diskret. Analiz., 3,* 25-30.

Weichsel, P. M. (1962). The Kronecker product of graphs. *Proc. Amer. Math. Soc., 13,* 47-52. http://dx.doi.org/10.2307/2033769

11

Computing of Z- valued Characters for the Projective Special Linear Group L_2 (2^m) and the Conway Group Co_3

Ali Moghani

Department of Computer Science, William Paterson University, Wayne, NJ. E-mail: moghania@wpunj.edu

Abstract

According to the main result of W. Feit and G. M. Seitz (see, Illinois J. Math. 33 (1), 103-131, 1988), the projective special linear group L_2 (2^m) for m = 3, 4, 5 and the smallest Conway group Co_3 are unmatured groups. In this paper, we continue our study on special finite groups (see Int. J. Theo. Physics, Group Theory, and Nonlinear Optics (17)1, 57-62, 2013) and the dominant classes and Q- conjugacy characters for the above groups are derived.

MSC Mathematics Subject Classification (2010): 20D05, 20C15

Keywords: projective special linear group, Sporadic Conway groups, Conjugacy class, Q-conjugacy character

1. Introduction

In recent years, the problems over group theory have drawn the wide attention of researchers in mathematics, physics and chemistry. Many problems of the computational group theory have been researched, such as the classification, the symmetry, the topological cycle index, etc. It is not only on the property of finite group, but also its wide-ranging connection with many applied sciences, such as Nanoscience, Chemical Physics and Quantum Chemistry, for instant see [Moghani, 2010].

S. Fujita suggested a new concept called the markaracter table, which enables us to discuss marks and characters for a finite group on a common basis, and then introduced tables of integer-valued characters and dominant classes, which are acquired for such groups. A dominant class is defined as a disjoint union of conjugacy classes corresponding the same cyclic subgroups, which is selected as a representative of conjugate cyclic subgroups. Moreover, the cyclic (dominant) subgroup selected from a non-redundant set of cyclic subgroups of G is used to compute the Q-conjugacy characters of G, as demonstrated in [Fujita, 1998].

The projective special linear groups L_2 (8), L_2 (16), L_2 (32) and the smallest Conway group Co_3 with orders 540, 4080, 32736 and 495766656000 respectively, are unmatured groups according to the main result of W. Feit and G. M. Seitz in [Feit et al., 1988]. The motivation for this study is outlined in [Safarisabet et al., 2013; Fujita, 1998; Moghani, 2009&2010; Aschbacher, 1997; Feit et al., 1988; Conway et al., 1985] and the reader is encouraged to consult these papers and [Moghani, 2009&2010; Aschbacher, 1997; Feit et al., 1988; Conway et al., 1985; GAP, 1995; Kerbe et al., 1982; Kerber, 1999] for background material as well as basic computational techniques.

This paper is organized as follows: In Section 2, we introduce some necessary concepts, such as the maturity and Q-conjugacy character of a finit group. In Section 3, we provide all the dominant classes and Q- conjugacy characters for the projective special linear group L_2 (2^m) for m = 3, 4, 5 and the Conway groups Co_3.

2. Preliminaries

Throughout this paper we adopt the same notations as in [Safarisabet et al., 2013; Conway, 1985]. For instance, we will use the ATLAS notations for conjugacy classes. Thus, nx, n is an integer and x = a, b, c…denotes an arbitrary conjugacy class of G of elements of order n.

Definition 2.1: Let G be an arbitrary finite group and $h_1, h_2 \in G$, we say h_1 and h_2 are Q-conjugate if $t \in G$ exists such that $t^{-1} < h_1 > t = < h_2 >$ which is an equivalence relation on group G and generates equivalence classes that are called dominant classes. Therefore, G is partitioned into dominant classes [Fujita, 1998].

Definition 2.2: Suppose H be a cyclic subgroup of order n of a finite group G. Then, the maturity discriminant of H denoted by m(H), is an integer number delineated by $|N_G(H): C_G(H)|$ in addition, the dominant class of $K \cap H$ in the normalizer $N_G(H)$ is the union of $t = \frac{m(H)}{\phi(|H|)}$ conjugacy classes of G where φ is Euler function, i.e. the maturity of G is clearly defined by examining how a dominant class corresponding to H contains conjugacy classes. The group G should be matured group if $t = 1$, but if $t \geq 2$, the group G is an unmatured concerning subgroup H, see [Safarisabet et al., 2013; Fujita, 1998; Moghani, 2009&2010]. For some properties of the maturity see the following theorem which is introduced by the author in [Moghani, 2009]:

Theorem 2.3: The wreath products of the matured groups again is a matured group, but the wreath products of at least one unmatured group is an unmatured group.

Definition 2.4: Let $C_{u\times u}$ be a matrix of the character table for an arbitrary finite group G. Then, C is transformed into a more concise form called the Q-Conjugacy character table denoted by C_G^Q containing integer-valued characters. By Theorem 4 in [Fujita, 1998], the dimension of a Q-conjugacy character table C_G^Q is equal to its corresponding markaracter table denoted by M_G^C, i.e. C_G^Q is a $m \times m$ –matrix where $m \leq u$ is the number of dominant classes or equivalently the number of non-conjugate cyclic subgroups denoted by denoted by SCS_G, see [Safarisabet et al., 2013; Fujita, 1998; Moghani, 2009&2010].

Definition 2.5: If $\chi_1, \ldots, \chi_k$ are all the irreducible characters of a finite group H, let $Q(H) = Q(\chi_1,\ldots, \chi_k)$ be the field generated by all χ_i (x), $x \in H$, $1 \leq I \leq k$.

A character χ is rational if $Q(\chi) = Q$. A group H is a rational group if $Q(H) = Q$ (e.g. every Weyl group is a rational group [Feit et al., 1988]).

Theorem 2.6 [Feit et al., 1988]: Let G be a non cyclic finite simple group. Then G is a composition factor of a rational group if and only if G is isomorphic to an alternating group or one of the following groups: $PSp_4(3)$, $Sp_6(2)$, $O_8^+(2)'$, $PSL_3(4)$, $PSU_4(3)$.

3. Conclusion

According to the Theorem 2.6, the projective special linear groups L_2 (8), L_2 (16), L_2 (32) and the Conway group Co_3 are unmatured groups. Now we are equipped to compute all the dominant classes and Q-conjugacy characters for the above groups with aid GAP program [GAP, 1995], http://www.gap–system.org.

Theorem 3.1

(i) The projective special linear group L_2 (8) has two unmatured dominant classes with $t = 3$ in definition 2.2. Furthermore, there are five Q- conjugacy characters for L_2 (8) with the following degrees: 1, 7, 8, 21 and 27.

(ii) The projective special linear group L_2 (16) has three unmatured dominant classes with t= 2, 4 and 8. Furthermore, there are eight Q- conjugacy characters for L_2 (16) with the following degrees: 1, 16, 17, 34, 68 and 120.

(iii) The projective special linear group L_2 (32) has three unmatured dominant classes with t= 5, 15 and 10. Furthermore, there are six Q- conjugacy characters for L_2 (32) with the following degrees: 1, 31, 32, 155, 310 and 495.

Proof: Here, because of similar discussions we verify via full discussions just (ii) for L_2(16) of order 4050. To find all the number of dominant classes for L_2(16) at first, we calculate the markaracter table for L_2(16) via GAP system, see definition 2.2 and GAP programs in [Safarisabet et al., 2013; GAP, 1995] for more details.

Hence, see the markaracter table for L_2 (16) (i.e. $M_{L2(16)}^C$) in Table 1, corresponding to five non-conjugate cyclic subgroups (i.e. $G_i \in SCS_{L2(16)}$) of orders 1, 2, 3, 5, 15 and 17 respectively, as follow:

G_1 = id, G_2 = < (2, 3)(4, 5)(6, 9)(7, 12)(8, 17)(10, 16)(11, 13)(14, 15) >, G_3 = < (3, 4, 5)(6, 10, 14)(7, 11, 15)(8, 12, 16)(9, 13, 17) >, G_4 = < (3, 8, 10, 13, 15)(4, 12, 14, 17, 7)(5, 16, 6, 9, 11) >, G_5 = < (3, 4, 5)(6, 10, 14)(7, 11, 15)(8, 12, 16)(9, 13, 17), (3, 8, 10, 13, 15)(4, 12, 14, 17, 7)(5, 16, 6, 9, 11) >, and G_6 = < (1, 2, 3, 6, 17, 11, 5, 13, 9, 10, 12, 8, 7, 4, 16, 14, 15) >.

Therefore, $|SCS_{L2(16)}| = 6$ and its dominant classes are 1a, 2a, 3a, K_5= 5a ∪ 5b, K_{15}= 15a ∪ 15b ∪ 15c ∪ 15d and K_{17} = 17a ∪ 17b ∪ 17c ∪ 17d ∪ 17e ∪ 17f ∪ 17g ∪ 17h, thus L_2 (16) has three unmatured dominant classes with t = 2, 4 and 8.

Furthermore, L_2(16) has three unmatured Q-conjugacy characters φ_2, φ_5 and φ_6 which are the sum of eight, two and four irreducible characters respectively. Therefore, there are eight, two and four column-reductions respectively (similarly row-reductions) in the character table of L_2(16). There are eight Q- conjugacy characters for L_2 (16) with the following degrees: 1, 16, 17, 34, 68 and 120, see Table 2.

Table 1. The markaracter Table of the projective special linear group L_2 (16)

$M^C_{L2(16)}$	G_1	G_2	G_3	G_4	G_5	G_6
$(L_2(16)/G_1)$	4080	0	0	0	0	0
$(L_2(16)/G_2)$	2040	8	0	0	0	0
$(L_2(16)/G_3)$	1360	0	10	0	0	0
$(L_2(16)/G_4)$	816	0	0	6	0	0
$(L_2(16)/G_5)$	272	0	2	2	2	0
$(L_2(16)/G_6)$	240	0	0	0	0	2

Besides, the dominant classes of L_2 (8) are 1a, 2a, 3a, D_7 = 7a ∪ 7b ∪ 7c and D_9 = 9a ∪ 9b ∪ 9c which has two unmatured dominant classes with t = 3. Similar discussions show that there are five Q- conjugacy characters for L_2 (8) with the following degrees: 1, 7, 8, 21 and 27.

L_2 (8) has two unmatured Q-conjugacy characters μ_3 and μ_5 which are the sum of three irreducible characters respectively, see Table 3.

Table 2. The Q-Conjugacy Character of the projective special linear group L_2 (16)

$C^Q_{L2(16)}$	1a	2a	3a	K_5	K_{15}	K_{17}
ϕ_1	1	1	1	1	1	1
ϕ_2	120	-8	0	0	0	1
ϕ_3	16	0	1	1	1	-1
ϕ_4	17	1	-1	2	-1	0
ϕ_5	34	2	4	-1	-1	0
ϕ_6	68	4	-4	-2	1	0

wherein K_5= 5a ∪ 5b, K_{15}= 15a ∪ 15b ∪ 15c ∪ 15d and K_{17} = 17a ∪ 17b ∪ 17c ∪ 17d ∪ 17e ∪ 17f ∪ 17g ∪ 17h

The dominant classes of L_2 (32) are 1a, 2a, 3a, L_{11} = 11a ∪ 11b ∪ 11c ∪ 11d ∪ 11e, L_{31} = 31a ∪ 31b ∪ 31c ∪ 31d ∪ 31e ∪ 31f ∪ 31g ∪ 31h ∪ 31i ∪ 31j ∪ 31k ∪ 31l ∪ 31m ∪ 31n ∪ 31o and L_{33} = 33a ∪ 33b ∪ 33c ∪ 33d ∪ 33e ∪ 33f ∪ 33g ∪ 33h ∪ 33i ∪ 33j which has three unmatured dominant classes with t = 5, 15 and 10.

Table 3. The Q-Conjugacy Character of the projective special linear group L_2 (8)

$C^Q_{L2(8)}$	1a	2a	3a	D_7	D_9
μ_1	1	1	1	1	1
μ_2	7	-1	-2	0	1
μ_3	21	-3	3	0	0
μ_4	8	0	-1	1	-1
μ_5	27	3	0	-1	0

Wherein, D_7 = 7a ∪ 7b ∪ 7c and D_9 = 9a ∪ 9b ∪ 9c

Table 4. The Q-Conjugacy Character of the projective special linear group L_2 (32)

$C^Q_{L2(32)}$	1a	2a	3a	L_{11}	L_{31}	L_{33}
ς_1	1	1	1	1	1	1
ς_2	120	-8	0	0	0	1
ς_3	16	0	1	1	1	-1
ς_4	17	1	-1	2	-1	0
ς_5	34	2	4	-1	-1	0
ς_6	68	4	-4	-2	1	0

Wherein L_{11} = 11a ∪ 11b ∪ 11c ∪ 11d ∪ 11e, L_{31} = 31a ∪ 31b ∪ 31c ∪ 31d ∪ 31e ∪ 31f ∪ 31g ∪ 31h ∪ 31i ∪ 31j ∪ 31k ∪ 31l ∪ 31m ∪ 31n ∪ 31o and L_{33} = 33a ∪ 33b ∪ 33c ∪ 33d ∪ 33e ∪ 33f ∪ 33g ∪ 33h ∪ 33i ∪ 33j.

We afford all the Q-conjugacy characters of $L_2(2^m)$ for m = 3, 4, 5 in Tables 2-4.□

Theorem 3.2

The Conway groups Co_3 has six unmatured dominant classes with the $t = 2$.

Furthermore, there are thirty eight Q- conjugacy characters for Co_3 with the following degrees: 1, 23, 253, 275, 1771, 1792, 2024, 4025, 5544, 7040, 7084, 8855, 19250, 23000, 26082, 31625, 31878, 40250, 41216, 57960, 63250, 73600, 80960, 91125, 93312, 129536, 177100, 184437, 221375, 226688, 246400, 249480, 253000 and 255024.

Proof: According to similar discussion in the previous theorem, it is enough to report the dominant classes of Co_3 as follow:

1a, 2a, 2b, 3a, 3b, 3c, 4a, 4b, 5a, 5b, 6a, 6b, 6c, 6d, 6e, 7a, 8a, 8b, 8c, 9a, 9b, 10a, 10b, M_{11} = 11a ∪ 11b, 12a, 12b, 12c, 14a, 15a, 15b, 18a, M_{20} = 20a ∪ 20b, 21a, M_{22} = 22a ∪ 22b, M_{23} = 23a ∪ 23b, 24a, 24b, 30a which has four unmatured dominant classes with t = 2.

Table 5. The Q-Conjugacy Character Table of the Conway group Co_3

C^Q_{Co3}	1a	2a	2b	3a	3b	3c	4a	4b	5a	5b	6a	6b
π_1	1	1	1	1	1	1	1	1	1	1	1	1
π_2	23	7	-1	-4	5	-1	-5	3	-2	3	4	-2
π_3	253	13	-11	10	10	1	9	1	3	3	10	4
π_4	253	29	-11	10	10	1	-11	5	3	3	2	2
π_5	275	35	11	5	14	-1	15	7	0	5	5	-1
π_6	1792	0	32	64	-8	-14	0	0	-8	2	0	0
π_7	1771	-21	11	-11	16	7	-5	-5	-4	1	21	-3
π_8	2024	104	0	-1	26	8	-24	8	-1	4	-1	5
π_9	7040	-128	0	-88	20	-16	0	0	-10	0	-8	16
π_{10}	4025	105	1	-25	29	-7	-35	5	0	5	15	-3
π_{11}	5544	168	0	-45	36	0	40	8	-6	4	3	-3
π_{12}	7084	-84	44	10	19	-14	-4	-4	9	-1	18	6
π_{13}	8855	231	55	-1	35	-7	19	11	5	0	-9	-3
π_{14}	19250	210	-110	80	-10	14	10	-6	0	0	0	12
π_{15}	41216	0	-32	-256	-40	14	0	0	16	6	0	0
π_{16}	23000	280	120	50	5	8	40	8	0	0	10	10
π_{17}	26082	-126	-54	81	0	0	-6	10	7	-3	9	9
π_{18}	31625	265	-55	35	35	-1	-55	9	0	0	-5	-5
π_{19}	31625	-55	-55	35	35	-1	25	-7	0	0	35	-1
π_{20}	31625	505	-55	35	35	-1	-35	5	0	0	-5	1
π_{21}	31878	294	-66	45	45	0	46	-2	3	3	-3	-3
π_{22}	40250	-70	10	-115	-25	14	10	10	0	0	5	-7
π_{23}	57960	168	120	126	45	0	-40	-8	10	0	6	6
π_{24}	63250	210	-110	-65	-20	22	-30	2	0	0	15	3
π_{25}	73600	0	144	160	16	13	0	0	0	-5	0	0
π_{26}	80960	-448	0	176	50	8	0	0	10	0	-16	-16
π_{27}	91125	405	45	0	0	27	45	-3	0	0	0	0
π_{28}	93312	0	-144	0	0	27	0	0	12	-3	0	0
π_{29}	129536	-512	0	-64	44	8	0	0	-14	-4	16	-8
π_{30}	129536	512	0	-64	44	8	0	0	-14	4	-16	8
π_{31}	177100	140	44	-20	-29	-14	-20	12	0	-5	20	-4
π_{32}	184437	405	-99	0	0	-27	45	-3	12	-3	0	0
π_{33}	221375	735	55	-160	-25	-7	-25	-9	0	0	0	-12
π_{34}	226688	0	-176	320	-40	-7	0	0	-12	3	0	0
π_{35}	246400	0	176	160	-56	7	0	0	0	5	0	0
π_{36}	249480	-504	0	-81	0	0	-24	8	5	0	-9	9
π_{37}	253000	-440	0	-125	10	-8	40	8	0	0	-5	1
π_{38}	255024	-336	0	-126	36	0	-16	-16	-1	4	-6	6

Table 5 (continued); wherein M_{11}= 11a ∪ 11b.

C^{Q}_{Co3}	6c	6d	6e	7a	8a	8b	8c	9a	9b	10a	10b	M_{11}	12a
π_1	1	1	1	1	1	1	1	1	1	1	1	1	1
π_2	1	-1	-1	2	1	-3	1	-1	2	2	-1	1	-2
π_3	-2	-2	1	1	-1	3	-1	1	1	3	-1	0	0
π_4	2	-2	1	1	-3	-3	1	1	1	-1	-1	0	-2
π_5	2	2	-1	2	1	5	1	-1	2	0	1	0	3
π_6	0	-4	2	0	0	0	0	4	-2	0	2	-1	0
π_7	0	2	-1	0	-1	-1	-1	-2	-2	4	1	0	1
π_8	2	0	0	1	4	-4	0	-1	-1	-1	0	0	-3
π_9	4	0	0	-2	0	0	0	2	2	2	0	0	0
π_{10}	-3	1	1	0	-1	-5	-1	2	2	0	1	-1	1
π_{11}	0	0	0	0	-4	4	0	0	0	-2	0	0	1
π_{12}	3	-1	2	0	0	0	0	4	-2	1	-1	0	2
π_{13}	3	1	1	0	5	1	1	2	-4	1	0	0	1
π_{14}	6	-2	-2	0	6	-2	-2	-4	2	0	0	0	4
π_{15}	0	4	-2	0	0	0	0	-4	-4	0	-2	-1	0
π_{16}	1	3	0	-2	0	0	0	-1	2	0	0	-1	-2
π_{17}	0	0	0	0	2	2	-2	0	0	-1	1	1	-3
π_{18}	-5	-1	-1	-1	1	1	1	-1	-1	0	0	0	-1
π_{19}	-1	-1	-1	-1	1	1	1	-1	-1	0	0	0	-5
π_{20}	7	-1	-1	-1	-5	-1	-1	-1	-1	0	0	0	1
π_{21}	-3	-3	0	0	2	2	-2	0	0	-1	-1	0	1
π_{22}	-1	1	-2	0	-2	-2	-2	5	-1	0	0	1	1
π_{23}	-3	3	0	0	0	0	0	0	0	-2	0	1	2
π_{24}	0	-2	-2	-2	2	2	2	4	1	0	0	0	3
π_{25}	0	0	-3	2	0	0	0	4	1	0	-1	-1	0
π_{26}	2	0	0	-2	0	0	0	-1	2	2	0	0	0
π_{27}	0	0	3	-1	-3	-3	1	0	0	0	0	1	0
π_{28}	0	0	3	2	0	0	0	0	0	0	1	-1	0
π_{29}	4	0	0	1	0	0	0	-1	-1	-2	0	0	0
π_{30}	-4	0	0	1	0	0	0	-1	-1	2	0	0	0
π_{31}	-1	-1	2	0	0	0	0	-5	1	0	-1	0	4
π_{32}	0	0	-3	1	-3	-3	1	0	0	0	1	0	0
π_{33}	3	1	1	0	3	3	-1	2	2	0	0	0	-4
π_{34}	0	4	1	0	0	0	0	2	-1	0	-1	0	0
π_{35}	0	-4	1	0	0	0	0	-2	-2	0	1	0	0
π_{36}	0	0	0	0	-4	4	0	0	0	1	0	0	-3
π_{37}	-2	0	0	-1	4	-4	0	1	1	0	0	0	1
π_{38}	0	0	0	0	0	0	0	0	0	-1	0	0	2

Table 5 (continued); wherein M_n= na ∪ nb, for n= 20, 22, 23.

C^{Q}_{Co3}	12b	12c	14a	15a	15b	18a	M_{20}	21a	M_{22}	M_{23}	24a	24b	30a
π_1	1	1	1	1	1	1	1	1	1	1	1	1	1
π_2	0	1	0	1	0	1	0	-1	-1	0	-2	0	-1
π_3	-2	0	-1	0	0	1	-1	1	0	0	2	0	0
π_4	2	-2	1	0	0	-1	-1	1	0	0	0	0	2
π_5	1	0	0	0	-1	-1	0	-1	0	-1	1	-1	0
π_6	0	0	0	4	2	0	0	0	-1	-2	0	0	0
π_7	1	-2	0	-1	1	0	0	0	0	0	-1	-1	1
π_8	-1	0	-1	-1	1	-1	1	1	0	0	1	-1	-1
π_9	0	0	-2	2	0	-2	0	-2	0	2	0	0	2
π_{10}	-1	1	0	0	-1	0	0	0	1	0	-1	1	0
π_{11}	-1	-2	0	0	1	0	0	0	0	1	-1	1	-2
π_{12}	2	-1	0	0	-1	0	1	0	0	0	0	0	-2
π_{13}	-1	1	0	-1	0	0	-1	0	0	0	-1	1	1
π_{14}	0	-2	0	0	0	0	0	0	0	-1	0	4	0
π_{15}	0	0	0	4	0	0	0	0	1	0	0	0	0
π_{16}	2	1	0	0	0	1	0	1	-1	0	0	0	0
π_{17}	1	0	0	1	0	0	-1	0	1	0	-1	-1	-1
π_{18}	3	-1	-1	0	0	1	0	-1	0	0	1	1	0
π_{19}	-1	1	1	0	0	-1	0	-1	0	0	1	1	0
π_{20}	-1	1	1	0	0	1	0	-1	0	0	1	-1	0
π_{21}	1	1	0	0	0	0	1	0	0	0	-1	-1	2
π_{22}	1	1	0	0	0	-1	0	0	-1	0	1	1	0
π_{23}	-2	1	0	1	0	0	0	0	-1	0	0	0	1
π_{24}	-1	0	0	0	0	0	0	1	0	0	-1	-1	0
π_{25}	0	0	0	0	1	0	0	-1	1	0	0	0	0
π_{26}	0	0	0	1	0	-1	0	1	0	0	0	0	-1
π_{27}	0	0	-1	0	0	0	0	-1	1	-1	0	0	0
π_{28}	0	0	0	0	0	0	0	-1	-1	1	0	0	0
π_{29}	0	0	-1	1	-1	1	0	1	0	0	0	0	1
π_{30}	0	0	1	1	-1	-1	0	1	0	0	0	0	-1
π_{31}	0	1	0	0	1	-1	0	0	0	0	0	0	0
π_{32}	0	0	-1	0	0	0	0	1	0	0	0	0	0
π_{33}	0	-1	0	0	0	0	0	0	0	0	0	0	0
π_{34}	0	0	0	0	0	0	0	0	0	0	0	0	0
π_{35}	0	0	0	0	0	0	0	0	0	1	0	0	0
π_{36}	-1	0	0	-1	0	0	1	0	0	-1	-1	1	1
π_{37}	-1	-2	1	0	0	1	0	-1	0	0	1	-1	0
π_{38}	2	2	0	-1	1	0	-1	0	0	0	0	0	-1

Furthermore, Co_3 has four unmatured Q-conjugacy characters π_6, π_9, π_{14} and π_{15} which are the sum of two irreducible characters respectively. Therefore, there are two column-reductions (similarly row-reductions) in the character table of Co_3.

There are thirty eight Q- conjugacy characters for Co_3 with the following degrees: 1, 23, 253, 275, 1771, 1792, 2024, 4025, 5544, 7040, 7084, 8855, 19250, 23000, 26082, 31625, 31878, 40250, 41216, 57960, 63250, 73600, 80960, 91125, 93312, 129536, 177100, 184437, 221375, 226688, 246400, 249480, 253000 and 255024, see all the Q-conjugacy characters of Co_3 which are stored in Table5.□

Acknowlagement

The author is indebted to dear Dr. John P. Najarian, Chairman of Department Computer Science William Paterson University, for his useful helps and partial support from

References

Aschbacher, M. (1997). *Sporadic Groups*. Cambridge, Cambridge University Press.

Conway, J. H., Curtis, R. T., Norton, S. P., Parker, R. A., & Wilson, R. A. (1985). *ATLAS of Finite Groups*. Oxford, Oxford Univ. Press.

Feit, W., & Seitz, G. M. (1988). On finite rational groups and related topics. *Illinois J. Math.*, *33*, 103-131.

Fujita, S. (1998). Inherent Automorphism and Q-Conjugacy Character Tables of Finite Groups. An Application to Combinatorial Enumeration of Isomers. *Bull. Chem. Soc. Jpn.*, *71*, 2309-2321. http://dx.doi.org/10.1246/bcsj.71.2309

GAP. (1995). *Groups, Algorithms and Programming*. Lehrstuhl De für Mathematik, RWTH, Aachen.

Kerbe, A., & Thurlings, K. (1982). *Combinatorial Theory*. Berlin Springer.

Kerber, A. (1999). *Applied Finite Group Actions*. Berlin, Springer-Verlag. http://dx.doi.org/10.1007/978-3-662-11167-3

Moghani, A. (2009). A New Simple Method for Maturity of Finite Groups and Application to Fullerenes and Fluxional Molecules. *Bull. Chem. Soc. Jpn.*, *82*, 1103-1106. http://dx.doi.org/10.1246/bcsj.82.1103

Moghani, A. (2010). Study of Symmetries on some Chemical Nanostructures. *J. Nano Res.*, *11*, 7-11. http://dx.doi.org/10.4028/www.scientific.net/JNanoR.11.7

Safarisabet, Sh. A., Moghani, A., & Ghaforiadl, N. (2013). A Study on The Q-Conjugacy Characters of Some Finite Groups. *Int. J. Theoretical Physics, Group Theory, and Nonlinear Optics*, *17*(1), 57-62.

12

Mathematical Models of Refugee Immigration and Recommendations of Policies

Qilong Cheng[1], Tiancheng Yu[2], Jingkai Yan[2], & Ru Wang[3]

[1] Department of Mechanical Engineering, Tsinghua University, 100084, China

[2] Department of Electronic Engineering, Tsinghua University, 100084, China

[3] School of Foreign Languages, Jiangxi Normal University, 330022, China

Correspondence: Qilong Cheng, Department of Mechanical Engineering, Tsinghua University, 100084, China. E-mail: chengqlthu@hotmail.com

The team was designated as Meritorious Winner in Interdisciplinary Contest in Modeling in 2016.

Abstract

Over the past two years, the refugee crisis resulted from the racial conflict, persecution, generalized violence and violations of human rights has forced an enormous number of refugees to flee to Europe. Aiming to address the problem caused by the flow of refugees, we analyzed the actual procedure of their movement and divide it into three major stages. We designed the gathering model, the entering model, the transferring model, even the health and safety model. Finally, we used the models described above to complete our assigned tasks. Also we put forward seven major policy recommendations to the committee. We accompanied every policy with a straightforward explanation so that people without any technical background can easily understand our insights. The main strength of our model is that it can forecast the flow of immigration and provide meaningful suggestions policies for refugees. With the help of modern computing software, we can track the current tendency and make judges efficiently.

Keywords: mathematical models, refugee immigration, policies

1. Introduction

In order to illustrate the problems caused by the influx of refugees inside or outside of their source country into Europe, the following background is worth mentioning.

1.1 Crisis Overview

With the serious and turbulent situation in the west of Asia and north of Africa in 2015, numerous people are forced to leave their homes, especially in Syria and Afghanistan, fleeing their countries through the sea and land route into Europe to find security, support and sufficient food and water to live on. This is the second biggest refugee immigration crisis that happened in Europe since the World War Two. A picture about a young baby huddling up his little body died on the beach is the epitome of lack of the safe transportation vehicles and protection. Attention also has been aroused by public that the refugee crisis has brought a series of social problems, for instance a sexual abuse in Germany and other crime cases from stealing, robbery and even murder.

1.2 International Response

The UN High Commissioner for refugees (UNHCA) has reached out hands to help these displaced people, calling up some countries to provide them with safe shelters and daily life necessities. Germany showed the most positive attitudes towards refugees and received the highest numbers of new asylum applications worldwide. On the contrast, some counties among the east of Europe displayed opposite opinions towards the flow of refugees. However, despite the harsh winter in 2016, there are still a large number of people making their dangerous journey continually across the Aegean Sea to Europe. They are provided with some blankets, pillows, and bed linens to survive the winter even below 24°C. And ambulances are also equipped to help take people with health problems to hospitals. But attention should also be focused that European countries may tighten asylum rules as refugee waves continue.

1.3 Our Faith

Our team, the ICM-RUN is very interested in refugee problems not only because of its large scale and fast expending trend, but also because we hold the belief that man was born to be equal, which rooted on our mind deeply and that man has his freedom to pursue his right to live and enjoy the esteem of humanity so that a better solution must be found with the aid of some scientific methods to help them.

2. The Description of the Problem

2.1 How Do Refugees Gather at Gathering Points of Travel Routes?

As a result of major political and social unrest and warfare, refugees will come at assembling places of six travel routes at a specified speed, which will be given in the model. We presume that warfare will burst in all places at the same time, so the refugees rush to the six gathering spots simultaneously. The gathering spots become models of which inputs are refugees that pour into and outputs are refugees that come into Europe through travel routes.

2.2 How Do Refugees Travel Through Routes?

Take the six travel routes as example, there are three by sea and three by land. The model is also suitable if there are more than six travel routes. What's more, routes by land is no different from routes by sea except some parameters like danger coefficient and transport capacity, so we shall restrict our discussion here to sea transport, and train transport can be analyzed just the same with some minor adjustment to values of parameters.

2.2.1 From the Perspective of Shipping Business

Obviously, all the refugees are traveling across the sea in the ships of local business illegally. The shipping business tends to load more refugees in ships and earn more money, but suffers from more danger of shipwreck and more financial loss. The shipping business, which aims at money, should have a plan on how to balance these two things.

2.2.2 From the Perspective of the Government

What is different is that the government cares more about people's lives rather than money. If the government does not supervise the shipping business at all, the shipping business will load as many refugees as they can and cause serious shipwreck. If the government do supervise the shipping business strictly, there will be many refugees waiting to be transported, who still suffer from warfare. The government should supervise to a certain extent where the number of endangered refugees is least.

2.2.3 The Compromise

Although refugees are not legal citizens of Europe so far, the government should impose a fine on the shipping business out of humanitarianism if they cause a shipwreck and make refugees die. Thus, the shipping business faces another condition of balance.

2.3 How Do Refugees Travel into Countries after Entry Points?

Considering limited speed of transportation and adaption to the local environment, we assume that refugees can only travel into adjacent countries in a time unit. We find that how they travel has something in common with Markov Random Field. We regard countries as minimum units and we define each country a parameter that stands for potential. We think that the potential relies on two sides, the national power and the number of refugees left in the nation. The potential of each country determines its energy. The energy of a country is proportionate to how attractive it is to refugees. Meanwhile, we consider each country to have a parameter called inertia probability which means the extent that refugees in one country are not willing to move. And the inertia probability will increase over time, which means refugees are more and more unwilling to move with time passing. Thus, we have a model that can simulate the flow of refugees through energy difference.

2.4 How Do Countries Grant Asylum Applications?

We know that each country will have many illegal refugees that will apply for asylum, but the country cannot grant all the applications. On the one hand, the refugees who are granted asylum will consume the country's resource. On the other hand, the refugees who are not granted asylum will be a kind of threat to the country bring with discontent, impatience, disturbance or even crimes to the country. Therefore, the country should decide how many applications will be granted in a time unit due to the maximum of its benefit.

2.5 How Do We Describe Healthcare and Security?

As for healthcare, the refugees' chronic diseases, acute diseases, infectious diseases should be considered first. Taking the poorer environment into account, refugees are more likely to suffer from the diseases mentioned above and become more susceptible to dying. Besides, both the security problems when refugees travel across the sea and the threat when they have many illegal refugees wandering about should also be considered. The most important, the extremely serious and sudden events such as the terrorist attack in Paris should be paid more attention to.

3. Models

3.1 Gathering Model

First of all, when warfare burst, refugees will choose which route to escape from their homes into Europe.

3.1.1 Assumptions

1) Six routes are considered.
2) Each refugee is independent and has his own choice of routes.
3) Refugees choose the route of the least danger he/she faces on the route.
4) The danger is proportional to the distance refugees have covered.
5) The density of refugees in the effected region is the same.

3.1.2 The Foundation of Model

1) Preparation

a. We number the routes on the map (Baronett, 2008) and set up a plane coordinate system as the following Figure 1.

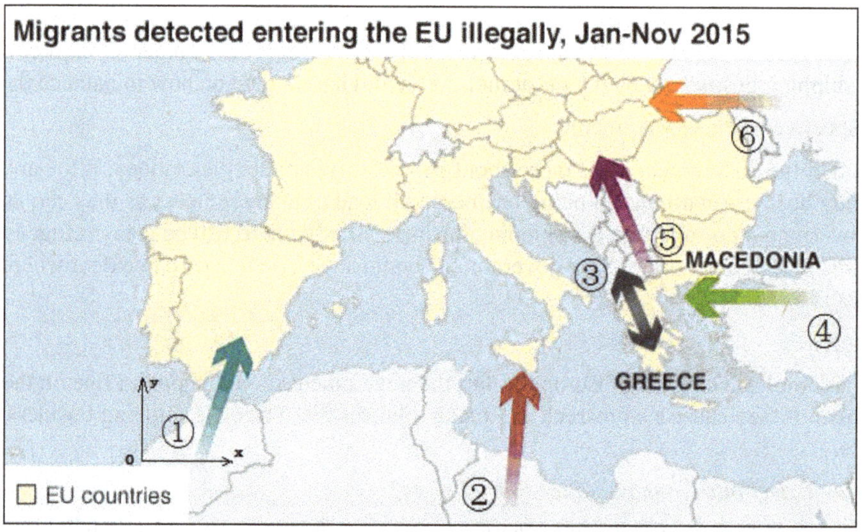

Figure 1. Map of the six routes

b. We measure the x-coordinate x_i and y-coordinate y_i of the gathering point and the length of the route on the map (under a certain proportion).

We used data retrieved from charts online (Baronett, 2008) to calculate the mortality rate l on each route. Note that the data of eastern Mediterranean covers four routes of the six, therefore we take the average of the eastern Mediterranean data to be the mortality rate of all four eastern routes.

Refugees' route from their homes to the gathering point of the travel routes must be on the land. But their travel routes may not be on the land, 1, 2, 4 on the sea and 3, 5, 6 on the land. We use the parameter ε to convert the route by sea into route by land. Because shipping is often quicker than walking on foot, we propose that ε should be a figure close to 0. ($\varepsilon = 0.1$)

Route #	x_i	y_i	d	ε	l
1	25	15	20	0.1	0.39%
2	130	0	35	0.1	1.93%
3	165	65	30	1	0.0668%
4	220	55	20	0.1	0.0668%
5	175	65	40	1	0.0668%
6	210	120	40	1	0.0668%

Chart 1. Route parameters of the six routes

2) The danger

The danger contains danger on the land and danger on the sea, but we use ε to convert the sea route into land route, so the danger contains the total distance and the danger parameter l. So we have:

$$f_i = l_i(\sqrt{(x-x_i)^2+(y-y_i)^2}+\varepsilon_i d_i) \tag{1}$$

As we can see from the Fig 1, Route 1 and 2 are associated, Route 3 and 5 are associated and Route 4 and 6 are associated. Take Route 1 and 2 for an example.

$$f_1 = f_2 \tag{2}$$

$$l_1\left[\sqrt{(x-x_1)^2+(y-y_1)^2}+\varepsilon_1 d_1\right] = l_2\left[\sqrt{(x-x_2)^2+(y-y_2)^2}+\varepsilon_2 d_2\right] \tag{3}$$

Thus, we have an equation which stands for the border where refugees can either choose Route 1 or 2. It is the same with Route 4 and 6. Note that Route 4 and 6 are naturally separated because of Black Sea. Note that Route 3 and 5 are different because they are inside of Europe and have convenient transportation, so we do not consider them here.

3.1.3 Solution and Result

The solution of the equations is showed as curves in the Figure 2, as follows

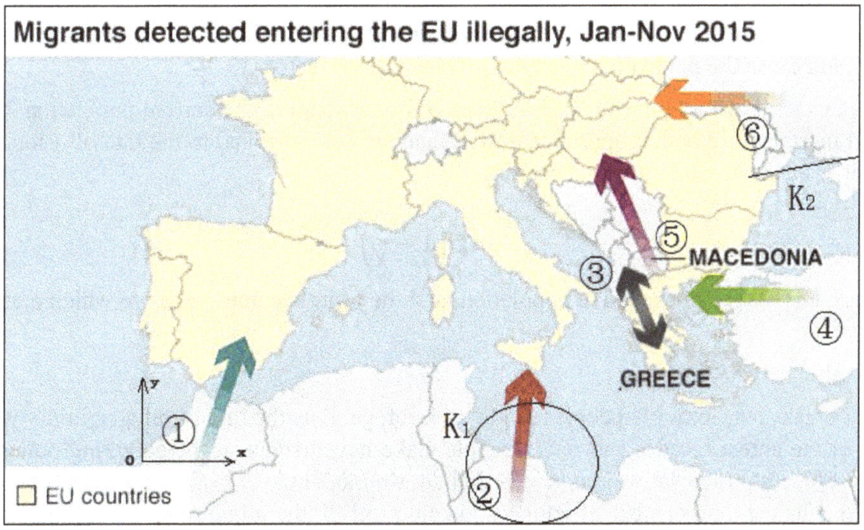

Figure 1. Map partition of the source of refugees

Figure 2 give us a partition of map, which indicates that refugees inside of K_1 should choose Route 2 and the others of North Africa should choose Route 1.

The border K_2 shows that the refugees north of K_2 should choose Route 6 and the refugees south of K_2 should choose Route 4.

The result is understandable because Route 2 is so dangerous that people should not choose it. But the fact is that there are quantities of refugees risking their lives across Central Mediterranean. We do not advocate the choice of routes. Besides, the border K_2 is basically consistent with the natural trend of terrain.

Our result is not sensitive to the parameter ε, so the model is stable itself.

3.2 Entering Model

After arriving at the gathering points, refugee wait for boats or trains to make the vital step of entering the Europe continent.

3.2.1 Assumptions

1) All refugees enter Europe via one of the six routes, setting off from a certain gathering point of each route.

2) The number of refugees to arrive at a assembling place is limited.

3) Each route has a limited capacity. The traffic amount of refugees cannot exceed the capacity.

4) Refugees either leave in batches by boat or train or leave on foot, such that the number of refugees leaving per batch is relatively small versus the frequency of their departure. Therefore the flow of refugees leaving for Europe can

be regarded as a constant continuous flow.

5) When a train ride is required, refugees manage to sneak onto trains without the normal procedure of buying tickets. Therefore the trains may be overloaded and thus brings a chance for chaos on trains.

6) When crossing a body of water (e.g. the Mediterranean) is required, refugees have to take boats owned by individuals who make a profit out of this. Therefore the boat owners may risk overloading the boat in order to make more profits.

7) Refugees are not well-informed about how overpopulated their destinations are, and are willing to go even if the destinations are overpopulated.

8) Each assembling place has limited resources for refugees to live on.

9) Refugees are endangered by local unrest, malnutrition and deadly diseases in the assembling place.

10) Refugees can reproduce while held up in the assembling place.

3.2.2 The Foundation of Model

1) Influence of resources

As assumed, the gathering point possesses resources enough to support S refugees. At some point, if the number of refugees exceeds S, then adverse conditions and lack of resources would cause the population to decrease. On the other hand, if the number of refugees is less than S, this would mark a surplus of resources and reproduction would become the dominant factor and cause the population to expand. This makes a factor $1-\frac{S}{X}$.

Also in either case, this term of the influence of resources is proportional to the current population X (because the rates of reproduction and mortality are both proportionate to it), and also proportional to the rate of resource insufficiency or surplus. Set the coefficient as α, and the term becomes

$$-\alpha X\left(1-\frac{S}{X}\right)$$

If X is greater than S, this marks an excessive population, and the term becomes negative which marks a decrease term for X.

2) Influence of arrival

When a local unrest breaks out, instantly people rush to seek refuge, thus the rate of refugee influx would soar. As time passes, the impact of the unrest recedes and fewer people make new arrivals at the gathering points. Considering that the total number of refugees produced by an unrest is limited, we model the arrival of refugees and an exponential term including time t. Let A be the total number of refugees to come and β the initial rate of incoming refuges, we have the expression of the rate of arrival

$$\beta e^{-\beta t/A}$$

3) Influence of departure

As assumed, we model the departing flow of refugees as a constant continuous flow L. In terms of designed load capacity and overload rate, this decrease term becomes

$$-L=-L_0(1+\theta)$$

4) The differential equation

Therefore, we have the differential equation for X, the number of refugee at the gathering point as follows:

$$\frac{\mathrm{d}X}{\mathrm{d}t}=-\alpha X\left(1-\frac{S}{X}\right)+\beta e^{-\beta t/A}-L \quad (4)$$

The equation has a natural initial condition:

$$X|_{t=0}=0 \quad (4')$$

5) Accidents caused by overloading

Due to overloading of boats and trains, there exists a positive correlation between the overload rate θ and the accident rate R.

Since a very minimal overload hardly makes any difference, and the accident rate is has a natural upper bound 1, therefore the curve of P versus θ has to be an S-curve. Here we apply and modify the most widely adopted S-curve, the logistic function. Considering that the function should be a mapping from $[0,\infty)$ to $[0,1]$, we therefore modify the logistic function by translation and stretching to be

$$R=\frac{1}{2}\left(1+\frac{1-e^{-k(\theta-m)}}{1+e^{-k(\theta-m)}}\right) \tag{5}$$

Here k is a parameter indicating the slope of the curve, and m is a parameter indicating the point at which the curve attains the steepest increase.

For our purpose of describing the shipwreck probability, we set $m=3$ and $k=5$. This corresponds well to statistics from the marine databases. (Vickers, 2001; Kosko, 1990) This set of parameters gives the following graph by MATLAB:

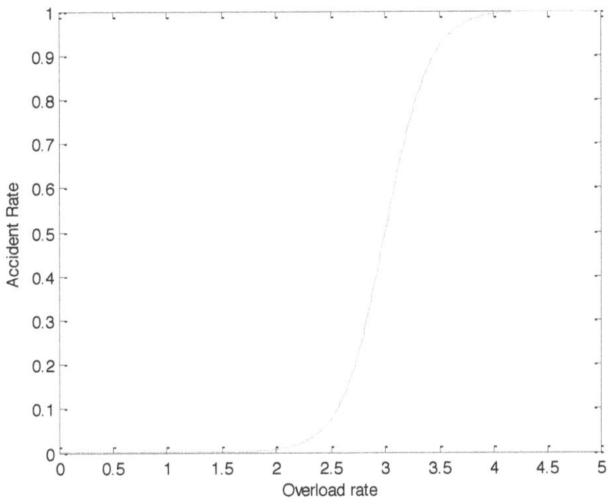

Figure 3. S-curve of the accident rate

6) Analyzing the motive of boat owners

Considering that the mortality rate is much higher on the routes overseas according to statistics (Baronett, 2008), we focus particularly on the overloading on sea, where the boat owners are driven by a desire for profit, and where the surroundings is much more hostile.

The boat owners commonly charges a fixed sum a for each passenger. When unfortunately a boat sinks, the loss for the boat owner comes from two sources – the value of the boat itself b, and less future income due to reduced boat number. Ironically he does not have to compensate for the drown refugees, since there'll be nobody to ask for the compensation.

Hence, there exists a balance between higher boat trip revenues and the risk of losing the boat. The profit function for the boat owners is

$$P(\theta)=La-Rb$$

that is

$$P(\theta)=L_0(1+\theta)a-\frac{1}{2}\left(1+\frac{1-e^{-k(\theta-m)}}{1+e^{-k(\theta-m)}}\right)b \tag{6}$$

The boat owner would seek an optimum overload rate θ to maximize his profit.

3.2.3 Solution and Result

1) Solution to the differential equation

Solve equation (4) with (4'), and we get:

$$X(t)=\left(S-\frac{L}{\alpha}\right)(1-e^{-\alpha t})+\frac{\beta}{A(\alpha-\beta)}(e^{-\beta t}-e^{-\alpha t}) \tag{7}$$

This is a composite of exponential decay functions. After the time period of 5τ, where $\tau = \max\left(\frac{1}{\alpha}, \frac{1}{\beta}\right)$ is the time constant , we can regard X as stable henceforth.

2) Optimal choice for boat owners

To find the maximum of $P(\theta)$, we set values for each parameter and use MATLAB for our calculation.

Set $a = 1$ as the fare as the unit, and $b = 1000$ as the loss of a sunken boat.

Set $L_0 = 20$ as the rated load capacity of a boat.

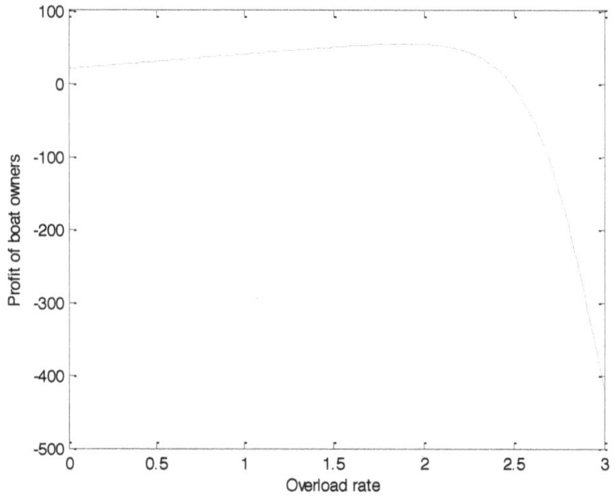

Figure 4. Optimum overload rate at 1.90

Numerical calculations give $\theta = 1.90$ as the maximum point.

3.2.4 Analysis of the Result

1) The pattern of population at gathering points

From the solution(7), we can see a pattern of exponential decay in the population, with a limit of $S - \frac{L}{\alpha}$ when $t \to \infty$, namely

$$\lim_{t \to \infty} X = S - \frac{L}{\alpha}$$

Since all parameters above are positive, this naturally yields $\lim_{t\to\infty} X < S$. This result corresponds to reality, because under the assumption that gathering points have the capacity for a certain number of refugees to live on, the stable condition must be a value less than S. Also in this particular case, combining the departure rate L and reproduction rate index α gives the stable size of the population.

2) Optimal choice for boat owners

The optimal overload rate solved above $\theta = 1.90$ indicates very crowded boats for the refugees. Boats on average would carry twice its capacity, which poses considerable risks for refugees on the vast seas.

3) Policy for less fatality over sea

However, we cannot let the lives of human beings be grasped in the hands of boat owners. International organizations ought to set up laws and regulations to reverse this tragic trend.

A simple way is to follow the logic of boat owners and give them penalties for every sinking accident that happens on their hand. This would add a weight on the scales against overloading.

Formally, an occurrence of sinking accident would bring a loss of b for boat owners. Adding a penalty is equivalent to raising the loss value. Assume penalty makes the loss ten times its original value, at $b = 10000$. Repeat the previous steps.

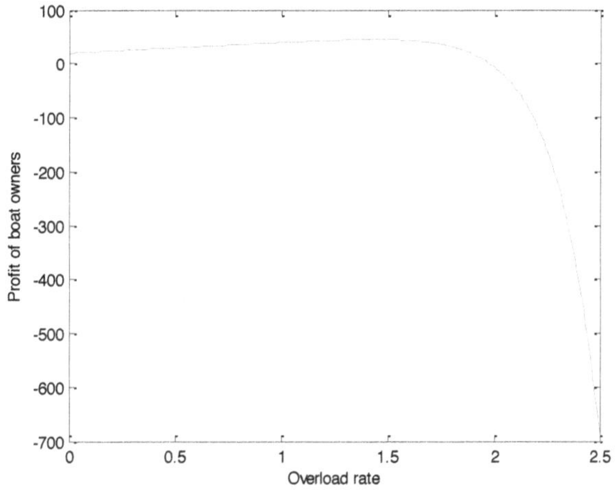

Figure 5. Optimum overload rate at 1.44 with penalty

Comparing Figure 4 (with penalty) with Figure 3 (without penalty), we see that the pattern of the figure is similar but the optimal overload rate becomes remarkably smaller. Besides, the slope of the initial increasing part becomes smaller in the model with penalty, which also means that boat owners may be less willing to risk overloading since there is no notable increase in profits. Therefore, applying penalty to boat owners can indeed lower the average overload rate of refugee boats and help save their souls.

3.3 Transferring Model

As we have already modeled the process of gathering and entering Europe, we now focus on the immigration flow inside Europe. We only focus on the main accepting countries in the below diagram and the transferring among them.

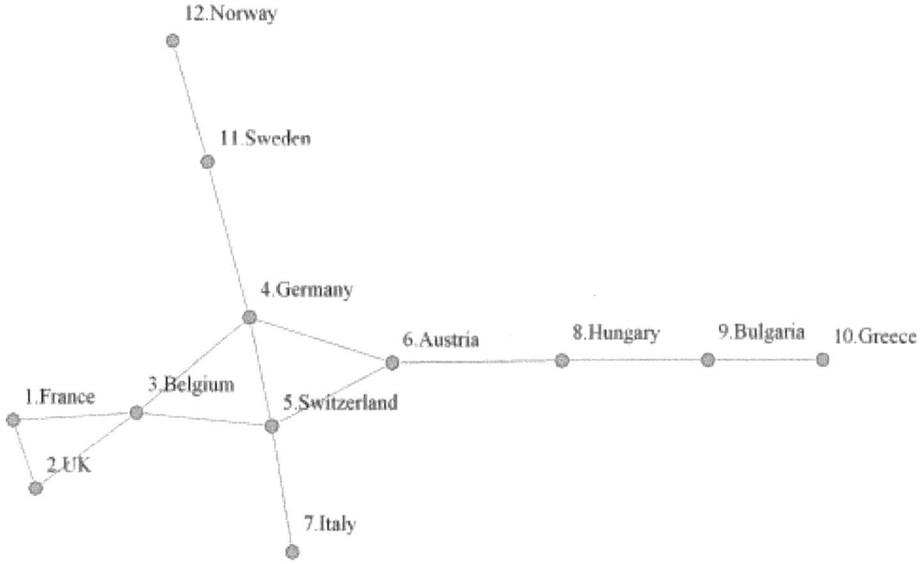

Figure 6. Transferring graph in Europe

3.3.1 Assumptions

1) Refugees can either come into Europe through customs or illegally.

2) In a single period, the refugees can only move among nearby countries. This is because the velocity of immigration is relatively slow.

3) The refugees are attracted by nearby countries considering both whether it is livable and how many refugees are already there. The attraction will increase as refugees accumulate first but finally decrease because of limited environmental capacity. The flow is also constrained by the transportation capacity.

4) The flow of immigration is radical at first but will finally become gentle. Refugees not accepted by any country will just stray in the country.

5) The EU will force each country to accept a certain number of refugees every period and EU will provide an amount of subsidy. The accumulation of unaccepted refugees will lead to crime and even riot so every country will accept an optimal number of refugees to keep a balance.

6) Based on the above assumptions, we model the flow by a Markov Random Field.

3.3.2 The Foundation of Model

1) The output of entry points

Refugees can either come into Europe through customs or illegally, so

$$O = O_1 + O_2$$

Illegal refugees into Europe has a proportion μ in the refugees that have not been accepted at entry points, so

$$O_2 = \mu(y - O_2)$$

Thus, we have a differential equation

$$\frac{dy}{dt} = L - O \tag{8}$$

$$y|_{t=0} = 0 \tag{8'}$$

Solving this equation gives the following

$$O = (1 - e^{-\mu t})L + (1 - \mu)e^{-\mu t}O_1 \tag{9}$$

We can apparently see that as time flies or as L changes, we can change O_1 to guarantee O at a relatively stable value, so European countries will not suffer from intense variation of refugees that pour in.

2) The transfer equation

Based on the assumptions above, we can construct an equivalent potential for each country to describe the related attraction. A proper option is the logistic:

$$\phi_i = -(\lambda_i + x_i - b_i x_i^2)$$

According to the statistical mechanics principle in physics, the partition of is proportional to the exponential of potential. Considering the time factor and inertia of immigration, we also include an inert parameter

$$\alpha = 1 - e^{-\mu\tau}$$

Thus we can compute the flow from the i_{th} country to the j_{th} country

$$f_{ij} = x_i(1 - \alpha)\frac{e^{-\phi_j}}{\sum_k e^{-\phi_k}} \tag{10}$$

$$f_{ii} = x_i\left[\alpha + (1 - \alpha)\frac{e^{-\phi_i}}{\sum_k e^{-\phi_k}}\right] \tag{11}$$

Finally, considering the limited transportation capacity, we must compute the minimum of the transportation capacity and the above flow

$$f_{ij} = \min\left\{x_i(1 - \alpha)\frac{e^{-\phi_j}}{\sum_k e^{-\phi_k}}, l_{ij}\right\} \tag{12}$$

$$f_{ii} = x_i - \sum_{j \neq i} f_{ij} \tag{13}$$

Finally we can compute the recursive relation of the current period and the next period

$$x'_i = \sum_j f_{ji} \tag{14}$$

3) The acceptance strategy

As refugees accumulates, they bring risk of crime and riot. The EU force each country to accept a certain numbers of refugees. At the same time, every country will find an optimal strategy to accept refugees and it is our focus in this section.

We conquer the problem by minimize a cost function, namely

$$C_i = (q - r)n_i + s(x_i - n_i)^2 + tn_i(x_i - n_i) \tag{15}$$

Here we give a brief explanation of the meaning of the loss function. The linear form is the resettlement cost and the EU subsidy. The quadratic form is a metric of the risk of crime because both the probability and the destructive power is proportional to the accepted refugees. The last cross term is a metric of the riot because the dissatisfaction of the accepted refugee is proportional to the number of refugee accepted so the sum is a proper metric.

Not surprisingly, the cost of the every country is independent so they will do the best decision for themselves. Finding the minimum of the cost function gives

$$n_i = \frac{t - 2s}{2t - 2s} x_i - \frac{q - r}{2t - 2s} \tag{16}$$

Finally, considering the quota

$$n_i = \max\left\{\frac{t - 2s}{2t - 2s} x_i - \frac{q - r}{2t - 2s}, quota_i\right\} \tag{17}$$

The accepted number is a linear function of the total refugee. Taking this in account, we can give a better recursive equation of the refugees:

$$f_{ij} = \min\left\{(x_i - n_i)(1 - \alpha)\frac{e^{-\phi_j}}{\sum_k e^{-\phi_k}}, l_{ij}\right\} \tag{12'}$$

$$f_{ii} = x_i - \sum_{j \neq i} f_{ij}$$

$$x'_i = \sum_j f_{ji}$$

3.3.3 Solution and Result

1) The asymptotic solution

Unfortunately, because of the appearance the minimum function, it is impossible for us to give an explicit solution of the recursive equation. We will utilize numerical simulation software later we will first use a physical principle to give an asymptotic solution and thus get some insight into the essence of the problem.

From the recursive equation, we know that the total potential of the system will always decrease therefore after a long time it will definitely reach a minimum point of potential(at least a local minimum, but as we prove later, the extreme point is unique so it is also the global minimum).

The minimum problem can be transformed into a constrained optimization problem

$$\text{minimize} \sum_i \phi_i$$

$$\text{s.t.} \sum_i x_i = X$$

This problem can be tackled by the Lagrange multiplier technique and the corresponding unique solution is

$$x_i = \frac{X}{\left(\sum_k \frac{1}{b_k}\right) b_i}$$

2) The simulation result

Using the current statistics, a quintessential diagram of the refugee in each country and the accumulating number of refugees accepted is as below.

The first row describe the refugee in the twelve major countries and each number represent a country as in Figure 6 above. The second row describe the accumulating number of refugee accepted in the twelve countries. The unit of the horizontal coordinate is a week and the vertical coordinate is a thousand persons.

We can easily find that the countries in north Europe only accept very few refugees and it is just that case. In the contrary, countries in the middle like Germany accepted a lot of refugees.

Another insight we can give is that the tendency of the flow. In the boundary countries, the number of refugee accumulates very fast at first but later decreases quickly. For most other countries, there is at least one extreme point.

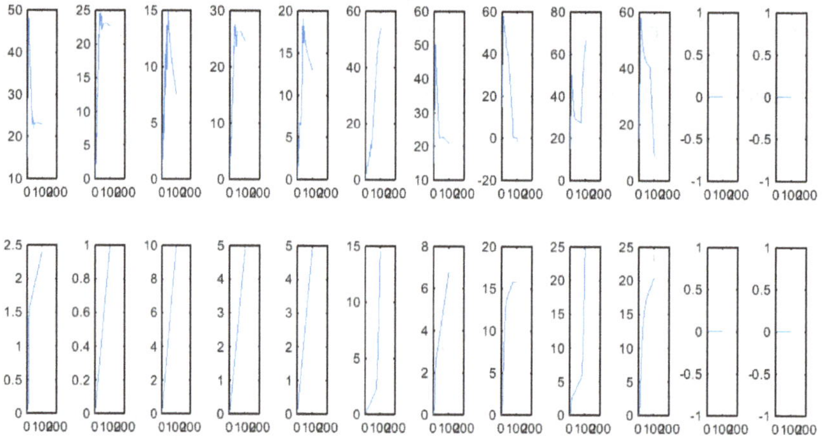

Figure 7. A quintessential diagram of the refugee in each country

3.4 Health/Security Model

In this section, we model the two crucial elements, namely the disease and crime.

3.4.1 Assumptions

1) Disease are divided into infectious diseases and non-infectious diseases. For both kind, we consider the different curing rate and lethal rate between permanent residents and refugees. For the first kind, we also consider the infectious rate

2) The riot is a small probability event. The probability is uniform in time and is proportional to the number of unaccepted refugees.

3.4.2 *The Foundation of Model*

1) The infectious diseases model

In this section, as there are so many qualities to compute, we use a short-hand form. We corporate the same quality with only different i into a vector, for instance,

$$u = [u_1, u_2, \dots \dots, u_n]$$

When we use this notation, the multiply operation should be regarded as operated between the corresponding terms between the two vectors.

Another notation we will utilize is when we add to any variable, that it means the value in the next period. Regarding this notation, we can write the evolutionary equation in a very explicit manner. We first list the equations below and then explain the meaning of each term.

$$\begin{gathered} u' = u - h_1 v_1 \\ v_1' = (1 - h_1 - f_1)v_1 + g(u - v_1)(v_1 + v_2) \\ w_1' = w_1 + h_1 v_1 \\ x' = x - h_2 v_2 \\ v_2' = (1 - h_2 - f_2)v_2 + g(x - v_2)(v_1 + v_2) \\ w_2' = w_2 + h_2 v_1 \end{gathered} \tag{18}$$

For x and u, we need only minus the dead people in the current period. For v_1 *and* v_2, we minus the dead and cured people and add the infectious people in the current period. Finally, for w_1 *and* w_2, we accumulate the dead people.

2) The non-infectious disease model

Different from infectious disease, non-infectious disease will not propagate, so we only need to calculate the dead rate. Similar notation gives

$$u' = u - h_1 v_1$$

$$\begin{aligned} v_1' &= (1 - h_1 - f_1)v_1 \\ w_1' &= w_1 + h_1 v_1 \\ x' &= x - h_2 v_2 \\ v_2' &= (1 - h_2 - f_2)v_2 \\ w_2' &= w_2 + h_2 v_1 \end{aligned} \tag{19}$$

A major difference is that the flow of refugee's influence here is not as crucial as in the previous situation.

3) The riot model

Different from disease, riots are small probability events and must be modeled stochastically. Based on the motivations stated in the assumption, we model the riots by a Compound Poisson Process.

In a period of time t, the probability that happened k riots is

$$P(N_i = k) = \frac{(\xi_i x_i t)^k}{k!} e^{-\xi_i x_i t} \tag{20}$$

Where we have already used the fact that the coefficient is proportional to the scale of refugees.

When a riot happen, we model the destructive power the scale of refugee multiplying a standard Rayleigh random variable R

$$R_i = x_i R$$

According to the Wald equation, the expectation

$$E_i = \xi_i {x_i}^2$$

Therefore, again, our problem is transformed into an optimization problem:

$$\text{minimize} \sum_i \xi_i {x_i}^2$$

$$\text{s.t.} \sum_i x_i = X$$

3.4.3 Solution and Result

1) The simulation of the infectious and non-infectious diseases model

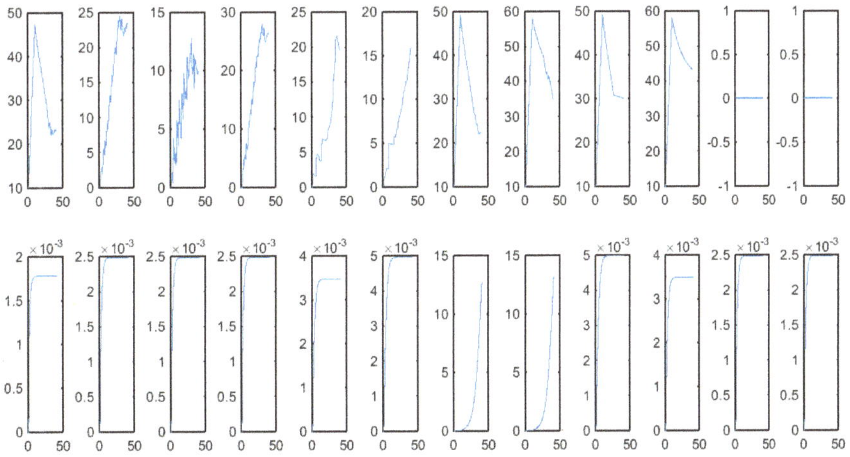

Figure 8. Simulation results with infectious disease

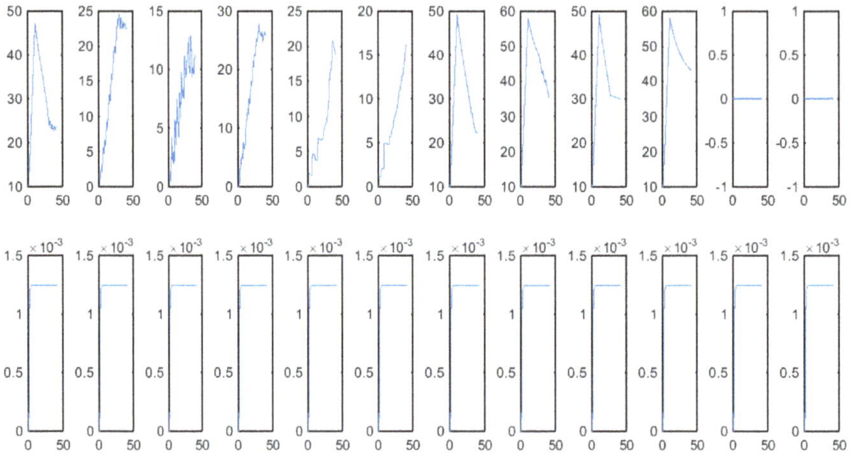

Figure 9. Simulation results with noninfectious disease

The two diagrams above describe the simulation results when there is an infectious disease or a noninfectious disease. The detailed information can be found in the appendix.

Comparing with infectious disease, the influence of non-infectious disease is much weaker. In fact, the major difference is the infectious rate. Therefore, it is a very crucial parameter for the system and we will argue later efficient policy is needed to limit the infectious disease in order not to let the disease outbreak.

2) The minimization of average destructive power

Fortunately, this time we can compute the minimum explicitly. By Cauchy's inequality

$$\sum_i \xi_i {x_i}^2 \sum_i \frac{1}{\xi_i} \geq \left(\sum_i x_i\right)^2 = X^2 \tag{21}$$

$$\sum_i \xi_i {x_i}^2 \geq \frac{X^2}{\sum_i \frac{1}{\xi_i}} \tag{22}$$

When the equality holds

$$x_i = \frac{X}{\xi_i \sum_i \frac{1}{\xi_i}} \tag{23}$$

4. Solutions to Tasks

4.1 Metrics of Refugee Crises

We classify the factors relevant to a refugee crisis into four categories: demographic data, route information, traffic condition, and resource capacity.

The specific measures and parameters are as follows:

- Total numbers of incoming refugee – *A (the total number of refugees that suffered from an outbreak of crisis)*

This measures the size of impact of a refugee crisis, which is a basic factor of crisis intensity.

- Rate of the refugee influx – β *(the initial speed of incoming refugees when a crisis breaks out)*

This measures the intensity of a refugee crisis. An abrupt crisis when people flood out to seek refugee is more troublesome than a chronic one provided the total number of refugees is the same.

- Constitution of the refugee population, including gender distribution, age distribution, etc. – *(a series of variables indicating proportions)*

This measures the refugees' ability to withstand exogenous adversities, such as diseases and lack of food and resources. Besides, there could be some factors that specially arise from the population's constitution.

- Load capacity of each route – *L (rated transport capacity of a necessary route for refugees to escape to safety),* l_{ij} *(transport capacity for refugees to move around between countries)*

This measures rate at which refugees are able to reach safe countries. This is a vital index about the outward flow of refugees.

- Danger on each route – *ε (index for the specific danger on a route), k (indicating the danger of overload of vehicles)*

This measures the safety of refugees on their way of migration. This factor is directly related to the well-being of the refugees. This can also significantly affect the condition and choices of the refugees.

- Resource capacity – *S (the amount of resource at a gathering point for refugees to cross over to safe regions), b (index for the capacity of a country to accommodate refugees)*

This measures the amount of environmental and social resource available to the refugees. Similar to the danger index, the resource capacity index strongly influences the refugees' survival and well-being.

4.2 Flow of Refugees

We divided the flow of refugees into 2 parts. One is the route from the refugees' effected homes to the gathering point. The other is the route from the gathering point to Europe.

- From Homes to the Gathering Point

We have made best options of route for refugees according to the model in 3.1. Taking the danger that refugees will meet on the route to Europe into account, our model assumes that refugees will choose the route with the least danger. After calculation, we suggest that refugees from North Africa choose Route 1 or 2 according to the border K1, and that refugees from West Asia choose Route 4 or 6 according to the border K2.

- From the Gathering Point to Europe

We have built a model of how refugees arrive in Europe in 3.2. We estimate the optimal option of boat owners considering their profits. Also, we estimate the minimum fatality of refugees.

4.3 Dynamics of the Crisis

A crucial character which have been considered thoroughly in our model is that the environmental change over time. As refugees move across Europe, the attraction of each country will definitely change, thus influence the flow. More detailed information can be found in the modeling section above.

- Forecasting

An essential advantage of our model is that it is rather easy to forecast the move the refugees. According to the results of simulation, we can preposition all kind of resources. According to the simulation results, the crucial resource include living necessities (clean water, food and clothes), medicine and medical care.

- Capacity and availability

The parameter b in our model is proportional to the reciprocal of the capacity. As the refugees accumulate in certain place, the situation will deteriorate exponentially which incorporate the limit of capacity. Although we do not include an absolute capacity which can never be exceed, but our assumptions are more realistic and easy to compute. Also, the availability of other typical resource is tackled similarly, like the transportation capacity and the availability of living necessities (already incorporated into b).

- The role of the government and the NGOs

We have not shown explicitly how government and NGOs work in the model before, so we will give an in-depth discussion here. As the government, it should give necessary resources to refugee unaccepted and incorporate the accepted refugee into the society. This is exactly what we do when finding the optimistic strategy for governments by minimizing the cost function.

For NGOs, their work can be considered global optimization. For NGOs who need to allocate their resource in advance, they can utilize the result of minimize the total potential and the results says that the asymptotic distribution is proportional to the capacity in each country so we recommend the NGOs to allocate their resource according to the current capacity in each country.

- Extensibility

Our assumptions above are fairly general that it can be easily utilized by non-European countries like Canada and China. We only need the remove the influence of quota because there is no such powerful local union.

4.4 Policy to support refugee model

We list our recommendations to both government, UN and NGOs below and then explain our reasons. As instructed, we prioritize the security and body health of both refugees and permanent residents in Europe.

- Set up regulations to control the phenomenon of overloading refugee boats, which poses a major threat to refugees' lives.

While it is totally understandable that refugees are extremely eager to reach Europe, even putting aside the danger of drowning themselves, efforts still need to be done to control the excessive overload. We call for some international organizations to set up a regulation that sets the maximum overload rate. We note that slight overloading actually meets the demand of refugees, so total elimination of overloading is unpractical, but certain restrictions on the upper limit would be beneficial nonetheless. An alternative choice would be to cooperate with some local boat owners, offering money to them to let them keep the overload rate at a relatively rational level.

- Supervise the output of refugees that come past customs at the entry points into Europe to supervise the total refugees into Europe.

From our model of entry points in 3.3.3, the number of refugees that enter Europe via entry points is influenced by the intensity of the inflow, the rate of illegally sneaking, and time. In order to control the rate of the entrance of refugees at a relatively stable level, customs need to frequently adjust the number of applications approved. For example, when the number of the incoming refugees sees a sharp rise, customs should not rise the number of approved applications along with it, otherwise the rate of refugees entering Europe would rise notably. Also, governments of those countries should devise more effective ways to settle the refugees held up at entry points, since it is unreasonable to admit too many refugees into the country at a time.

- The government and NGOs should pay enough attention to decrease the infectious rate, especially between refugee and permanent residents.

Based on our infectious disease model, we conducts over 100 simulation in different conditions and find the result is really sensitive to the infectious rate g. In fact, when g increases 10 times, the infectious and dead people in a year increase 10000 times. Comparing with the infectious rate, the model is rather robust to the change of curing and lethal rate of the disease. So we strongly recommend the government and NGOs to provide enough facilities to limit the infectious rate.

- The EU and NGOs should guide the refugees to come to countries less crowded like those in North Europe for better stability and less riot destruction.

Both the analysis of expectation of riot destruction and global stability shows that the best distribution of the refugee is uniform in some sense. Our first result is the most stable (least potential) distribution is proportional to the capacity and the second result shows the least potential riot destruction happens when the distribution is proportional to the square of population in each country. Our simulation also found that the refugee in North Europe is a lot less than in the middle and west part of Europe. Guiding refugee to these districts will moderate the stress in other parts.

- The EU should adjust the refugee subsidy, weaken the quota system and convince the European countries there will be harmful if not accepting any refugee.

According to our model and the corresponding simulation results, the best subsidy which make the global situation most stable and decrease more radical immigration is a little higher than the resettlement cost of each refugee. The current subsidy is proper in this regard. Also, we find the performance of the quota policy is rather weak, especially in the long run. Rather, they should convince the European countries of the harm and guide them to use the optimal strategies.

4.5 Exogenous Events

In general, our following procedure can deal with all kinds of exogenous events but to be concrete, we assume like in the task that Belgium was placed in a lockdown after the Paris raids in attempts to capture possible terrorists. We assume a protest campaign happens in Belgium and Belgium forbids all the refugee to enter the country. The following are our simulation, analysis and policy recommendation.

- Parameters shifts

The livable degree and environmental capacity in Belgium is completely changed. It can be considered as if λ decreases and b increases both dramatically. A rational simplification is to consider $\lambda \to 0$ while $b \to \infty$.

- Cascading effects

Again, we simulate the resulting refugee flow in the following diagram.

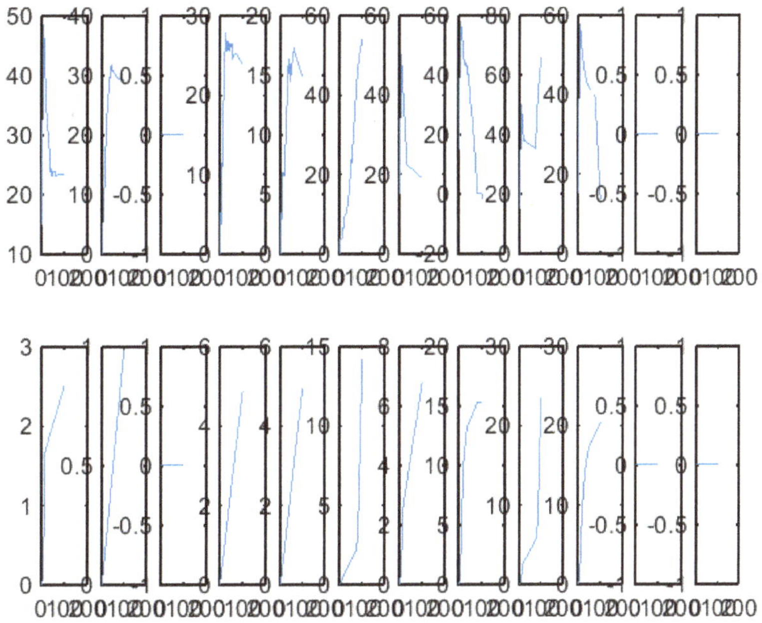

Figure 10. Simulation results with exogenous event - shutdown of city

Comparing with the diagram when Belgium functions well, we can see that the stress of the neighboring countries, namely, the accumulated refugee and refugee accepted, increased a lot and the extreme points come earlier.

- Policy recommendation

As we can see, if an important country is paralyzed, the result is serious. As EU or NGOs, we suggest they allocate the resources more uniformly, which we have already suggested before. This can make the policy more resilient. Also, we suggest the EU to moderately increase the subsidy for the neighboring countries to moderate the stress.

4.6 Scalability

When the scale have increased by ten times, the framework of our model do not need any further major adjustments. Only the number of refugees will increase. We simulate the new scenario and the result is as below

Comparing the diagram with the diagram before, we can see that the peak have been dramatically increased and the number of refugee in the boundary countries have increased more than 10 times. This is the result of relatively limited transportation capacity. The time for the unaccepted refugee scale to be decreased to the same scale has been obviously prolonged and this bring us new challenges in health and security, just like we discussed before. Finally, we suggest the EU give some extra support to help facilitate the flow. If the transportation capacity also increases by 20 times, we can get the new simulation results below.

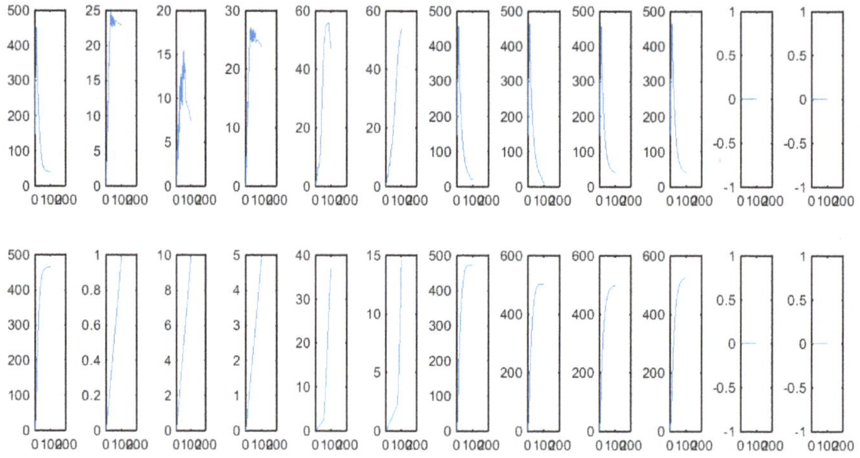

Figure 11. Simulation result of refugee population expanded by 10 times

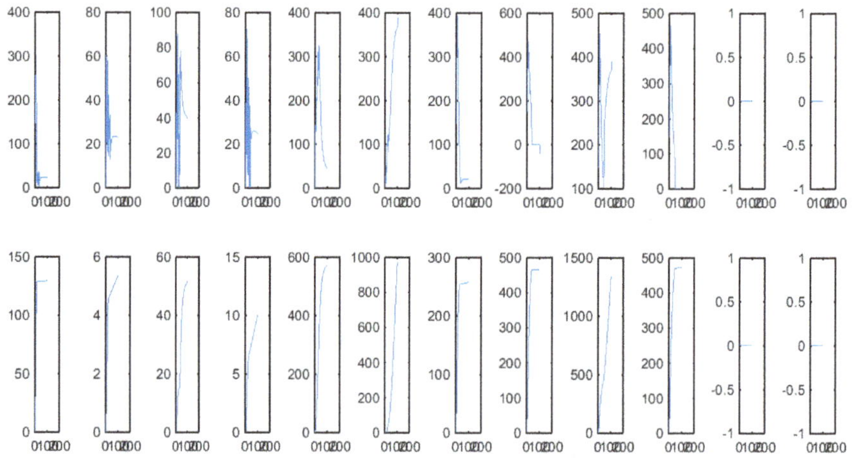

Figure 12. Simulation result with traffic capacity expanded along with population

We can see that this time, the refugee distribution is comparatively more uniform and more controllable.

5. Evaluations of Models

5.1 Sensitivity

5.1.1 Assembling Model

The partition of North Africa and West Asia relies on the parameter ε . And our result is not sensitive to the parameter ε , so the model is stable itself.

5.1.2 Entering Model

This model of population size includes many contributing factors and none of them is sensitive. The calculation of optimum overload rate is quite sensitive to the parameter k which indicates the shape of the S-curve and is not sensitive to load capacity L_0.

5.1.3 Transferring Model

The flow of immigration relies on the parameter b asymptotically but not too sensitive to the parameter. The flow of immigration is relatively more sensitive to the transportation capacity and the other parameter's influence is rather weak.

5.1.4 Health/Security Model

The most sensitive parameter is the infectious rate g and the other parameters' influence is a lot weaker than g.

5.2 Strength/Weakness

5.2.1 Assembling Model

The model is prioritize refugees' lives. The model only considers about six routes and we can use the same way to solve it if there are more routes. But we do not consider about population distribution of different regions.

5.2.2 Entering Model

The model is inclusive of many aspects, including resource capacity and internal attributes of refugees. This model well explains the overloading phenomenon. A point to be improved is that we consider the sinking accidents as evenly distributed instead of discrete incidents.

5.2.3 Transferring Model

The model's principal advantage is that we can easily forecast the flow of the immigration and get many insights into the nature of the problem. The weakness should be attributed to the abstract and simplified nature of the model and thus ignore some potential influential effects like the inherent structure of the refugee.

5.2.4 Health/Security Model

The model's major advantage is that we can find the propagation tendency of certain disease and thus we can preposition our resource to moderate the stress. The weakness is that we do not include the correlation between different diseases and riots, which is crucial when considering the global situation.

6. Conclusion

Aiming to address the problem caused by the flow of refugees, we analyzed the actual procedure of their movement and divide it into three major stages. In addition, health and security and threat of refugees and local people should also be considered.

In the assembling model, under the basic assumption to minimize risk, our model proposes a partition of Northern Africa and Western Asia, which proposes the choice of the six main routes for the refugees.

In the entering model, we modeled the gathering points with input of refugees from nearby regions, output of refugees to Europe and the influence of danger and resource limitations. In particular, we studied the pervasive phenomenon of boat overload, and found that moderate overloading could both yield higher profits for boat owners and meet the refugees' demand. We determined the optimal value of overload rate from the perspective of boat owners. Also our model gives advice that international organizations should cooperate with boat owners and lower the overload rate to a safe level via contractual means.

In the transferring model, upon arriving at entry places, refugees seek to go through customs, whether with granted asylum applications or with stealth. We modeled this process and gave an insight into the flow control policy. Then we considered the transfer of refugees between European countries. This is modeled as a Markov Random Field to forecast the immigration flow together with a constrained optimization technique to choose the optimal strategy for countries to accept refugees.

In the health and security model, we modeled two major concerns about refugees as well as permanent residents in Europe, namely, diseases and riots. We evaluated the lethal rate, curing rate and infectious rate of the diseases and gave a policy recommendation. For the riots, we modeled them stochastically and used an optimization technique to give the most stable distribution of unaccepted refugees.

Finally, we used the models described above to complete our assigned tasks. Also we put forward seven major policy recommendations to the committee. We accompanied every policy with a straight forward explanation so that people without any technical background can easily understand our insights.

The main strength of our model is that it can forecast the flow of immigration and provide meaningful suggestions policies for refugees. With the help of modern computing software, we can track the current tendency and make judges efficiently. Our major weakness, due to the abstract and simplified nature of our model, is that it does not give enough attention to some less essential factors like inherent structure of the refugees.

References

Baronett, S. (2008). Logic. Upper Saddle River, NJ: Pearson Prentice Hall.[M] p. 321–325.

Copi, I. M., Cohen, C., & Flage, D. E. (2007). Essentials of Logic (Second ed.). Upper Saddle River, NJ: Pearson Education.

Gray, P. (2001). *Psychology* (Sixth ed.).[M] New York: Worth.

Herms, D. (2010). Logical Basis of Hypothesis Testing in Scientific Research. Stanford Encyclopedia of Philosophy: Kant's account of reason.

Vickers, J. (2011). *The Problem of Induction*.[M] The Stanford Encyclopedia of Philosophy.

Kosko, B. (1990). Fuzziness vs. Probability. *International Journal of General Systems*. p: 211–240.

Popper, K. R., & Miller, D. W. (1983). A proof of the impossibility of inductive probability. *Nature*. 687–688.

Rathmanner, S., & Hutter, M. (2011). A Philosophical Treatise of Universal Induction. *Entropy*.

Tymoczko, T. (1980).The Four-Color Problem and its Mathematical Significance.

Trans. R.G. Bury. (1933). Harvard University Press, Cambridge, Massachusetts, p. 283.

Vickers, J. (2010). The Problem of Induction (Section 2). Stanford Encyclopedia of Philosophy.

Appendix

A chart of symbols and definitions

1. Gathering model

Symbol	Definition
d	Distance of the i-th route into Europe
ε	Parameter that stands for the route by sea or by land
l	Parameter that stands for the probability of meeting danger
f	Parameter that stands for the danger of different routes

2. Entering model

Symbol	Definition / Explanation
X	The number of refugees at a certain gathering point
t	Time elapsed since the beginning of the refugee crisis
S	The amount of resources at the gathering point, measured by the number of refugees it can stably support
A	The total number of refugees to come to a gathering point because of a crisis outbreak
α	A parameter indicating the impact of resources on the number of refugees
β	The initial rate of incoming refugees when a crisis breaks out
L_0	The designed load capacity of vehicles
L	The actual load of vehicles
θ	The overload rate of vehicles
R	The accident rate of vehicles, as caused by overloading
k	A parameter indicating the correlation between fatality rate and overload rate
a	The price the boat owners charges per person
b	The loss of a boat owner when a boat sinks
P	The profit of boat owners in a time unit

3. Transferring model

Symbol	Definition
y	The number of refugees at entry points

O	The output of refugees who come into Europe from entry points
O_1	The output of refugees who come into Europe through customs
O_2	The output of refugees who come into Europe illegally
μ	The proportion of illegal refugees into Europe in the refugees that have not been accepted at entry points
λ_i	The livable degree of the *i*-th country
x_i	Refugee in the *i*-th country in the current period
x'_i	Refugee in the *i*-th country in the next period
b_i	The decay coefficient of attraction in the *i*-th country
ϕ_i	The equivalent potential of the *i*-th country
α	The inert parameter of the refugee flow
l_{ij}	The transportation capacity between the *i*-th and the *j*-th country
f_{ij}	The refugee flow between the *i*-th and the j-th country
n_i	The number of refugee the *i*-th country accept in the current period
C_i	The loss function of the *i*-th country
q	The subsidy EU provide for each refugee accepted
r	The resettlement cost for each refugee accepted
s	The crime parameter for unaccepted refugees
t	The riot parameter for unaccepted refugees
$quota_i$	The quota for the *i*-th country

4. Health/Security Model

Symbol	Definition
$h_{1,2}$	The lethal rate for permanent resident and refugee
g	The infectious rate
$f_{1,2}$	The curing rate for permanent resident and refugee
u_i	The population of the *i*-th country
$v_{1,2i}$	The number of infected permanent resident/ refugee in the *i*-th country
$w_{1,2i}$	The number of permanent resident/ refugee dead in the *i*-th country
N_i	Riots happened in a certain period of time in the *i*-th country
ξ_i	The risk coefficient in the *i*-th country
R	Standard Rayleigh random variable
R_i	The destructive power for each riot in the *i*-th country

13

Mixed and Hybrid Finite Element Methods for Convection-Diffusion Problems and Their Relationships with Finite Volume: The Multi-Dimensional Case

Michel Fortin[1] & Abdellatif Serghini Mounim[2]

[1] Département de Mathématiques et Statistique , Université Laval, Québec, G1K 7P4, Canada

[2] Department of Mathematics and Computer Science, Laurentian University, Sudbury, Ontario, P3E 2C6, Canada

Correspondence: Abdellatif Serghini Mounim, Department of Mathematics and Computer Science, Laurentian University, Sudbury, Ontario, P3E 2C6, Canada. E-mail: aserghini@cs.laurentian.ca

Abstract

We introduced in (Fortin & Serghini Mounim, 2005) a new method which allows us to extend the connection between the finite volume and dual mixed hybrid (DMH) methods to advection-diffusion problems in the one-dimensional case. In the present work we propose to extend the results of (Fortin & Serghini Mounim, 2005) to multidimensional hyperbolic and parabolic problems. The numerical approximation is achieved using the Raviart-Thomas (Raviart & Thomas, 1977) finite elements of lowest degree on triangular or rectangular partitions. We show the link with numerous finite volume schemes by use of appropriate numerical integrations. This will permit a better understanding of these finite volume schemes and the large number of DMH results available could carry out their analysis in a unified fashion. Furthermore, a stabilized method is proposed. We end with some discussion on possible extensions of our schemes.

Keywords: convection-diffusion, dual mixed and hybrid finite elements, finite volume, upwinding

1. Introduction

The aim of this paper is to extend to multidimensional hyperbolic and parabolic problems the results of (Fortin & Serghini Mounim, 2005). In (Fortin & Serghini Mounim, 2005) we presented in a mixed finite element framework, two non-standard formulations in one space dimension allowing, among other things, to recover in a unique fashion, numerous finite volume schemes. We shall present in this paper a two-dimensional extension. To achieve the discretization we shall use mixed finite elements to approximate the convective and diffusive fluxes. We concentrate on the case of the lowest order Raviart-Thomas spaces (Raviart & Thomas, 1977), though many of our results will be more general. In particular, to obtain higher-order accuracy finite volume schemes less sensitive to partitions of domain, we can follow the same procedure but utilize others usual mixed finite element approximation subspaces of $H(\text{div}, \Omega)$ such as BDM spaces (Brezzi, Douglas & Marini, 1986).

This paper is organized as follows. In the next section, we systematically build the extension of the formulations of (Fortin & Serghini Mounim, 2005) to the multi-dimensional case. Results of existence and uniqueness are presented. In Section 3, we build the numerical schemes on triangular or rectangular partitions of the domain Ω using the Raviart-Thomas finite elements of lowest degree (Rarviart & Thomas, 1977) to approximate the fluxes. In Section 4, we show the link between the finite volume methods and our formulations using the suitable quadrature formulas suggested in (Baranger, Maitre, & Oudin, 1996) (triangular case) or trapezoidal quadrature (rectangular case) to diagonalize the element mass matrix and numerical integration formulas to approximate the term $\int_K \underline{f}(u).q\,dx$ (here $\underline{f}(u)$ is the convective flux). Moreover, we focus our attention to the multidimensional extension of some recent schemes. Finally, in Section 5 we discuss how we can also establish the extension of the second method to both convection-diffusion equations and system of equations. We end with some discussion on possible extensions of our schemes. The DMH formulation requires the solution of a linear system; the approach adopted here was suggested by Arnold and Brezzi (Arnold & Brezzi, 1985). In their method, one eliminates the velocity and the flux unknowns to leave a system for the Lagrange multipliers alone. The lowest order Raviart-Thomas spaces have one Lagrange multiplier unknown per edge.

2. Multidimensional Extension

2.1 Preliminaries

In this section, we start by presenting a few properties relative to the decomposition of Ω. The proofs can be found in Thomas (Thomas, 1977) or Brezzi-Fortin (Brezzi & Fortin, 1991). Let us consider the following partial differential equation for u:

$$\frac{\partial u}{\partial t} + \mathrm{div}(\underline{a}u) - \mathrm{div}(\nu\,\mathrm{grad}\,u) = f \quad \text{in}\,\Omega, \tag{1}$$

with given initial data $u(x,0) = u_0(x)$ in Ω, and corresponding suitable boundary conditions on $\Gamma = \partial\Omega$. Here the velocity field and the diffusion coefficient (which is positive) are denoted by $\underline{a}(x,t)$ and ν respectively. Moreover, for simplicity and without loss of generality, we restrict our attention to the case where $\mathrm{div}(\underline{a}) = 0$; taking into account $\underline{a}.\underline{n} = 0$ on $\Gamma\times]0,T[$, and assuming that $\underline{a}(.,t) \in L^\infty(\Omega)$; also, that $u = 0$ on $\Gamma\times]0,T[$.
Let then $T_h = \{K_i\}_{i=1}^{NT}$ be the usual non-overlapping finite element triangulation of the domain $\Omega = \cup_{K\in T_h}K$. We also consider for $1 \le t < \infty$ the spaces

$$X^t(\Omega) = \{\underline{v} \in (L^t(\Omega))^2,\ \underline{v}|_K \in W_t(\mathrm{div};K) \quad \forall K \in T_h\}, \tag{2}$$

and

$$W_t(\mathrm{div};\,\Omega) = \{\underline{v} \in (L^t(\Omega))^n; \quad \mathrm{div}(\underline{v}) \in L^t(\Omega)\}, \tag{3}$$

which are both Banach spaces endowed with the usual norms

$$\|\underline{v}\|_{X^t(\Omega)} = \|\underline{v}\|_{0,t,\Omega} + \sum_{K\in T_h} \|\mathrm{div}(\underline{v})\|_{0,t,K}, \tag{4}$$

and

$$\|\underline{v}\|_{W_t(\mathrm{div};\,\Omega)} = \|\underline{v}\|_{0,t,\Omega} + \|\mathrm{div}(\underline{v})\|_{0,t,\Omega}. \tag{5}$$

The following proposition, showing that the space $W_t(\mathrm{div};\,\Omega)$ can be characterized as a subspace of $X^t(\Omega)$, is well known:

Proposition 1 *Let $s > 1$, a vector $\underline{q} \in X^t(\Omega)$ then $\underline{q} \in W_t(\mathrm{div};\Omega)$ if and only if*

$$\sum_{K\in T_h} \langle \underline{q}.\underline{n}, v\rangle_{W^{-1/t,t}(\partial K)\times W^{1/t,s}(\partial K)} = 0, \tag{6}$$

for all $v \in W_0^{1,s}(\Omega)$, where t is the conjugate of s.

Finally, let us recall that for all $\mu \in W^{1/t,s}(\Gamma)$ and $\mu^* \in W^{-1/t,t}(\Gamma)$ we have

$$\|\mu\|_{1/t,s,\Gamma} = \sup_{\underline{q}\in W_t(\mathrm{div};\Omega)} \frac{\langle \underline{q}.\underline{n}, \mu\rangle_\Gamma}{\|\underline{q}\|_{W_t(\mathrm{div};\Omega)}}, \tag{7}$$

and

$$\|\mu^*\|_{-1/t,t,\Gamma} = \sup_{v\in W^{1,s}(\Omega)} \frac{\langle \mu^*, v\rangle_\Gamma}{\|v\|_{1,s,\Omega}}. \tag{8}$$

Here $\langle .,.\rangle$ denotes the duality pairing between $W^{-1/t,t}(\Gamma)$ and $W^{1/t,s}(\Gamma)$.

Now, we are able to formulate precisely the parabolic problem (1).

2.2 Dual Mixed and Hybrid Finite Element Method DMH1

In this section, we deal with the first variational formulation of the convection-diffusion problem previously considered. The principal objective of the following is to extend to the 2D case the first formulation proposed in (Fortin & Serghini Mounim, 2005).

As in (Fortin & Serghini Mounim, 2005), we shall begin from equation (1). Next, we introduce convective and diffusion fluxes as auxiliary variables to obtain the following first order system:

$$\begin{cases} \dfrac{\partial u}{\partial t} - \nu\,\mathrm{div}(\underline{p}) + \mathrm{div}(\underline{\hat{p}}) = f & \text{in } \Omega\times]0,T[, \\ \underline{p} = \mathrm{grad}\,u \quad \text{in } \Omega\times]0,T[, \\ \underline{\hat{p}} = \underline{a}u \quad \text{in } \Omega\times]0,T[. \end{cases} \tag{9}$$

Boundary conditions and initial data are added to these equations.

Here the boundary of Ω is supposed polygonal. Introducing λ, a Lagrange multiplier to enforce the continuity constraint of $\underline{q}.\underline{n}$ at elements' interfaces, we obtain

$$\int_\Omega \underline{p}.\underline{q}\,dx = -\sum_{K_i}\int_{K_i} u\,\mathrm{div}(\underline{q})\,dx + \sum_{K_i}\int_{\partial K_i} \lambda\,\underline{q}.\underline{n}\,ds. \tag{10}$$

We shall evidently have to make a proper choice for the space of λ. This technique was first used by Fraejis DE Veubeke (Fraejis DE Veubeke, 1965; Fraejis DE Veubeke, 1975). In the same way, we shall impose the continuity of the convective flux $\hat{p}$ at the interfaces of elements K_i, with the aid of a Lagrange multiplier noted $\hat{\lambda}$. Hence, the last equation of the system (9) can now be rewritten as:

$$\int_\Omega \underline{\hat{p}}.\underline{q}\,dx = \sum_{K_i}\int_{K_i} \underline{au}.\underline{q}\,dx + \sum_{K_i}\int_{\partial K_i} \hat{\lambda}\,\underline{q}.\underline{n}\,ds. \tag{11}$$

Now, to make the variational formulation precise we have to define the functional spaces. Let us consider an exponent $\frac{4}{3} < s < 2$ and its conjugate t. In addition, we set $X = \{\underline{q} \in (L^t(\Omega))^2;\ \underline{q}|_K \in W_t(\mathrm{div};K)\,\forall K \in T_h\}$, $M_2 = \{\mu \in \prod_{K\in T_h} W^{1/t,s}(\partial K)\,;\mu|_{\partial K\cap\Gamma} = 0 \quad \forall K \in T_h\}$, and $M_1 = L^t(\Omega)$. Hence, the system (9) could be written as follows:

Find $(u(t), p(t), \hat{p}(t)) \in M_1 \times X^2$ and $(\lambda(t), \hat{\lambda}(t)) \in M_2^2$ such that

$$\left\{\begin{array}{ll}
\displaystyle\int_\Omega \frac{\partial u}{\partial t} v\,dx - \nu \sum_{K_i}\int_{K_i} \mathrm{div}(\underline{p})\,v\,dx + & \\
\displaystyle\sum_{K_i}\int_{K_i} \mathrm{div}(\underline{\hat{p}})\,v\,dx = \int_\Omega f\,v\,dx \qquad \forall v \in M_1, & \\
\displaystyle\int_\Omega \underline{p}.\underline{q}\,dx = -\sum_{K_i}\int_{K_i} u\,\mathrm{div}(\underline{q})\,dx + \sum_{K_i}\int_{\partial K_i} \lambda\,\underline{q}.\underline{n}\,ds & \forall \underline{q} \in X, \\
\displaystyle\int_\Omega \underline{\hat{p}}.\underline{q}\,dx = \sum_{K_i}\int_{K_i} (\underline{au}.\underline{q})\,dx + \sum_{K_i}\int_{K_i} \hat{\lambda}\,\underline{q}.\underline{n}\,ds & \forall \underline{q} \in X, \\
\displaystyle\sum_{K_i}\int_{\partial K_i} \underline{p}.\underline{n}\,\mu\,ds = 0 \quad \forall \mu \in M_2, & \\
\displaystyle\sum_{K_i}\int_{\partial K_i} \underline{\hat{p}}.\underline{n}\,\mu\,ds = 0 \quad \forall \mu \in M_2. &
\end{array}\right. \tag{12}$$

The last two equations of system (12) insure the continuity of $\underline{p}.\underline{n}$ and $\underline{\hat{p}}.\underline{n}$ at the interfaces of T_h.

Remark 1

The expressions $\int_{\partial K} \underline{q}.\underline{n}\,\mu\,ds$ and $\int_{\partial K} \underline{\hat{q}}.\underline{n}\,\mu\,ds$ are duality pairing between $W^{-1/t,t}(\partial K)$ and $W^{1/t,s}(\partial K)$. Indeed, $\underline{q}|_K \in (W_t(\mathrm{div};K))^2$ and its normal trace is defined in $W^{-1/t,t}(\partial K)$.

Remark 2

The proof of the existence of Lagrange multipliers λ and $\hat{\lambda}$, which insure the continuity of normal traces is similar to the one given in Brezzi-Fortin (Brezzi & Fortin, 1991).

Once the spaces are defined, we can establish the existence and uniqueness of problem (12) inspired from Girault-Raviart (Godlewski & Raviart, 1991) (for details we can see (Serghini Mounim, 2000)).

Theorem 1 *Let u be the solution of the convection-diffusion problem (1). If* $\underline{p} = \mathrm{grad}\,u$ *and* $\underline{\hat{p}} = \underline{au}$ *verify*

$$(\underline{p}(t), \underline{\hat{p}}(t)) \in (W_t(\mathrm{div};\Omega))^2, \tag{13}$$

then $((\underline{p}, \underline{\hat{p}}, \hat{\lambda});\ (u, u|_{\partial K}))$ *is the unique solution of (12).*

2.3 Dual Mixed and Hybrid Finite Element DMH2

Let us consider the following partial differential equation for u:

$$\frac{\partial u}{\partial t} + \operatorname{div}(\underline{a}u) = 0 \qquad \text{in } \Omega\times]0,T[, \tag{14}$$

where $\underline{a}(x,t)$ is the velocity.

The dual mixed and hybrid formulation of the problem (14) can be achieved proceeding in a similar way as in (Fortin & Serghini Mounim, 2005). First, we introduce again the convective flux as an auxiliary variable to obtain:

$$\begin{cases} \dfrac{\partial u}{\partial t} + \operatorname{div}(\underline{\hat{p}}) = 0 & \text{in } \Omega\times]0,T[, \\ \underline{\hat{p}} = \underline{a}u, & \text{in } \Omega\times]0,T[. \end{cases} \tag{15}$$

Boundary conditions and initial data are added to these equations.

Next, we assume that the solution has discontinuities at each boundary of every K_i. Finally, we suggest that the "Rankine-Hugoniot" jump condition should be imposed on the numerical flux through the elements' interfaces. With this aim, we employ a Lagrange multiplier to relax these jump conditions at the interelements.

$$\int_\Omega \underline{\hat{p}}.\underline{q}\, dx = \sum_{K_i}\int_{K_i} \underline{a}u.\underline{q}\, dx + \sum_{K_i}\int_{\partial K_i} \hat{\lambda}\, \underline{q}.\underline{n}\, ds \qquad \forall \underline{q} \in X, \tag{16}$$

and

$$\sum_{K_i}\int_{\partial K_i} [\underline{\hat{p}}.\underline{n}]\mu\, ds = \sum_{K_i}\int_{\partial K_i} sign(\underline{a}.\underline{n})a_i\,[u]\mu\, ds \quad \forall \mu \in M_2. \tag{17}$$

We denote by $\hat{\lambda}$ the Lagrange multiplier and by $a_i \geq 0$ some approximation of the local propagation speeds. As we will see, it plays a stabilization parameter role, and for an appropriate choice of a_i, we can recover in particular the well-known upwind numerical fluxes (approximate Riemann solvers). Here, we have set $[\underline{\hat{p}}.\underline{n}] = \underline{\hat{p}}^i.\underline{n} - \underline{\hat{p}}^e.\underline{n}$, where "the interior normal trace " $\underline{\hat{p}}^i.\underline{n}$ (resp. "the exterior normal trace" $\underline{\hat{p}}^e.\underline{n}$) is defined as the normal trace of the restriction $\underline{\hat{p}}|_K$ (resp. $\underline{\hat{p}}|_{(\Omega\setminus K)}$) on ∂K, and with $[u] = u^i - u^e$ stands for the jump of u, where "the interior trace" u^i and "the exterior trace" u^e of u are defined from the trace of u as above. In order to obtain the non-standard mixed hybrid finite element formulation, we keep the same notation for functional spaces as in the previous section. Thus, the variational formulation of (14) is to find $(u(t), \hat{\lambda}(t)) \in M_1 \times M_2$ and $\underline{\hat{p}}(t) \in X$ such that:

$$\begin{cases} \displaystyle\int_\Omega \frac{\partial u}{\partial t} v\, dx + \sum_{K_i}\int_{K_i} \operatorname{div}(\underline{\hat{p}})\, v\, dx = 0 & \forall v \in M_1, \\ \displaystyle\int_\Omega \underline{\hat{p}}.\underline{q}\, dx = \sum_{K_i}\int_{K_i} (\underline{a}u.\underline{q})\, dx + \sum_{K_i}\int_{\partial K_i} \hat{\lambda}\, \underline{q}.\underline{n}\, ds & \forall \underline{q} \in X, \\ \displaystyle\sum_{K_i}\int_{\partial K_i} [\underline{\hat{p}}.\underline{n}]\mu\, ds = \sum_{K_i}\int_{\partial K_i} sign(\underline{a}.\underline{n})a_i\,[u]\mu\, ds & \forall \mu \in M_2. \end{cases} \tag{18}$$

The last equation of (18) imposes jump conditions to the flux $\underline{\hat{p}}.\underline{n}$ at the interelement interfaces of T_h. We are now able to obtain a result of existence and uniqueness of the solution for problem (18), using an argument of regularity as in Theorem 2.23.

Theorem 2 *Consider u the solution of the problem (14). If* $\underline{\hat{p}} = \underline{a}u$ *satisfies*

$$\underline{\hat{p}}(t) \in W_t(\operatorname{div};\Omega), \tag{19}$$

then $((\underline{\hat{p}}, \hat{\lambda}); (u, u|_{\partial K}))$ *is the unique solution of (18).*

3. Spatial Approximation

Let us now deal with the finite element approximation of the above dual variational problem. To this aim, we assume that Ω is a polygonal bounded domain of R. Let T_h be a standard finite element partition of Ω.

3.1 Spatial Approximation of DMH1

To define a discretization of problem (12), we first define the approximation spaces X_h, M_{1h} and M_{2h} of X, M_1 and M_2, respectively. We thus build an appropriate approximation space for $\underline{q} \in X$ in the standard way as follows:

$$X_h = \left\{ \underline{q}_h \in X; \quad \underline{q}_h|_K \in RT_0(K) \qquad \forall K \in T_h \right\}, \tag{20}$$

where $RT_0(K)$ is the local Raviart-Thomas space of lowest degree.

To obtain the approximation of the spaces M_1 and M_2, we define

$$M_{1h} = \{ v_h \in M_1; \quad v_h|_K \in \mathrm{P}_0(K) \qquad \forall K \in T_h \}, \tag{21}$$

and

$$M_{2h} = \{ \mu_h \in M_2; \quad \mu_h|_{\partial K} \in R_0(\partial K) \qquad \forall K \in T_h \}, \tag{22}$$

where

$$R_0(\partial K) = \{ \varphi; \quad \varphi|_e \in \mathrm{P}_0(e) \qquad \forall e \in \partial K \}.$$

Finally, the discrete version of (12) is given by:

Find $(u_h(t), \lambda_h(t), \hat{\lambda}_h(t)) \in M_{1h} \times M_{2h}^2$ and $(p_h(t), \hat{p}_h(t)) \in X_h^2$ such that:

$$\left\{ \begin{array}{l} \displaystyle\int_\Omega \frac{\partial u_h}{\partial t} v_h \, dx - \nu \sum_{K_i} \int_{K_i} \mathrm{div}(\underline{p}_h) v_h \, dx + \\[2ex] \displaystyle\sum_{K_i} \int_{K_i} \mathrm{div}(\underline{\hat{p}}_h) v_h \, dx = \int_\Omega f v_h \, dx \qquad \forall v_h \in M_{1h}, \\[2ex] \displaystyle\int_\Omega \underline{p}_h . \underline{q}_h \, dx = - \sum_{K_i} \int_{K_i} u_h \, \mathrm{div}(\underline{q}_h) \, dx + \sum_{K_i} \int_{\partial K_i} \lambda_h \underline{q}_h . \underline{n} \, ds \quad \forall \underline{q}_h \in X_h, \\[2ex] \displaystyle\int_\Omega \underline{\hat{p}}_h . \underline{q}_h \, dx = \sum_{K_i} \int_{K_i} (\underline{a} u_h . \underline{q}_h) \, dx + \sum_{K_i} \int_{\partial K_i} \hat{\lambda}_h \underline{q}_h . \underline{n} \, ds \qquad \forall \underline{q}_h \in X_h, \\[2ex] \displaystyle\sum_{K_i} \int_{\partial K_i} \underline{p}_h . \underline{n} \mu_h \, ds = 0 \quad \forall \mu_h \in M_{2h}, \\[2ex] \displaystyle\sum_{K_i} \int_{\partial K_i} \underline{\hat{p}}_h . \underline{n} \mu_h \, ds = 0 \quad \forall \mu_h \in M_{2h}. \end{array} \right. \tag{23}$$

To these equations we have to add given initial data, and corresponding suitable boundary conditions.

Theorem 3 *Problem (23) has a unique solution.*

Proof. Solving the problem (23) leads to solve a square finite-dimensional system of linear algebraic equations. Hence, we only have to prove the uniqueness of the solution.

If we consider the homogeneous problem associated to (23), f=0, and if the boundary conditions and initial value vanish, the proof is similar to that of the Theorem 1, in regard to u_h, $\underline{p}_h$ and $\underline{\hat{p}}_h$. For the uniqueness of λ_h and $\hat{\lambda}_h$, we consider

$$\sum_{K_i} \int_{\partial K_i} \underline{q}_h . \underline{n} \lambda_h \, ds = 0 \quad \forall \underline{q}_h \in M_{1h},$$

let

$$\int_{\partial K_i} \underline{q}_h . \underline{n} \lambda_h \, ds = 0 \quad \forall \underline{q}_h \in RT_0(K_i),$$

and we then deduce that $\lambda_h = 0$, and similarly we have $\hat{\lambda}_h = 0$.

3.2 *Spatial Approximation of DMH2*

To define the associated discrete problem of (18), we may now use the same suitable subspaces as above, and we also proceed in the same manner. Hence, we obtain the following discrete problem:

Find $(u_h(t), \hat{\lambda}_h(t)) \in M_{1h} \times M_{2h}$ and $\hat{p}_h(t) \in X_h$ such that

$$\left\{ \begin{array}{ll} \int_\Omega \frac{\partial u_h}{\partial t} v_h \, dx + \sum_{K_i} \int_{K_i} \mathrm{div}(\underline{\hat{p}}_h) v_h \, dx = 0 & \forall v_h \in M_{1h}, \\ \int_\Omega \underline{\hat{p}}_h . \underline{q}_h \, dx = \sum_{K_i} \int_{K_i} (\underline{a} u_h . \underline{q}_h) \, dx + \sum_{K_i} \int_{\partial K_i} \hat{\lambda}_h \underline{q}_h . \underline{n} \, ds & \forall \underline{q}_h \in X_h, \\ \sum_{K_i} \int_{\partial K_i} [\underline{\hat{p}}_h . \underline{n}] \mu_h \, ds = \sum_{K_i} \int_{\partial K_i} sign(\underline{a}.\underline{n}) a_i [u_h] \mu_h \, ds & \forall \mu_h \in M_{2h}, \end{array} \right. \tag{24}$$

a_i is again the approximation of the local velocity at ∂K_i, $(a_i \geq 0)$. In a similar way as we did for Theorem 3, we can prove the following result:

Theorem 4 *The problem (24) has a unique solution.*

4. Analogy with the Finite Volume Method

The contents of this section are discussed in more details in (Serghini Mounim, 2000). Notice that the most attractive property of finite volume methods is that the conservation of quantities such as mass, momentum and energy is exactly satisfied on each finite volume and also on the whole domain. To make the link explicit with finite volume methods, we use the *RT* vector finite elements to approximate the diffusion and convective fluxes. We point out this relationship, using a suitable quadrature formula to approximate the terms $\int_K \underline{p}.\underline{q} \, dx$ and $\int_K \underline{f}(u).\underline{q} \, dx$, $\underline{f}(u)$ denotes here the convective flux.

4.1 *DMH1 and scheme1*

4.1.1 Rectangular Case

We consider a quadrangulation T_h of Ω, and we denote by K the generic rectangle of T_h defined by $[x^K, x^K + \Delta x^K] \times [y^K, y^K + \Delta y^K]$, and the degrees of freedom introduced by Raviart-Thomas. The usual *RT* basis functions of K are defined by:

$$\left\{ \begin{array}{lcl} p_S & = & \frac{1}{|K|}(0, \, y - y^K - \Delta y^K), \\ p_E & = & \frac{1}{|K|}(x - x^K, \, 0), \\ p_N & = & \frac{1}{|K|}(0, \, y - y^K), \\ p_W & = & \frac{1}{|K|}(x - x^K - \Delta x^K, \, 0), \end{array} \right. \tag{25}$$

where $|K|$ is the area of K.

In order to diagonalize the mass matrix $\int_K \underline{p}.\underline{q} \, dx$, still retaining the good numerical properties of the mixed approximation, we employ the trapezoidal quadrature rule to obtain the following diagonal matrix:

$$\frac{1}{2} \begin{pmatrix} \frac{\Delta y}{\Delta x} & 0 & 0 & 0 \\ 0 & \frac{\Delta x}{\Delta y} & 0 & 0 \\ 0 & 0 & \frac{\Delta y}{\Delta x} & 0 \\ 0 & 0 & 0 & \frac{\Delta x}{\Delta y} \end{pmatrix}.$$

For simplicity, we suppose $f = 0$, and in what follows we shall write $u, p, \hat{p}$ instead $u_h, p_h, \hat{p}_h, \ldots$ and omit the dependance of index sets $u, p, \hat{p}, \ldots$ on h. Now, let us fix our attention on two adjacent rectangles K and $K_i, i \in \{S, E, N, W\}$, and denote the values of u in K and K_i by u_K and u_i respectively. Next, as in the 1D case (see (Fortin & Serghini Mounim, 2005)) a direct computation provides:

$$\lambda_i = \frac{u_K + u_i}{2}, \tag{26}$$

and

$$\hat{\lambda}_i = \frac{|K|}{4} \frac{\underline{f}(u_i) - \underline{f}(u_K)}{|e_i|} .\underline{n}_i, \tag{27}$$

from which we can deduce the approximation of the convective flux at the faces of K

$$\hat{\alpha}_i = \frac{|e_i|}{2}(\underline{f}(u_K) + \underline{f}(u_i)).\underline{n}_i, \tag{28}$$

where the outward normal to the faces e_i of K is denoted by $\underline{n}_i$, with $|e_i|$ stands for the length of e_i for $i \in \{S, E, N, W\}$.

Then, we observe that the expressions of $\hat{\alpha}_i$ are equivalent to a finite volume method. Indeed, we obtain a "central difference scheme" (CDS). This scheme is not spatially stable; it does not prevent oscillations near discontinuities and at shocks, (for the properties of CDS see Lazarov *et al* (Lazarov, & Vassilevski, 1996)).

4.1.2 Other Formulation and Upwind Schemes

If we impose the continuity of convective flux only at inflow boundary of each control volume K, we can obtain an upwind scheme. For this purpose, we utilize the following form:

$$\begin{aligned} c((\hat{p}_h, \hat{\lambda}); q_h) &= \int_\Omega \hat{p}_h.q_h\, dx - \sum_{K_i} \int_{K_i} f(u_h).q_h\, dx + \sum_{K_i} \int_{\partial K_-} \hat{\lambda}\, \underline{q}.\underline{n}\, ds, \\ &= 0 \qquad \forall\, q_h \in X_h. \end{aligned}$$

∂K_- denotes the inflow boundary of K, which is defined by

$$\partial K_- = \left\{ x \in \partial K \quad : \quad \underline{a}.\underline{n} \le 0 \right\}.$$

Using the same techniques as in the previous subsection, we can derive easily the numerical convective flux at the interface between adjacent cells:

$$\hat{\alpha}_i = |e_i| \left[\left(\frac{\underline{f}(u_K) + \underline{f}(u_i)}{2} \right) .\underline{n}_i - \frac{1}{2} | \left(\frac{\partial \underline{f}}{\partial u} \right)_{\underline{u}} .\underline{n}_i | (u_i - u_K) \right], \tag{29}$$

since

$$\underline{f}(u_i) - \underline{f}(u_K) = \left(\frac{\partial \underline{f}}{\partial u} \right)_{\underline{u}} (u_i - u_K),$$

where $\underline{u}$ represents some average state between u_i and u_K.
For the linear convection case ($\underline{f}(u) = \underline{a}u$ where $\underline{a}$ denotes again the convective field), the above expression then reads:

$$\hat{\alpha}_i = |e_i| \left[\frac{\underline{f}(u_K).\underline{n}_i + |\underline{f}(u_K).\underline{n}_i|}{2} + \frac{\underline{f}(u_i).\underline{n}_i - |\underline{f}(u_i).\underline{n}_i|}{2} \right], \tag{30}$$

where $i \in \{S, E, N, W\}$, and $|e_S| = |e_N| = \Delta x$, $|e_E| = |e_W| = \Delta y$. This scheme coincides with the standard UDS (upwind difference scheme) see for example (LeVeque, 2002).
Before considering the generalization of the previous schemes, we note that one advantage of the use of $\hat{\lambda}$ is that no (approximate) Riemann solvers are required.

Remark 3

Since u is discretized using piecewise polynomial functions and our formulation involves the boundary integrals, we can introduce an upwinding in the formulation. To achieve the generalization of the schemes above, we propose the following form

$$\sum_K \left\{ \frac{1+\beta}{2} \int_{\partial K_-} \hat{\lambda}\, \underline{q}.\underline{n}\, ds + \frac{1-\beta}{2} \int_{\partial K_+} \hat{\lambda}\, \underline{q}.\underline{n}\, ds \right\}, \tag{31}$$

to replace the form already used $\sum_K \int_{\partial K} \hat{\lambda}\, \underline{q}.\underline{n}\, ds$ in the system (23). The following notation is used:

$$\partial K_+ = \left\{ x \in \partial K \quad : \quad \underline{a}.\underline{n} > 0 \right\}.$$

Taking the same approach as above, this results in the generalized upwind scheme:

$$\sum_{i=1}^{4} \hat{\alpha}_i = \Delta x \left((1-\beta)\frac{f_2(u_N) - f_2(u_S)}{2} + 2\beta\frac{f_2(u_K) - f_2(u_S)}{2} \right) +$$
$$\Delta y \left((1-\beta)\frac{f_1(u_E) - f_1(u_W)}{2} + 2\beta\frac{f_1(u_K) - f_1(u_W)}{2} \right).$$

Here $\underline{f} = (f_1, f_2)$, and β is the dissipation parameter ($0 \leq \beta \leq 1$) which controls the combination of fully upwind and centered scheme. We have supposed without loss of generality that the inflow boundary of K is $e_S \cup e_W$. Notice that for $\beta = 1$, we recover the scheme (29), (30).

4.1.3 Triangular Case

We consider now the case where T_h is a standard triangulation of Ω (here, we suppose again that Ω is a polygonal bounded domain of R). We begin with a brief description of the notation used here. For each triangle K, we have:

- $|K|$: area of K;
- $\underline{a}_r$: r^{th} vertex of K with coordinates (x_r, y_r);
- e_r: face opposite to vertex a_r, with length $|e_r|$;
- h_r: the length of the perpendicular dropped from the vertex $\underline{a}_r$;
- $\underline{n}_r$: unit exterior normal to face e_r;
- θ_r: angle at vertex a_r;

for $r \in \{i, j, k\}$.

We also consider the local shape functions $\{\underline{p}_r\}_{r\in\{i,j,k\}}$ of $RT_0(K)$ for every triangle K. The degrees of freedom are the fluxes through the faces $\{e_r\}_{r\in\{i,j,k\}}$, and are defined by:

$$\underline{p}_r(\underline{x}) = \frac{\underline{x} - \underline{a}_r}{2|K|} \equiv \frac{\underline{x} - \underline{a}_r}{h_r|e_r|} \qquad \forall\, \underline{x} = (x_1, x_2) \in K, \tag{32}$$

from which, for every local vectorial function $\underline{p}_r$ of $RT_0(K)$ the well known properties follow:

$$\begin{cases} \operatorname{supp}(\underline{p}_r) = K \cup K', \quad \text{with } K \cap K' = e_r, \\ \operatorname{div}(\underline{p}_r) = \dfrac{1}{|K|}, \\ \underline{p}_r.\underline{n}_s|_{e_s} = \dfrac{\delta_{rs}}{|e_r|}, \end{cases} \tag{33}$$

where δ_{rs} denotes the Kronecker symbol $\delta_{rs} = \begin{cases} 1 & \text{if } r = s, \\ 0 & \text{otherwise.} \end{cases}$

We introduce the bilinear form $a(\underline{p}, \underline{q}) \equiv \int_K \underline{p}.\underline{q}\, dx$ for all $\underline{p}, \underline{q}$ in $RT_0(K)$. Then, $a(\underline{p}_r, \underline{p}_s)$ is a 3×3 symmetric, positive definite matrix,

$$a(\underline{p}_r, \underline{p}_s) \equiv a_{rs} = \frac{1}{12}\begin{pmatrix} 3c_i + c_j + c_k & c_k - (c_i + c_j) & c_j - (c_k + c_i) \\ c_k - (c_i + c_j) & 3c_j + c_k + c_i & c_i - (c_j + c_k) \\ c_j - (c_k + c_i) & c_i - (c_j + c_k) & 3c_k + c_i + c_j \end{pmatrix}, \tag{34}$$

with $c_r \equiv \cot\theta_r$, for $r, s \in \{i, j, k\}$.

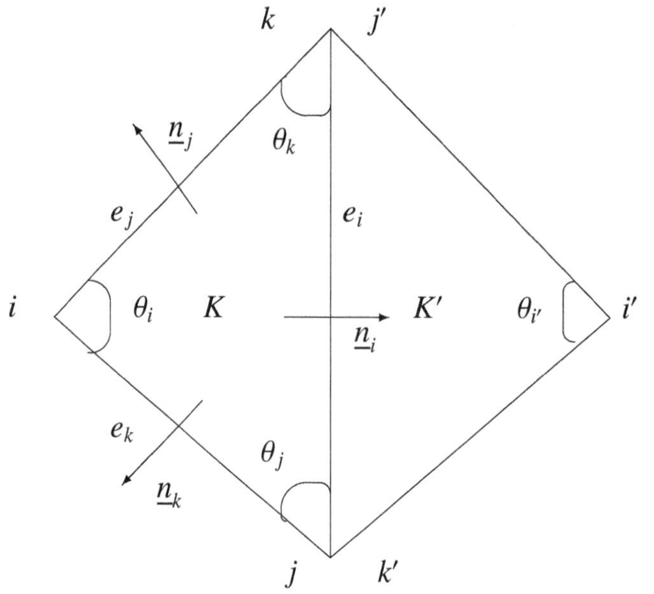

Figure 1. Notations for triangles K and K'

We employ the quadrature formula proposed in (Barranger, Maitre, & Oudin, 1996), to diagonalize the local mass matrix (34). We obtain the matrix:

$$a^h_{rs} \equiv \frac{1}{2}\delta_{rs} cot\theta_r, \quad \text{for } r, s \in \{i, j, k\}. \tag{35}$$

From this, after temporal approximation, and considering the neighbor elements of K which appear in the formula, the equation expressing the discrete conservation law becomes:

$$|K|\frac{u^{n+1}_K - u^n_K}{\delta t} - \nu \sum_{r=i,j,k} \alpha_r + \sum_{r=i,j,k} \hat{\alpha}_r = 0, \tag{36}$$

and thanks to the following relation,

$$\int_K \underline{p}.\underline{q}\, dx = -\int_K u\, \text{div}(\underline{q})\, dx + \int_{\partial K} \lambda\, \underline{q}.\underline{n}\, ds,$$

we obtain

$$\beta_r \alpha_r = -u_K + \lambda_r, \tag{37}$$

where we have set $\beta_r = \frac{1}{2}cot\theta_r$, with $r = i, j, k$.
From the above equation, in conjunction with the continuity of the normal trace of the diffusive flux, we deduce

$$\lambda_r = \frac{\beta_{r'}}{\beta_r + \beta_{r'}} u_K + \frac{\beta_r}{\beta_r + \beta_{r'}} u_{K_r}, \quad r = i, j, k. \tag{38}$$

Using (37) and (38), the numerical diffusion flux across edge e_r reads

$$\alpha_r = \frac{u_{K_r} - u_K}{\beta_r + \beta_{r'}}. \tag{39}$$

We note that $\beta_r = \frac{d_r}{|e_r|}$, $r = i, j, k$ (see the annexe in (Bank & Rose, 1987; Kerkhoven-Jerome, 1990)), where d_r denotes the distance between the circumscribed circle of K and the center m_r on sides e_r (see previous and below figures). We mention that β_r is also given by $\beta_r = -\frac{|e_s|\underline{t}_s.|e_{s'}|\underline{t}_{s'}}{4|K|}$, where we have set $s = I(r)$, $s' = I(s)$, with I defined as $I(i) = j, I(j) = k$ and $I(k) = i$, and $\underline{t}_r$ stands for the unit tangent to e_r for $r = i, j, k$.

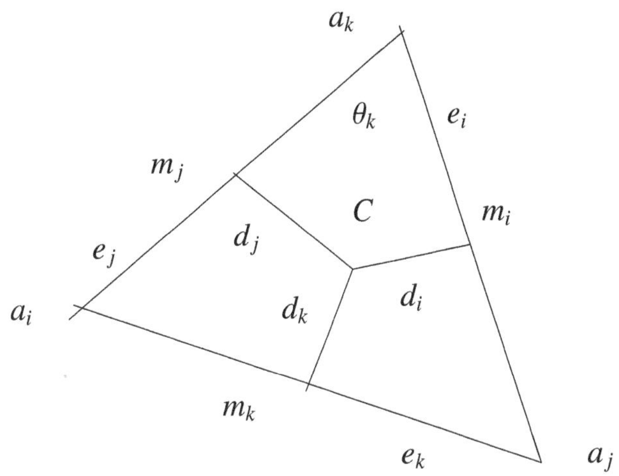

Figure 2. Circumscribed circle and distances d_i.

With the aid of the new expression of β_r the system (39) becomes

$$\alpha_r = |e_r|\frac{u_{K_r} - u_K}{d_r + d_{r'}}, \qquad r = i, j, k, \tag{40}$$

where $d_r + d_{r'}$ is the distance between the circumcenters C, C_r of two adjoining triangles K and K_r. The above distance does not vanish, in the case of a strictly acute triangle.

We now deal with the convective flux using the same notation and procedure as above. We thus start from the equation

$$\int_K \hat{\underline{p}}.\underline{q}\, dx = \int_K \underline{f}(u).\underline{q}\, dx + \int_{\partial K} \hat{\lambda}\, \underline{q}.\underline{n}\, ds.$$

We approximate the mass matrix with the same quadrature formula as in the diffusive case, in addition, we use the following numerical integration which is exact for polynomials of degree 0:

$$\int_K p(x, y)\, dxdy \simeq |K| p(H_K), \tag{41}$$

where H_K is the orthocenter of the triangle K which is considered acute, and $|K|$ denotes the area of K as illustrated in Figure 3.

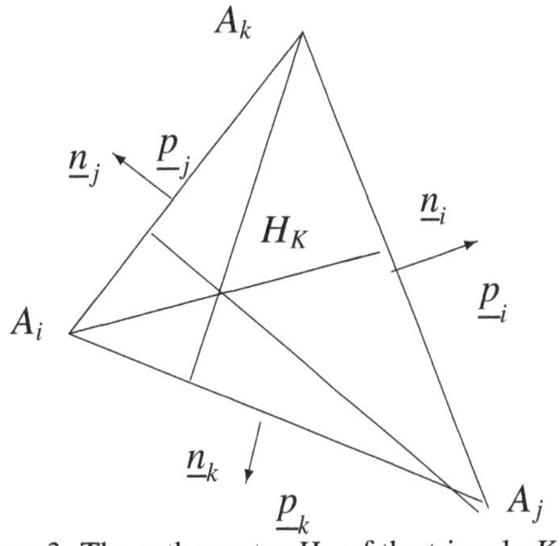

Figure 3. The orthocenter H_K of the triangle K.

Thanks to the above formula (41), the integral $\int_K \underline{f}(u)\underline{p}_r\, dx$ is approximated by

$$\int_K \underline{f}(u).\underline{p}_r\, dx \simeq \frac{\underline{f}(u)}{2}.\overrightarrow{A_r H_K}, \qquad r = i, j, k.$$

Finally, these approximate relations allow us to write the numerical convective flux as

$$\hat{\alpha}_r = \frac{|e_r|}{2} \frac{\underline{f}(u_K)\,||\overrightarrow{A_r H_K}|| + \underline{f}(u_{K_r})\,||\overrightarrow{A_{r'} H_{K_r}}||}{d_r + d_{r'}} .\underline{n}_r, \qquad r = i, j, k, \tag{42}$$

since $\beta_r = \dfrac{d_r}{|e_r|}$, for all $r \in \{i, j, k\}$.

If we only use equilateral triangles, we get $||\overrightarrow{A_r H_K}|| = ||\overrightarrow{A_{r'} H_{K_r}}|| = 2d$, with $d_r = d_{r'} = d$, and the approximation formula (42), becomes

$$\hat{\alpha}_r = |e_r| \left(\frac{\underline{f}(u_K) + \underline{f}(u_{K_r})}{2} \right) .\underline{n}_r, \qquad r = i, j, k. \tag{43}$$

It is precisely the central finite volume scheme "CDS".

Remark 4

In view of the linear case $\underline{f}(u) = \underline{a}u$, we can define $\underline{a}$ as a linear combination of the local shape functions $\underline{p}_r$ of $RT_0(K)$, after some manipulation, we can exhibit the numerical flux through the face e_r

$$\hat{\alpha}_r = \frac{|e_r|}{d_r + d_{r'}} \frac{u_K + u_{K_r}}{2} a_{h_r}, \qquad r = i, j, k,$$

where $\underline{a}_h = \sum_{r=i}^{k} a_{h_r} \underline{p}_r$. For other approximations the reader can refer to (Serghini Mounim, 2000).

Remark 5

If we utilize the bilinear form (31) and the quadrature formula (41), the numerical convective flux (43) becomes

$$\hat{\alpha}_r = |e_r| \left[\frac{(1+\beta)\underline{f}(u_K) + (1-\beta)\underline{f}(u_{K_r})}{2} + \beta \Xi \left(\underline{f}(u_K) - \underline{f}(u_{K_r}) \right) \right] .\underline{n}_r,$$

for $r \in \{i, j, k\}$, with $\Xi = min\left(sign(\underline{a}.\underline{n}_r), 0 \right)$ and β being again the upwinding parameter ($0 \leq \beta \leq 1$). Note that for $\beta = 1$ (or the continuity constraint imposed to $\hat{p}$ only at the inflow boundary of K), we get the following upwind finite volume scheme:

$$\hat{\alpha}_r = |e_r| \left[\underline{f}(u_K) + \Xi \left(\underline{f}(u_K) - \underline{f}(u_{K_r}) \right) \right] .\underline{n}_r, \quad r = i, j, k,$$

for the stability and error estimates of CDS and UDS on Voronoi meshes we refer to (Mishev, 1998).

4.2 *DMH2 and Scheme2*

The results of this section are derived in a similar fashion to DMH1 case.

4.2.1 Rectangular Case

Consider again the case of a rectangular mesh T_h of Ω, and the local degrees of freedom associated to $RT_0(K)$. Here we denote by $\underline{n}_r^e$ (resp. $\underline{n}_r^i$) the exterior unit normal vector, i.e., oriented from K to K_r (resp. interior unit normal vector, i.e., oriented from K_r to K) along each edge $e_r = \partial K \cap \partial K_r$; clearly, the property $\underline{n}_r^e = -\underline{n}_r^i$ holds. Now, following the same lines as in previous sections, we can state the result:

Proposition 2 *If we use the trapezoidal quadrature formula to diagonalize the element mass matrix, and exact computation of the spatial integral* $\int_K \underline{f}(u).\underline{q}\,dx$; *then, the exterior* $\hat{\alpha}_r^e$ *and the interior* $\hat{\alpha}_r^i$ *fluxes approximating* $\int_{e_r} \hat{\underline{p}}^e.\underline{n}_r^e\,ds$ *and* $\int_{e_r} \hat{\underline{p}}^i.\underline{n}_r^e\,ds$ *respectively across edge* e_r *are given by:*

$$\hat{\alpha}_r^e = \frac{|e_r|}{2} \left((\underline{f}(u_K) + \underline{f}(u_r)).\underline{n}_r^e + sign(\underline{a}.\underline{n}_r^e)\, a_r|_{e_r} (u_r - u_K) \right),$$

and

$$\hat{\alpha}_r^i = \frac{|e_r|}{2} \left((\underline{f}(u_K) + \underline{f}(u_r)).\underline{n}_r^e + sign(\underline{a}.\underline{n}_r^e)\, a_r|_{e_r} (u_K - u_r) \right),$$

where $\hat{\lambda}_r$ *is given by:*

$$\hat{\lambda}_r = \frac{|K|}{4} \frac{(\underline{f}(u_r) - \underline{f}(u_K)).\underline{n}_r^e + sign(\underline{a}.\underline{n}_r^e)\, a_r|_{e_r} (u_K - u_r)}{|e_r|},$$

for $r \in \{S, E, N, W\}$, *and where* $a_r|_{e_r}$ *stands for an approximation of the propagation speed at the edge* e_r.

Hence, we can state that the Rankine-Hugoniot jump condition at e_r is recovered in the expression of Lagrange multipliers. It is also important to notice that assuming that e_r is the upstream boundary of ∂K, *i.e.*, $sign(\underline{a}.\underline{n}_r^e)|_{e_r} < 0$ (resp. e_r is the downstream boundary of ∂K), then the numerical fluxes used in the FV methods at e_r are our exterior fluxes: $F_i = \alpha_r^e$ (resp. our interior fluxes $\hat{\alpha}_r^i$). Then we can generalize the above formulation taking the same approach used in (Fortin & Serghini Mounim, 2005). To this end, we consider the values of the trace normal of the fluxes $\hat{p}.\underline{n}$ at the boundary of element ∂K as a convex combination of the exterior fluxes $\underline{\hat{p}}^e.\underline{n}$ and the interior fluxes $\underline{\hat{p}}^i.\underline{n}$:

$$\underline{\hat{p}}.\underline{n} = \begin{cases} \frac{1+\beta}{2}\underline{\hat{p}}^e.\underline{n} + \frac{1-\beta}{2}\underline{\hat{p}}^i.\underline{n} & \text{on } \partial K^-, \\ \frac{1-\beta}{2}\underline{\hat{p}}^e.\underline{n} + \frac{1+\beta}{2}\underline{\hat{p}}^i.\underline{n} & \text{on } \partial K^+, \end{cases} \tag{44}$$

where $\beta \in [0, 1]$ is the upwinding parameter, and $\underline{n}$ is again the exterior unit normal vector oriented from K to neighboring triangles K_r. Consequently, by means of (44), we also obtain the local conservativity of the scheme (the numerical fluxes are obviously consistent). Hence, we replace in each K_i, $\int_{K_i} \text{div}(\underline{p})\, v_h\, dx$ by

$$\int_{\partial K_i^-} (\frac{1+\beta}{2}\underline{\hat{p}}^e.\underline{n} + \frac{1-\beta}{2}\underline{\hat{p}}^i.\underline{n})v_h\, ds + \int_{\partial K_i^+} (\frac{1+\beta}{2}\underline{\hat{p}}^i.\underline{n}) + \frac{1-\beta}{2}\underline{\hat{p}}^e.\underline{n})v_h\, ds.$$

We mention that the usual expressions of the finite volume numerical fluxes (see for example (LeVeque, 2002; Beux *et al*, 1993) are obtained via our formulation, taking $\beta = 1$ and for the choice $\beta = 0$ we recover the centered scheme. Now, it is clear that the appropriate choice of the speed a_i allows us to recover the standard volume schemes. Next, we focus our attention to giving the extensions of some recent schemes.

Remark 6

In the following we shall use in each K the piecewise polynomial interpolant at time t^n, $P^K(., t^n)$, and the local velocities of propagation a_r^n, proposed in (Kurganov & Levy, 2000). Next, taking a dimension by dimension approach, we can recover the two-dimensional, third order, non oscillatory scheme of Kurganov-Levy scheme, computing the interior and exterior traces of u, $u_r^{i,e}$ at the faces $e_r = \partial K \cap \partial K_r$, $r \in \{S, E, N, W\}$ following the recipe. The recipe is as follows (the computation of the exterior values u_r^e can be carried out in the similar way, using $P^r(., t^n)$ in the neighboring rectangles)

$$u_r^i = w_L P_L^K(x_r; t^n) + w_R P_R^K(x_r; t^n) + w_C P_C^K(x_r; t^n),\ r \subset \{W, E\},$$

and

$$u_r^i = w_L P_L^K(y_r; t^n) + w_R P_R^K(y_r; t^n) + w_C P_C^K(y_r; t^n),\ r \in \{S, N\},$$

where the polynomials P_L^K, P_R^K, P_C^K, and the weights w_L, w_R, w_C are also introduced in (Kurganov & Levy, 2000). Now, if we utilize u_r^i and u_r^e in (24), and thanks to Proposition 4.21, and the help to (44), this will result in the 2D extension Kurganov-Levy semi-discrete scheme.

4.2.2 Triangular Case

We consider now the case where T_h is a triangular mesh. For each triangle K, we use the same notations as DMH1 case and we summarize the result obtained as follows:

Proposition 3 *If we use the quadrature formula proposed in (Baranger, Maitre, & Oudin, 1996) to diagonalize the element mass matrix and we employ the numerical integration formula (41) then, the fluxes* $\hat{\alpha}_r^e$ *and* α_r^i *through the face* e_r *are approximated by*

$$\begin{aligned} \hat{\alpha}_r^e &= \frac{|e_r|}{d_r + d_{r'}}(\frac{||\overrightarrow{A_r H_K}||\underline{f}(u_K).\underline{n}_r^e + ||\overrightarrow{A_{r'} H_{K_r}}||\underline{f}(u_{K_r}).\underline{n}_r^e}{2} \\ &+ sign(\underline{a}.\underline{n}_r^e) d_r a_r|_{e_r}(u_{K_r} - u_K)), \end{aligned}$$

and

$$\begin{aligned} \hat{\alpha}_r^i &= \frac{|e_r|}{d_r + d_{r'}}(\frac{||\overrightarrow{A_r H_K}||\underline{f}(u_K).\underline{n}_r^e + ||\overrightarrow{A_{r'} H_{K_r}}||\underline{f}(u_{K_r}).\underline{n}_r^e}{2} \\ &+ sign(\underline{a}.\underline{n}_r^e) d_{r'} a_r|_{e_r}(u_K - u_{K_r})), \end{aligned}$$

where $\hat{\lambda}_r$ is given by:

$$\hat{\lambda}_r = -\frac{d_{r'}}{d_r + d_{r'}}\|\overrightarrow{A_r H_K}\|\frac{\underline{f}(u_K)}{2}.\underline{n}_r^e + \frac{d_r}{d_r + d_{r'}}\|\overrightarrow{A_r H_{K_r}}\|\frac{\underline{f}(u_{K_r})}{2}.\underline{n}_r^e,$$

$$+ \quad sign(\underline{a}.\underline{n}_r^e)\frac{d_r d_{r'}}{d_r + d_{r'}} a_r|_{e_r}(u_K - u_{K_r}),$$

for $r = i, j, k$.

If we restrain to equilateral triangles, one then gets obviously $\|\overrightarrow{A_r H_K}\| = \|\overrightarrow{A_{r'} H_{K_r}}\| = 2d$, with $d_r = d_{r'} = d$, and, in particular, the approximation of the exterior flux becomes:

$$\hat{\alpha}_r^e = |e_r|\left(\frac{\underline{f}(u_K) + \underline{f}(u_{K_r})}{2}.\underline{n}_r^e + sign(\underline{a}.\underline{n}_r^e)\frac{a_r|_{e_r}}{2}(u_{K_r} - u_K)\right), \tag{45}$$

for $r = i, j, k$.

Remark 7

In view of the linear case $\underline{f}(u) = \underline{a}u$, we can express $\underline{a}$ as linear combination of the local shape functions $\underline{p}_r$ of $RT_0(K)$. After some manipulations, we can exhibit the exterior numerical flux through the face e_r:

$$\hat{\alpha}_r^e = \frac{d_r u_K + d_{r'} u_{K_r}}{d_r + d_{r'}} a_{h_r} + sign(\underline{a}.\underline{n}_r^e)\frac{|e_r| d_{r'}}{d_r + d_{r'}} a_r|_{e_r}(u_{K_r} - u_K),$$

for $r = i, j, k$, and the approximation a_h of $\underline{a}$ is defined by $\underline{a}_h = \sum_{r=i}^{k} a_{h_r}\underline{p}_r$.

Remark 8

To simplify, let K be an equilateral triangle with centroid G, and let K_r, $r = i, j, k$ be the neighbouring equilateral triangles, with centroids G_r $(r = i, j, k)$. If we assume $u|_K$ linear, and the least-squares gradient gradu(t)$|_K = (a_K(t), b_K(t))$ for the triangle K is then chosen in order to minimize the functional

$$F(a_K(t), b_K(t)) = \sum_{r=i}^{k}\left\{u_K(t) + \overrightarrow{GG_r}.gradu(t)|_K - u_{K_r}(t)\right\}^2.$$

Of course, the minimum is obtained when

$$\frac{\partial F}{\partial a_K(t)} = \frac{\partial F}{\partial b_K(t)} = 0,$$

for details and the expressions of $a_K(t)$ and $b_K(t)$ see (Champier, 1992).
Next, we define over all K

$$P_K(x, y, t) = u_G(t) + (x - x_G)a_K(t) + (y - y_G)b_K(t),$$

and the exterior and interior traces of u at the face e_r are now given by:

$$u_r^e(t^n) = P_{K_r}(x_{m_r}, t^n) \quad \text{and} \quad u_r^i(t^n) = P_K(x_{m_r}, t^n),$$

where x_{m_r} denote the coordinates of the center on sides e_r (see Fig. 2). The resulting exterior flux is

$$\hat{\alpha}_r^e = |e_r|\left(\frac{\underline{f}(u_K) + \underline{f}(u_{K_r})}{2}\right).\underline{n}_r^e + sign(\underline{a}.\underline{n}_r^e)\frac{|e_r|}{2} a_r|_{e_r}(u_r^e(t^n) - u_r^i(t^n)), \tag{46}$$

for $r = i, j, k$. The cell average flux $\underline{f}(u_K)$, (also $\underline{f}(u_{K_r})$, $r = i, j, k$) can be approximated by a second-order quadrature in time (such as the mid-point rule). The resulting scheme can be viewed as a natural new version of multidimensional extension of the second order central NT scheme (Nessyahu & Tadmor, 1990). For other extensions of the NT scheme, see (Arminjon & Viallon, 1995; Jiang & Tadmor, 1998).

We close by noting that we can get a 2D extension of the family of semi-discrete central schemes of Kurganov and Tadmor (Kurganov & Tadmor, 2000). Our recipe is similar to that above, taking into account the quadrature formula given below:

$$\int_K \underline{f}(u(t))\underline{p}_r\,dx \simeq \frac{\underline{f}(u_r^i)(t)}{2}\overrightarrow{A_r H_K} \qquad r = i, j, k.$$

From this, we can express the exterior numerical flux in the form:

$$\begin{aligned}\hat{\alpha}_r^e = & |e_r|\left(\frac{\underline{f}(u_r^i(t)) + \underline{f}(u_r^e(t))}{2}\right).\underline{n}_r^e \\ & + sign(\underline{a}.\underline{n}_r^e)\frac{|e_r|}{2}a_r(t)|_{e_r}(u_r^e(t) - u_r^i(t)),\end{aligned} \tag{47}$$

for $r = i, j, k$, we then obtain the extension to the triangular grid case of the semi-discrete KT scheme (Kurganov & Tadmor, 2000), and where $a_r(t)$ are the maximal local speeds. For more details see (Kurganov & Tadmor, 2000).

5. Discussion

1- The extension to a system of equations can be carried out in the same fashion as (Fortin & Serghini Mounim, 2005), and again with the aid of Roe's construction (ROE, 1981).

2- The generalization to a convection-diffusion equation can also be performed using the extension to 2D of the variational formulation used in (Fortin & Serghini Mounim, 2005). We use, for this, the fact that the solutions to the general initial value problem of the viscous conservation laws converge, in the zero dissipation limit as ν goes zero, to the solutions of the hyperbolic conservation laws. Next, the "RH" jump conditions at the interelements are imposed again to the convective flux. Finally, we can establish the existence and uniqueness of the solution in this case taking the same approach to previous formulations, and using in particular an argument of regularity.

3- Note that the simplicity of the method makes it easy to extend to the three-dimensional case, in addition, if the triangulation of Ω is made of tetrahedrons, and the mesh is restricted to regular subdivisions of the domain, the quadrature formula given in Baranger *et al* (Baranger, Maitre, & Oudin, 1996) can be used to diagonalize the mass matrix. Moreover, the result of (Baranger, Maitre, & Oudin, 1996) cannot be extended to the general 3D case, for more details see (Thomas & Trujillo, 1997).

4- Notice that the methods presented here can also be viewed as an extension of the mortar finite elements method to transport problems. On the other hand, we mention that there is a possible link between our methods and the mimetic finite difference methods (see for example (Hyman & Shashkov, 1997)) that could be made by choosing an appropriate inner product.

Finally, the stability and the feasibility of the method had been successfully tested on numerical examples. For this, the dual mixed finite element method associated to the first-order accurate implicit Euler (for the time-discretization) has been experimented on structured and unstructured meshes see (Serghini Mounim, 2000).

References

Adams, R. A. (1975). *Sobolev Spaces.* Academic Press, New-York.

Agouzal, A., Baranger, J., Maitre, J. F., & Oudin, F. (1995). Connection Between Finite volume and mixed finite element methods for a diffusion problem with nonconstant coefficients. Application to a convection diffusion problem. *East-West J. Num. Math., 3*(4), 237-254.

Arminjon, P., & Viallon, M. C. (1995). Généralisation du schéma de Nessyahu-Tadmor pour une équation hyperbolique à deux dimensions d'espace. *Comptes Rendus de l'Acad. des Sciences, Paris, t.320, série I,* 85-88.

Arnold, D. N., Brezzi, F. (1985). Mixed & nonconforming finite element methods: implementation, postprocessing and error estimates. *RAIRO Modél. Math. Anal. Numér., 19,* 7-32.

Babuska, I. (1984). Error Bounds for Finite Elemnet Methods. *Math., 16,* 1-22.

Bank, R. E., & Rose, D. J. (1987). Some error estimates for the box method. *SIAM J. Num. An., 24,* 777-787.

Baranger, J., Maitre, J. F., & Oudin, F. (1993). Application de la théorie des éléments finis mixtes à l'étude d'une classe de schémas aux volumes finis pour les problèmes elliptiques. *C. R. Acad. Sci. Paris, 316,* série I, 509-512.

Baranger, J., Maitre, J. F., & Oudin, F. (1996). Connection Between Finite Volume And Mixed Finite Element Methods. *RAIRO M^2AN, 30*(4), 445-465.

Beux, F., Lanteri, S., Dervieux, A., & Larrouturou, B. (1993). Upwind Stabilization of Navier-Stokes Solvers. *Rapport de Recherche, INRIA N 1885.*

Brezzi, F. (1974). On the Existence, Uniqueness & Approximation of Saddle-Point Problems Arising from Lagrangian Multipliers. *RAIRO, Anal. Numer. Math., 2,* 129-151.

Brezzi, F., Douglas, J., Duran, R., & Fortin, M. (1987). Mixed finite elements for second order elliptic problems in three variables. *Numer. Math., 51*, 237-250,

Brezzi, F., Douglas, J., Fortin, M., Marini, L. D. (1987). Efficient rectangular mixed finite elements in two and three space variables. *Math. Model. Numer. Anal., 21*, 581-604.

Brezzi, F., Douglas, J., & Marini, L. D. (1986). *Recent results on mixed finite element methods for second order elliptic problems, in Vistas in Applied Math., Numerical Analysis, Atmospheric Sciences, Immunology (Balakrishanan, Dorodnitsyn, and Lions, eds.).* Optimization Software Publications, New York.

Brezzi, F., & Fortin, M. (1991). *Mixed and Hybrid Finite Element Methods.* Springer Series in Computational Mathematics. Springer-Verlag.

Brezzi, F., & Marini, L. D. (1975). On the numerical solution of palte bending problems by hybrid methods. *RAIRO, Anal.*, 5-50.

Champier, S. (1992). *Convergence de schémas numériques type volumes finis pour la résolution d'équations hyperboliques.* Thèse, Université de St-Etienne.

Ciarlet, P. G. (1978). *The Finite Element Method for Elliptic Problems.* North-Holland.

Ekeland, I., & Temam, R. (1974). *Analyse convexe et problèmes variationnels.* Dunod, Paris.

Farhloul, M. (1991). *Méthode d'éléments finis mixtes et volumes finis.* PhD thesis, Université Laval.

Fortin, M., (1976). *An analysis of the Convergence of Mixed Finite Element Methods. Rapport interne INRIA, 186.*

Fortin, M. (1977). An analysis of the Convergence of Mixed Finite Element Methods. *RAIRO, Anal. Numer., 11*, 341-354.

Fortin, M., & SERGHINI MOUNIM, A. (2005). Mixed and hybrid finite element methods for convection-diffusion equations and their relationships with finite volume, Part I. *Calcolo, 42*(1), 1–30.

Fraeijs de veubeke, B. X. (1965). Displacement and Equilibrium Models in the Finite Element Method. *Stress Analysis (O.C. Zienkiewicz and G. Holister, eds.), Wiley,* New-York.

Fraeijs de veubeke, B.X. (1975). *Stress Function Approach.* World Congress on the Finite Element Method in Structural Mechanics, Bournemouth.

Franca, L., & Hugues, T. J. R. (1988). Two classes of mixed finite element methods. *Comp. Meth. Appl. Mech. Eng., 69*, 89-129.

Girault, V., & Raviart, P. A. (1980). *Finite Elements Methods for Navier-Stokes Equations: Theory and Algorithms.* Springer-Verlag.

Godlewski, E., & Raviart, P. A. (1991). Hyperbolic Systems of Conservation Laws. *Mathématiques et Applications. Ellipse.*

Hansbo, P. (1994). Aspects of consevation in finite element flow computations. *Comput. Meth. Appl. Mech. Engrg., 117*, 423-437.

Hou, T. Y., & Lefloch, P. (1992). Why conservative schemes converge to wrong solutions : Error analysis. *Technical Report 255*, Ecole Polytechique de Paris.

Hughes, T. J. R. (1978). A simple finite element scheme for developping upwind finite elements. *Int. J. Num. Meth. Eng., 12*, 1359-1365.

Hughes, T. J. R., & BROOKS, A. (1979). A multidimensional upwind scheme with no crosswind diffusion. *Finite Element Methods for convection dominated flows,* AMD 34, ASME, New York.

Hyman, J. M., & Shashkov, M. (1997). Natural discretizations for the divergence, gradient, and curl on logically rectangular grids. *Int. J. Comp. Math. Appl., 33*(4), 81-104.

Jiang, G.-S., & Tadmor, E. (1998). Non-oscillatory central schemes for multidimensional hyperbolic conservation laws. *SIAM J. Sci. Comput., 19*, 1892-1917.

Kerkhoven, T., & Jerome, J. (1990). L^∞ stability of finite element approximations to elliptic gradient equations. *Numer. Math., 57,* 561-575.

Kruzkov, S. N. (1970). First order quasilinear equtions in several independant variables. *Numer. Math. USSR Sb., 10,* 217-243.

Kurganov, A., & levy, D. (2000). A Third-Order Semi-Discrete Central Scheme For Conservation Laws and Convection-

Diffusion Equations. *SIAM Journal on Scientific Computing, 22*(4), 1461-1488.

Kurganov, A., & Tadmor, E., (2000). New high-resolution central schemes for nonlinear conservation laws and convection-diffusion equations. *J. Comput. Phys., 160* (1), 214-282.

Ladyzhenskaya, O. A., Solonnikov, V. A., & Ouraltseva, N. N. (1967). Equations paraboliques linéaires et quasi linéaires. *MIR.*

Lazarov, R. D., Mishev, I. D., & Vassilevski, P. S. (1996). Finite volume methods for convection-diffusion problems. *SIAM J. Numer. Anal, 33,* 31-55,

Lesaint, P., & Raviart, P. A. (1974). On a finite element method for solving the neutron transport equation. *In Mathematical aspect of finite elements in PDE.C de Boor ed. Academic Press,* 89-123,

Leveque, R. J. (1990). Numerical Methods for Conservation Laws. *Lectures in Mathematics. Birkhauser.*

Leveque, R. J. (2002). Finite Volume Methods for Hyperbolic Problems. *Cambridge Texts In Applied Mathematics.*

Majda, A. (1984). Compressible Fluid Flow and Systems of Conservation Laws in Serval Space Variables. *Bull. Applied Math. Sciences, 53.* Springer-Verlag,

Mishev, I. D. (1998). Finite Volume Methods on Voronoi Meshes. *Numer. Meth. PDE, 14*,(2), 193-212,

Nedelec, J. C. (1986). A New Family of Mixed Finite Elements in R^3. *Numer. Math., 50*, 57-81.

Nessyahu, H., & Tadmor, E. (1990). Non-oscillatoy Central Differencing for Hyperbolic Conservation Laws. *JCP, 87*(2), 408-463.

Raviart, P. A., & Thomas, J. M. (1977). A Mixed Finite Element Method for 2-nd Order Elliptic Problems. *Lecture Notes in Mathematics, 606,* Springer-Verlag, New-York, 292-315.

Roe, P. L. (1985). Some contributions to the modeling of discontinuous flows. *Lect. Notes Appl. Math., 22,* 163-193.

Roe, P. L. (1981). Approximate Riemann solvers, parameter vectors, and difference schemes. *J. Comput. Phys., 43,* 357-372.

Sacco, R., & Saleri, F. (1997). Mixed Finite Volume Methods For Semiconductor Device Simulation. *Numer. Methods Partial Differential Equations, 13,* 215-236.

Serghini Mounim, A. (2000). *Méthode d'éléments finis mixtes: Application aux équations de convection-diffusion et Navier-Stokes.* Ph.D. thesis, Université Laval.

Smoller, J. (1994). *Shock Waves and Reaction-Diffusion Equations.* Spriger-Verlag, 2 edition.

Thomas, J. M. (1977). *Sur l'analyse numérique des méthodes d'éléments finis hybrides et mixtes.* Thèse, Université Pierre et Marie Curie, Paris 6.

Thomas, J. M., & Trujillo, D. (1997). Finite volume methods for elliptic problems; Convergence on unstructured meshes. *In. Numerical Methods in Mechanics, Conca C, Gatica G (eds). Pitman Research Notes in Mathematics Series, 371.* Addison-Wesley Longman: Reading MA, 163-174.

14

Useful Numerical Statistics of Some Response Surface Methodology Designs

Iwundu M. P.

Department of Mathematics and Statistics, Faculty of Science, University of Port Harcourt, Port Harcourt, Nigeria.
E-mail: mary.iwundu@uniport.edu.ng

Abstract

Useful numerical evaluations associated with three categories of Response Surface Methodology designs are presented with respect to five commonly encountered alphabetic optimality criteria. The first-order Plackett-Burman designs and the 2^k Factorial designs are examined for the main effects models and the complete first-order models respectively. The second-order Central Composite Designs are examined for second-order models. The A-, D-, E-, G- and T-optimality criteria are employed as commonly encountered optimality criteria summarizing how good the experimental designs are. Relationships among the optimality criteria are pointed out with regards to the designs and the models. Generally the designs do not show uniform preferences in terms of the considered optimality criteria. However, one interesting finding is that central composite designs defined on cubes and hypercubes with unit axial distances are uniformly preferred in terms of E-optimality and G-optimality criteria.

Keywords: Plackett-Burman design, 2^k Factorial design, Central Composite Design, A-optimality criterion, D-optimality criterion, E-optimality criterion, G-optimality criterion and T-optimality criterion.

1. Introduction

First- and second-order models are the most frequently utilized approximating models in response surface methodology. In most cases, they adequately represent the true unknown functional relationship between the response variable and the control or independent variables. Designs for fitting the first-order models and the second-order models are called the first-order designs and the second-order designs, respectively. The use of first-order designs is vital in preliminary stage of experimentation involving several factors. Commonly encountered first-order designs are the 2^k Factorial designs, 2^{k-p} Fractional factorial designs and the Plakett-Burman designs, where k is the number of control variables and p is the number of defining relations. The 2^k Factorial design involves k factors measured at 2 levels each. As the number of control variables increases, the use of full factorial designs becomes very demanding and burdensome. This then requires running only the 2^{k-p} Fractional factorial designs which are subsets of the full factorial designs, where each subset could be made up of one-half, one-quarter and so on, of the full factorial design.

Plackett-Burman designs, due to Plackett and Burman (1946), are also two level factorial designs but used for studying up to k = N-1 response variables in N runs, where N is a multiple of 4. Each of Plackett-Burman design, 2^k Factorial design and 2^{k-p} Fractional factorial design plays a vital role in experiments involving several factors and where the interest is to examine the joint effects of the factors on a response variable. They are very suitable in screening experiments for the purpose of identifying factors that contribute significantly to the process under study thereby screening out less influential factors. The 2^k factorial designs require large experimental runs when compared to 2^{k-p} fractional factorial designs and the Plackett-Burman designs in k variables. In fact, the Plackett-Burman designs are very economical as each requires a much smaller number of experimental runs, especially for large control variables k, where N=k+1. In this wise, the Plakett-Burman designs are considered saturated. Although Plackett-Burman designs exist for multiples of 4, they are equivalent to 2^{k-p} Fractional factorial designs for N that is powers of 2.

Whereas first-order Response Surface Methodology designs exhibit lack of fits in the presence of curvature and hence inadequate at such stage of experimentation, second-order designs are considered suitable for addressing the quadratic effects. One commonly encountered second-order response surface methodology design is the Central Composite Design due to Box and Wilson (1951). The Central Composite Designs have been extensively studied and there exist vast literature on the subject. For reference purpose, see Dette and Grigoriev (2014), Chigbu and Nduka (2006), Iwundu (2015), Lucas (1976), Myer *et al.*(1992), Giovannitti-Jensen and Myers (1989), Zahran *et al.* (2003), Chigbu *et al.* (2009) and Ukaegbu and Chigbu (2015). We investigate in this paper the optimality properties of three categories of

Response Surface Methodology designs namely, the Plackett-Burman designs, 2^k Factorial design and the Central Composite Designs. Rady *et al.* (2009) presented a survey on optimality criteria that are often encountered in the theory of designing experiments as well as the relationships among several optimality criteria. Although concise enough, no illustrative examples were presented. The aim of this paper is thus to present useful numerical values associated with the three categories of Response Surface Methodology designs, with respect to some commonly encountered alphabetic optimality criteria.

2. Methodology

Most early stages of experimentation assume the absence of curvature in the system and thus utilize the main effects models or complete first-order models to serve as some "good" representation of the response surface in a small region of experimentation. Plackett-Burman designs and the full Factorial designs shall be examined for the main effects models and the complete first-order models respectively, in k factors. The Plackett-Burman designs shall be examined for k = 3, 7, 11, 15. The 2^k Factorial designs shall be examined for k = 2, 3, 4. However when k exceeds 4 the Fractional factorial designs should be encouraged. In a similar fashion, the central composite designs shall be examined for second-order models in k factors (k = 2, 3, 4) and at varying α values. Five alphabetic optimality criteria, namely A-, D-, E-, G- and T-optimality criteria shall be employed in the numerical evaluations as optimality criteria summarize how good experimental designs are. As reported in Khuri and Mukhopadhyay (2010), optimal designs are constructed on the basis of optimality criteria. Rady *et al.* (2009) remarked that optimal designs are model dependent as designs are generally optimal only with respect to specific statistical models. As can be seen in most standard literatures on optimal designs such as Raymond *et al.* (2009) and Rady *et al.* (2009), D-optimality criterion focuses on good model parameter estimation. A D-optimal design is one in which the determinant of the moment matrix

$$M = \frac{X^T X}{N}$$

is maximized over all designs, where X represents the design matrix associated with the design and X^T represents its transpose. The criterion of D-optimality equivalently minimizes the determinant of M^{-1}. A-optimality criterion also focuses on good model parameter estimation. However unlike the D-optimality criterion, it does not take into account the covariances among model coefficients. A-optimality criterion is one in which the sum of the variances of the model coefficients is minimized. It is defined as

$$\text{Min tr}\,(M^{-1})$$

where Min implies that minimization is over all designs and tr represents trace.

E-optimality criterion aims at finding a design which maximizes the minimum eigen value of M or equivalently finds a design which minimizes the maximum eigen value of M^{-1}. By E-optimality, the maximum variance of all possible normalized linear combinations of parameter estimates is minimized. E-optimality criterion is defined by

$$\text{Max } \lambda_{min}(M) \equiv \text{Min } \lambda_{max}(M^{-1})$$

where λ_{min} and λ_{max} represent minimum eigen value and maximum eigen value, respectively. On the other hand, G-optimality criterion considers designs whose maximum scaled prediction variance, $v(\underline{x})$, in the region of the design is not too large. Hence, a G-optimal design minimizes the maximum scaled prediction variance and is defined by

$$\text{Min}\{\max_{\underline{x}\in R} v(\underline{x})\}$$

The concept of scaled prediction variance has been well explained in Myers *et al.* (2009). Finally, the T-optimality criterion aims at finding a design that maximizes the trace of M.

The first-order main effects model to be employed in this study is

$$y = \beta_0 + \sum_{i=1}^{k} \beta_i x_i + \varepsilon \quad ; -1 \le x_i \le 1 \tag{1}$$

where the model parameters retain their usual definitions as contained in standard textbooks on response surface methodology. The first-order complete model (main effects and interactions model) to be employed in this study is

$$y = \beta_0 + \sum_{i=1}^{k} \beta_i x_i + \{ \sum_{i=1}^{k-1} \sum_{j=i+1}^{k} \beta_{ij} (x_i x_j)\} + \{ \sum_i \sum_j \sum_k \beta_{ijk}(x_i x_j x_k)\} + \cdots + \varepsilon \,; -1 \le x_i \le 1 \tag{2}$$

The second-order complete model (main effects, interaction effects and quadratic effects model) to be employed in this study is

$$y = \beta_0 + \sum_{i=1}^{k} \beta_i x_i + \{ \sum_{i=1}^{k-1} \sum_{j=i+1}^{k} \beta_{ij} (x_i x_j)\} + \sum_{i=1}^{k} \beta_{ii} x_i^2 + \varepsilon \quad ; -\alpha \le x_i, x_j \le \alpha \tag{3}$$

The real value α represents the axial distance associated with central composite design. According to Myers *et al.*(2009; pg.298) the value of the axial distance generally lies between 1.0 and $\sqrt{k}$. An α value of 1.0 places the axial points on the cube or hypercube and an α value of $\sqrt{k}$ places all points on a common sphere. To achieve rotatability of designs, the axial distance is set at $\alpha = F^{\frac{1}{4}}$, where F is the number of factorial points contained in the design.

3. Results

We present in this section some useful numerical evaluations for the considered response surface methodology designs.

3.1 Main Effects Model and the Placket-burman Design

Case 1: K=3, N=4

The 4-point Plackett-Burman design in three controllable variables is

$$\xi = \begin{pmatrix} -1 & 1 & 1 \\ 1 & -1 & 1 \\ 1 & 1 & -1 \\ -1 & -1 & -1 \end{pmatrix}$$

Using the design and the model in (1) yields the normalized information matrix

$$M= \begin{pmatrix} 1.0000 & 0 & 0 & 0 \\ 0 & 1.0000 & 0 & 0 \\ 0 & 0 & 1.0000 & 0 \\ 0 & 0 & 0 & 1.0000 \end{pmatrix}$$

For the purpose of evaluating the optimality measures the following computations are made.

The Eigen values of M are

1.0000

1.0000

1.0000

1.0000

The determinant value and the trace of M are respectively 1.0000 and 4.0000.

The variance of prediction at each corresponding design point in ξ is respectively

4.0000

4.0000

4.0000

4.0000

$$M^{-1}= \begin{pmatrix} 1.0000 & 0 & 0 & 0 \\ 0 & 1.0000 & 0 & 0 \\ 0 & 0 & 1.0000 & 0 \\ 0 & 0 & 0 & 1.0000 \end{pmatrix}$$

The Eigen values of M^{-1} are

1.0000

1.0000

1.0000

1.0000

The determinant value and the trace of M^{-1} are respectively 1.0000 and 4.0000.

Case 2: K=7, N=8

$$\xi = \begin{pmatrix} 1 & 1 & 1 & -1 & 1 & -1 & -1 \\ -1 & 1 & 1 & 1 & -1 & 1 & -1 \\ -1 & -1 & 1 & 1 & 1 & -1 & 1 \\ 1 & -1 & -1 & 1 & 1 & 1 & -1 \\ -1 & 1 & -1 & -1 & 1 & 1 & 1 \\ 1 & -1 & 1 & -1 & -1 & 1 & 1 \\ 1 & 1 & -1 & 1 & -1 & -1 & 1 \\ -1 & -1 & -1 & -1 & -1 & -1 & -1 \end{pmatrix}$$

Using the design and the model in (1) yields the normalized information matrix

$$\mathbf{M}= \begin{pmatrix} 1.0000 & 0 & 0 & 0 & 0 & 0 & 0 & 0 \\ 0 & 1.0000 & 0 & 0 & 0 & 0 & 0 & 0 \\ 0 & 0 & 1.0000 & 0 & 0 & 0 & 0 & 0 \\ 0 & 0 & 0 & 1.0000 & 0 & 0 & 0 & 0 \\ 0 & 0 & 0 & 0 & 1.0000 & 0 & 0 & 0 \\ 0 & 0 & 0 & 0 & 0 & 1.0000 & 0 & 0 \\ 0 & 0 & 0 & 0 & 0 & 0 & 1.0000 & 0 \\ 0 & 0 & 0 & 0 & 0 & 0 & 0 & 1.0000 \end{pmatrix}$$

The Eigen values of M are

1.0000

1.0000

1.0000

1.0000

1.0000

1.0000

1.0000

1.0000

The determinant value and the trace of M are respectively 1.0000 and 8.0000

The variance of prediction at each corresponding design point in ξ is respectively

8.0000

8.0000

8.0000

8.0000

8.0000

8.0000

8.0000

8.0000

$$
\mathbf{M}^{-1}=\begin{pmatrix}
1.0000 & 0 & 0 & 0 & 0 & 0 & 0 & 0 \\
0 & 1.0000 & 0 & 0 & 0 & 0 & 0 & 0 \\
0 & 0 & 1.0000 & 0 & 0 & 0 & 0 & 0 \\
0 & 0 & 0 & 1.0000 & 0 & 0 & 0 & 0 \\
0 & 0 & 0 & 0 & 1.0000 & 0 & 0 & 0 \\
0 & 0 & 0 & 0 & 0 & 1.0000 & 0 & 0 \\
0 & 0 & 0 & 0 & 0 & 0 & 1.0000 & 0 \\
0 & 0 & 0 & 0 & 0 & 0 & 0 & 1.0000
\end{pmatrix}
$$

The Eigen values of M^{-1} are

1.0000

1.0000

1.0000

1.0000

1.0000

1.0000

1.0000

1.0000

The determinant value and the trace of M^{-1} are respectively 1.0000 and 8.0000.

Case 3: K=11, N=12

The 12-point Plackett-Burman design in eleven controllable variables is

$$
\xi=\begin{pmatrix}
1 & 1 & -1 & 1 & 1 & 1 & -1 & -1 & -1 & 1 & -1 \\
-1 & 1 & 1 & -1 & 1 & 1 & 1 & -1 & -1 & -1 & 1 \\
1 & -1 & 1 & 1 & -1 & 1 & 1 & 1 & -1 & -1 & -1 \\
-1 & 1 & -1 & 1 & 1 & -1 & 1 & 1 & 1 & -1 & -1 \\
-1 & -1 & 1 & -1 & 1 & 1 & -1 & 1 & 1 & 1 & -1 \\
-1 & -1 & -1 & 1 & -1 & 1 & 1 & -1 & 1 & 1 & 1 \\
1 & -1 & -1 & -1 & 1 & -1 & 1 & 1 & -1 & 1 & 1 \\
1 & 1 & -1 & -1 & -1 & 1 & -1 & 1 & 1 & -1 & 1 \\
1 & 1 & 1 & -1 & -1 & -1 & 1 & -1 & 1 & 1 & -1 \\
-1 & 1 & 1 & 1 & -1 & -1 & -1 & 1 & -1 & 1 & 1 \\
1 & -1 & 1 & 1 & 1 & -1 & -1 & -1 & 1 & -1 & 1 \\
-1 & -1 & -1 & -1 & -1 & -1 & -1 & -1 & -1 & -1 & -1
\end{pmatrix}
$$

Using the design and the model in (1) yields the normalized information matrix

$$
\mathbf{M}=\begin{pmatrix}
1.0000 & 0 & 0 & 0 & 0 & 0 & 0 & 0 & 0 & 0 & 0 & 0 \\
0 & 1.0000 & 0 & 0 & 0 & 0 & 0 & 0 & 0 & 0 & 0 & 0 \\
0 & 0 & 1.0000 & 0 & 0 & 0 & 0 & 0 & 0 & 0 & 0 & 0 \\
0 & 0 & 0 & 1.0000 & 0 & 0 & 0 & 0 & 0 & 0 & 0 & 0 \\
0 & 0 & 0 & 0 & 1.0000 & 0 & 0 & 0 & 0 & 0 & 0 & 0 \\
0 & 0 & 0 & 0 & 0 & 1.0000 & 0 & 0 & 0 & 0 & 0 & 0 \\
0 & 0 & 0 & 0 & 0 & 0 & 1.0000 & 0 & 0 & 0 & 0 & 0 \\
0 & 0 & 0 & 0 & 0 & 0 & 0 & 1.0000 & 0 & 0 & 0 & 0 \\
0 & 0 & 0 & 0 & 0 & 0 & 0 & 0 & 1.0000 & 0 & 0 & 0 \\
0 & 0 & 0 & 0 & 0 & 0 & 0 & 0 & 0 & 1.0000 & 0 & 0 \\
0 & 0 & 0 & 0 & 0 & 0 & 0 & 0 & 0 & 0 & 1.0000 & 0 \\
0 & 0 & 0 & 0 & 0 & 0 & 0 & 0 & 0 & 0 & 0 & 1.0000
\end{pmatrix}
$$

The Eigen values of M are 1.0000

1.0000

1.0000

1.0000

1.0000

1.0000

1.0000

1.0000

1.0000

1.0000

1.0000

1.0000

The determinant value and the trace of M are respectively 1.0000 and 12.0000.

The variance of prediction at each corresponding design point in ξ is respectively

12

12

12

12

12

12

12

12

12

12

12

12

$$M^{-1}=\begin{pmatrix}
1.0000 & 0 & 0 & 0 & 0 & 0 & 0 & 0 & 0 & 0 & 0 & 0 \\
0 & 1.0000 & 0 & 0 & 0 & 0 & 0 & 0 & 0 & 0 & 0 & 0 \\
0 & 0 & 1.0000 & 0 & 0 & 0 & 0 & 0 & 0 & 0 & 0 & 0 \\
0 & 0 & 0 & 1.0000 & 0 & 0 & 0 & 0 & 0 & 0 & 0 & 0 \\
0 & 0 & 0 & 0 & 1.0000 & 0 & 0 & 0 & 0 & 0 & 0 & 0 \\
0 & 0 & 0 & 0 & 0 & 1.0000 & 0 & 0 & 0 & 0 & 0 & 0 \\
0 & 0 & 0 & 0 & 0 & 0 & 1.0000 & 0 & 0 & 0 & 0 & 0 \\
0 & 0 & 0 & 0 & 0 & 0 & 0 & 1.0000 & 0 & 0 & 0 & 0 \\
0 & 0 & 0 & 0 & 0 & 0 & 0 & 0 & 1.0000 & 0 & 0 & 0 \\
0 & 0 & 0 & 0 & 0 & 0 & 0 & 0 & 0 & 1.0000 & 0 & 0 \\
0 & 0 & 0 & 0 & 0 & 0 & 0 & 0 & 0 & 0 & 1.0000 & 0 \\
0 & 0 & 0 & 0 & 0 & 0 & 0 & 0 & 0 & 0 & 0 & 1.0000
\end{pmatrix}$$

The Eigen values of M^{-1} are

1.0000

1.0000

1.0000

1.0000

1.0000

1.0000

1.0000

1.0000

1.0000

1.0000

1.0000

1.0000

The determinant value and the trace of M^{-1} are respectively 1.0000 and 12.0000.

Case 4: K=15, N=16

The 16-point Plackett-Burman design in fifteen controllable variables is

$$\xi = \begin{pmatrix}
1 & 1 & 1 & 1 & 1 & -1 & 1 & -1 & 1 & 1 & -1 & -1 & 1 & -1 & -1 & -1 \\
1 & -1 & 1 & 1 & 1 & 1 & -1 & 1 & -1 & 1 & 1 & -1 & -1 & 1 & -1 & -1 \\
1 & -1 & -1 & 1 & 1 & 1 & 1 & -1 & 1 & -1 & 1 & 1 & -1 & -1 & 1 & -1 \\
1 & -1 & -1 & -1 & 1 & 1 & 1 & 1 & -1 & 1 & -1 & 1 & 1 & -1 & -1 & 1 \\
1 & 1 & -1 & -1 & -1 & 1 & 1 & 1 & 1 & -1 & 1 & -1 & 1 & 1 & -1 & -1 \\
1 & -1 & 1 & -1 & -1 & -1 & 1 & 1 & 1 & 1 & -1 & 1 & -1 & 1 & 1 & -1 \\
1 & -1 & -1 & 1 & -1 & -1 & -1 & 1 & 1 & 1 & 1 & -1 & 1 & -1 & 1 & 1 \\
1 & 1 & -1 & -1 & 1 & -1 & -1 & -1 & 1 & 1 & 1 & 1 & -1 & 1 & -1 & 1 \\
1 & 1 & 1 & -1 & -1 & 1 & -1 & -1 & -1 & 1 & 1 & 1 & 1 & -1 & 1 & -1 \\
1 & -1 & 1 & 1 & -1 & -1 & 1 & -1 & -1 & -1 & 1 & 1 & 1 & 1 & -1 & 1 \\
1 & 1 & -1 & 1 & 1 & -1 & -1 & 1 & -1 & -1 & -1 & 1 & 1 & 1 & 1 & -1 \\
1 & -1 & 1 & -1 & 1 & 1 & -1 & -1 & 1 & -1 & -1 & -1 & 1 & 1 & 1 & 1 \\
1 & 1 & -1 & 1 & -1 & 1 & 1 & -1 & -1 & 1 & -1 & -1 & -1 & 1 & 1 & 1 \\
1 & 1 & 1 & -1 & 1 & -1 & 1 & 1 & -1 & -1 & 1 & -1 & -1 & -1 & 1 & 1 \\
1 & 1 & 1 & 1 & -1 & 1 & -1 & 1 & 1 & -1 & -1 & 1 & -1 & -1 & -1 & 1 \\
1 & -1 & -1 & -1 & -1 & -1 & -1 & -1 & -1 & -1 & -1 & -1 & -1 & -1 & -1 & -1
\end{pmatrix}$$

Using the design and the model in (1) yields the normalized information matrix

$$M = \begin{pmatrix}
1.0000 & 0 & 0 & 0 & 0 & 0 & 0 & 0 & 0 & 0 & 0 & 0 & 0 & 0 & 0 & 0 \\
0 & 1.0000 & 0 & 0 & 0 & 0 & 0 & 0 & 0 & 0 & 0 & 0 & 0 & 0 & 0 & 0 \\
0 & 0 & 1.0000 & 0 & 0 & 0 & 0 & 0 & 0 & 0 & 0 & 0 & 0 & 0 & 0 & 0 \\
0 & 0 & 0 & 1.0000 & 0 & 0 & 0 & 0 & 0 & 0 & 0 & 0 & 0 & 0 & 0 & 0 \\
0 & 0 & 0 & 0 & 1.0000 & 0 & 0 & 0 & 0 & 0 & 0 & 0 & 0 & 0 & 0 & 0 \\
0 & 0 & 0 & 0 & 0 & 1.0000 & 0 & 0 & 0 & 0 & 0 & 0 & 0 & 0 & 0 & 0 \\
0 & 0 & 0 & 0 & 0 & 0 & 1.0000 & 0 & 0 & 0 & 0 & 0 & 0 & 0 & 0 & 0 \\
0 & 0 & 0 & 0 & 0 & 0 & 0 & 1.0000 & 0 & 0 & 0 & 0 & 0 & 0 & 0 & 0 \\
0 & 0 & 0 & 0 & 0 & 0 & 0 & 0 & 1.0000 & 0 & 0 & 0 & 0 & 0 & 0 & 0 \\
0 & 0 & 0 & 0 & 0 & 0 & 0 & 0 & 0 & 1.0000 & 0 & 0 & 0 & 0 & 0 & 0 \\
0 & 0 & 0 & 0 & 0 & 0 & 0 & 0 & 0 & 0 & 1.0000 & 0 & 0 & 0 & 0 & 0 \\
0 & 0 & 0 & 0 & 0 & 0 & 0 & 0 & 0 & 0 & 0 & 1.0000 & 0 & 0 & 0 & 0 \\
0 & 0 & 0 & 0 & 0 & 0 & 0 & 0 & 0 & 0 & 0 & 0 & 1.0000 & 0 & 0 & 0 \\
0 & 0 & 0 & 0 & 0 & 0 & 0 & 0 & 0 & 0 & 0 & 0 & 0 & 1.0000 & 0 & 0 \\
0 & 0 & 0 & 0 & 0 & 0 & 0 & 0 & 0 & 0 & 0 & 0 & 0 & 0 & 1.0000 & 0 \\
0 & 0 & 0 & 0 & 0 & 0 & 0 & 0 & 0 & 0 & 0 & 0 & 0 & 0 & 0 & 1.0000
\end{pmatrix}$$

The Eigen values of M are

1.0000

1.0000

1.0000

1.0000

1.0000
1.0000
1.0000
1.0000
1.0000
1.0000
1.0000
1.0000
1.0000
1.0000
1.0000
1.0000

The determinant value and the trace of M are respectively 1.0000 and 16.0000

The variance of prediction at each corresponding design point in ξ is respectively

16
16
16
16
16
16
16
16
16
16
16
16
16
16
16
16

$$M^{-1} = \begin{pmatrix}
1.0000 & 0 & 0 & 0 & 0 & 0 & 0 & 0 & 0 & 0 & 0 & 0 & 0 & 0 & 0 & 0 \\
0 & 1.0000 & 0 & 0 & 0 & 0 & 0 & 0 & 0 & 0 & 0 & 0 & 0 & 0 & 0 & 0 \\
0 & 0 & 1.0000 & 0 & 0 & 0 & 0 & 0 & 0 & 0 & 0 & 0 & 0 & 0 & 0 & 0 \\
0 & 0 & 0 & 1.0000 & 0 & 0 & 0 & 0 & 0 & 0 & 0 & 0 & 0 & 0 & 0 & 0 \\
0 & 0 & 0 & 0 & 1.0000 & 0 & 0 & 0 & 0 & 0 & 0 & 0 & 0 & 0 & 0 & 0 \\
0 & 0 & 0 & 0 & 0 & 1.0000 & 0 & 0 & 0 & 0 & 0 & 0 & 0 & 0 & 0 & 0 \\
0 & 0 & 0 & 0 & 0 & 0 & 1.0000 & 0 & 0 & 0 & 0 & 0 & 0 & 0 & 0 & 0 \\
0 & 0 & 0 & 0 & 0 & 0 & 0 & 1.0000 & 0 & 0 & 0 & 0 & 0 & 0 & 0 & 0 \\
0 & 0 & 0 & 0 & 0 & 0 & 0 & 0 & 1.0000 & 0 & 0 & 0 & 0 & 0 & 0 & 0 \\
0 & 0 & 0 & 0 & 0 & 0 & 0 & 0 & 0 & 1.0000 & 0 & 0 & 0 & 0 & 0 & 0 \\
0 & 0 & 0 & 0 & 0 & 0 & 0 & 0 & 0 & 0 & 1.0000 & 0 & 0 & 0 & 0 & 0 \\
0 & 0 & 0 & 0 & 0 & 0 & 0 & 0 & 0 & 0 & 0 & 1.0000 & 0 & 0 & 0 & 0 \\
0 & 0 & 0 & 0 & 0 & 0 & 0 & 0 & 0 & 0 & 0 & 0 & 1.0000 & 0 & 0 & 0 \\
0 & 0 & 0 & 0 & 0 & 0 & 0 & 0 & 0 & 0 & 0 & 0 & 0 & 1.0000 & 0 & 0 \\
0 & 0 & 0 & 0 & 0 & 0 & 0 & 0 & 0 & 0 & 0 & 0 & 0 & 0 & 1.0000 & 0 \\
0 & 0 & 0 & 0 & 0 & 0 & 0 & 0 & 0 & 0 & 0 & 0 & 0 & 0 & 0 & 1.0000
\end{pmatrix}$$

The Eigen values of M^{-1} are

1.0000

1.0000

1.0000

1.0000

1.0000

1.0000

1.0000

1.0000

1.0000

1.0000

1.0000

1.0000

1.0000

1.0000

1.0000

1.0000

The determinant value and the trace of M^{-1} are respectively 1.0000 and 16.0000

3.2 Complete Model and the 2^k Factorial Design

Case 1: k=2, N=4

The 4-point full factorial design is

$$\xi = \begin{pmatrix} -1 & -1 \\ 1 & -1 \\ -1 & 1 \\ 1 & 1 \end{pmatrix}$$

Using the design and the model in (2) yields the normalized information matrix

$$\mathbf{M} = \begin{pmatrix} 1.0000 & 0 & 0 & 0 \\ 0 & 1.0000 & 0 & 0 \\ 0 & 0 & 1.0000 & 0 \\ 0 & 0 & 0 & 1.0000 \end{pmatrix}$$

The Eigen values of M are

1.0000

1.0000

1.0000

1.0000

The determinant value and the trace of M are respectively 1.0000 and 4.0000.

The variance of prediction at each corresponding design point in ξ is respectively

4.0000

4.0000

4.0000

4.0000

$$\mathbf{M}^{-1}=\begin{pmatrix} 1.0000 & 0 & 0 & 0 \\ 0 & 1.0000 & 0 & 0 \\ 0 & 0 & 1.0000 & 0 \\ 0 & 0 & 0 & 1.0000 \end{pmatrix}$$

The Eigen values of M^{-1} are

1.0000

1.0000

1.0000

1.0000

The determinant value and the trace of M^{-1} are respectively 1.0000 and 4.0000.

Case 2: k=3, N=8

The 8-point full factorial design is

$$\xi = \begin{pmatrix} -1 & -1 & -1 \\ 1 & -1 & -1 \\ -1 & 1 & -1 \\ 1 & 1 & -1 \\ -1 & -1 & 1 \\ 1 & -1 & 1 \\ -1 & 1 & 1 \\ 1 & 1 & 1 \end{pmatrix}$$

Using the design and the model in (2) yields the normalized information matrix

$$\mathbf{M}=\begin{pmatrix} 1.0000 & 0 & 0 & 0 & 0 & 0 & 0 & 0 \\ 0 & 1.0000 & 0 & 0 & 0 & 0 & 0 & 0 \\ 0 & 0 & 1.0000 & 0 & 0 & 0 & 0 & 0 \\ 0 & 0 & 0 & 1.0000 & 0 & 0 & 0 & 0 \\ 0 & 0 & 0 & 0 & 1.0000 & 0 & 0 & 0 \\ 0 & 0 & 0 & 0 & 0 & 1.0000 & 0 & 0 \\ 0 & 0 & 0 & 0 & 0 & 0 & 1.0000 & 0 \\ 0 & 0 & 0 & 0 & 0 & 0 & 0 & 1.0000 \end{pmatrix}$$

The Eigen values of M are

1.0000

1.0000

1.0000

1.0000

1.0000

1.0000

1.0000

1.0000

The determinant value and the trace of M are respectively 1.0000 and 8.0000.

The variance of prediction at each corresponding design point in ξ is respectively

8.0000

8.0000

8.0000

8.0000

8.0000

8.0000

8.0000

8.0000

$$\mathbf{M}^{-1}=\begin{pmatrix} 1.0000 & 0 & 0 & 0 & 0 & 0 & 0 & 0 \\ 0 & 1.0000 & 0 & 0 & 0 & 0 & 0 & 0 \\ 0 & 0 & 1.0000 & 0 & 0 & 0 & 0 & 0 \\ 0 & 0 & 0 & 1.0000 & 0 & 0 & 0 & 0 \\ 0 & 0 & 0 & 0 & 1.0000 & 0 & 0 & 0 \\ 0 & 0 & 0 & 0 & 0 & 1.0000 & 0 & 0 \\ 0 & 0 & 0 & 0 & 0 & 0 & 1.0000 & 0 \\ 0 & 0 & 0 & 0 & 0 & 0 & 0 & 1.0000 \end{pmatrix}$$

The Eigen values of M^{-1} are

1.0000

1.0000

1.0000

1.0000

1.0000

1.0000

1.0000

1.0000

The determinant value and the trace of M^{-1} are respectively 1.0000 and 8.0000.

Case 3: k=4, N=16

The 16-point full factorial design is

$$\xi=\begin{pmatrix} -1 & -1 & -1 & -1 \\ 1 & -1 & -1 & -1 \\ -1 & 1 & -1 & -1 \\ 1 & 1 & -1 & -1 \\ -1 & -1 & 1 & -1 \\ 1 & -1 & 1 & -1 \\ -1 & 1 & 1 & -1 \\ 1 & 1 & 1 & -1 \\ -1 & -1 & -1 & 1 \\ 1 & -1 & -1 & 1 \\ -1 & 1 & -1 & 1 \\ 1 & 1 & -1 & 1 \\ -1 & -1 & 1 & 1 \\ 1 & -1 & 1 & 1 \\ -1 & 1 & 1 & 1 \\ 1 & 1 & 1 & 1 \end{pmatrix}$$

Using the design and the model in (2) yields the normalized information matrix

M=

$$\begin{pmatrix}
1.0000 & 0 & 0 & 0 & 0 & 0 & 0 & 0 & 0 & 0 & 0 & 0 & 0 & 0 & 0 & 0 \\
0 & 1.0000 & 0 & 0 & 0 & 0 & 0 & 0 & 0 & 0 & 0 & 0 & 0 & 0 & 0 & 0 \\
0 & 0 & 1.0000 & 0 & 0 & 0 & 0 & 0 & 0 & 0 & 0 & 0 & 0 & 0 & 0 & 0 \\
0 & 0 & 0 & 1.0000 & 0 & 0 & 0 & 0 & 0 & 0 & 0 & 0 & 0 & 0 & 0 & 0 \\
0 & 0 & 0 & 0 & 1.0000 & 0 & 0 & 0 & 0 & 0 & 0 & 0 & 0 & 0 & 0 & 0 \\
0 & 0 & 0 & 0 & 0 & 1.0000 & 0 & 0 & 0 & 0 & 0 & 0 & 0 & 0 & 0 & 0 \\
0 & 0 & 0 & 0 & 0 & 0 & 1.0000 & 0 & 0 & 0 & 0 & 0 & 0 & 0 & 0 & 0 \\
0 & 0 & 0 & 0 & 0 & 0 & 0 & 1.0000 & 0 & 0 & 0 & 0 & 0 & 0 & 0 & 0 \\
0 & 0 & 0 & 0 & 0 & 0 & 0 & 0 & 1.0000 & 0 & 0 & 0 & 0 & 0 & 0 & 0 \\
0 & 0 & 0 & 0 & 0 & 0 & 0 & 0 & 0 & 1.0000 & 0 & 0 & 0 & 0 & 0 & 0 \\
0 & 0 & 0 & 0 & 0 & 0 & 0 & 0 & 0 & 0 & 1.0000 & 0 & 0 & 0 & 0 & 0 \\
0 & 0 & 0 & 0 & 0 & 0 & 0 & 0 & 0 & 0 & 0 & 1.0000 & 0 & 0 & 0 & 0 \\
0 & 0 & 0 & 0 & 0 & 0 & 0 & 0 & 0 & 0 & 0 & 0 & 1.0000 & 0 & 0 & 0 \\
0 & 0 & 0 & 0 & 0 & 0 & 0 & 0 & 0 & 0 & 0 & 0 & 0 & 1.0000 & 0 & 0 \\
0 & 0 & 0 & 0 & 0 & 0 & 0 & 0 & 0 & 0 & 0 & 0 & 0 & 0 & 1.0000 & 0 \\
0 & 0 & 0 & 0 & 0 & 0 & 0 & 0 & 0 & 0 & 0 & 0 & 0 & 0 & 0 & 1.0000
\end{pmatrix}$$

The Eigen values of M are

1.0000

1.0000

1.0000

1.0000

1.0000

1.0000

1.0000

1.0000

1.0000

1.0000

1.0000

1.0000

1.0000

1.0000

1.0000

1.0000

The determinant value and the trace of M are respectively 1.0000 and 16.0000.

The variance of prediction at each corresponding design point in ξ is respectively

16

16

16

16

16

16

16

16

16

16

16

16
16
16
16
16

$$M^{-1}= \begin{pmatrix}
1.0000 & 0 & 0 & 0 & 0 & 0 & 0 & 0 & 0 & 0 & 0 & 0 & 0 & 0 & 0 & 0 \\
0 & 1.0000 & 0 & 0 & 0 & 0 & 0 & 0 & 0 & 0 & 0 & 0 & 0 & 0 & 0 & 0 \\
0 & 0 & 1.0000 & 0 & 0 & 0 & 0 & 0 & 0 & 0 & 0 & 0 & 0 & 0 & 0 & 0 \\
0 & 0 & 0 & 1.0000 & 0 & 0 & 0 & 0 & 0 & 0 & 0 & 0 & 0 & 0 & 0 & 0 \\
0 & 0 & 0 & 0 & 1.0000 & 0 & 0 & 0 & 0 & 0 & 0 & 0 & 0 & 0 & 0 & 0 \\
0 & 0 & 0 & 0 & 0 & 1.0000 & 0 & 0 & 0 & 0 & 0 & 0 & 0 & 0 & 0 & 0 \\
0 & 0 & 0 & 0 & 0 & 0 & 1.0000 & 0 & 0 & 0 & 0 & 0 & 0 & 0 & 0 & 0 \\
0 & 0 & 0 & 0 & 0 & 0 & 0 & 1.0000 & 0 & 0 & 0 & 0 & 0 & 0 & 0 & 0 \\
0 & 0 & 0 & 0 & 0 & 0 & 0 & 0 & 1.0000 & 0 & 0 & 0 & 0 & 0 & 0 & 0 \\
0 & 0 & 0 & 0 & 0 & 0 & 0 & 0 & 0 & 1.0000 & 0 & 0 & 0 & 0 & 0 & 0 \\
0 & 0 & 0 & 0 & 0 & 0 & 0 & 0 & 0 & 0 & 1.0000 & 0 & 0 & 0 & 0 & 0 \\
0 & 0 & 0 & 0 & 0 & 0 & 0 & 0 & 0 & 0 & 0 & 1.0000 & 0 & 0 & 0 & 0 \\
0 & 0 & 0 & 0 & 0 & 0 & 0 & 0 & 0 & 0 & 0 & 0 & 1.0000 & 0 & 0 & 0 \\
0 & 0 & 0 & 0 & 0 & 0 & 0 & 0 & 0 & 0 & 0 & 0 & 0 & 1.0000 & 0 & 0 \\
0 & 0 & 0 & 0 & 0 & 0 & 0 & 0 & 0 & 0 & 0 & 0 & 0 & 0 & 1.0000 & 0 \\
0 & 0 & 0 & 0 & 0 & 0 & 0 & 0 & 0 & 0 & 0 & 0 & 0 & 0 & 0 & 1.0000
\end{pmatrix}$$

The Eigen values of M^{-1} are

1.0000
1.0000
1.0000
1.0000
1.0000
1.0000
1.0000
1.0000
1.0000
1.0000
1.0000
1.0000
1.0000
1.0000
1.0000
1.0000

The determinant value and the trace of M^{-1} are respectively 1.0000 and 16.0000.

3.3 Second Order Model and the Central Composite Design

Case 1: K=2, N=9; α=1

The 9-point central composite design is

$$\xi = \begin{pmatrix}
-1 & -1 \\
1 & -1 \\
-1 & 1 \\
1 & 1 \\
1 & 0 \\
-1 & 0 \\
0 & 0 \\
0 & -1 \\
0 & 0
\end{pmatrix}$$

Using the design and the model in (3) yields the normalized information matrix

$$M=\begin{pmatrix} 1.0000 & 0 & 0 & 0 & 0.6667 & 0.6667 \\ 0 & 0.6667 & 0 & 0 & 0 & 0 \\ 0 & 0 & 0.6667 & 0 & 0 & 0 \\ 0 & 0 & 0 & 0.4444 & 0 & 0 \\ 0.6667 & 0 & 0 & 0 & 0.6667 & 0.4444 \\ 0.6667 & 0 & 0 & 0 & 0.4444 & 0.6667 \end{pmatrix}$$

The Eigen values of M are

0.1111

0.2222

0.4444

0.6667

0.6667

2.0000

The determinant value and the trace of M are respectively 0.0098 and 4.1111.

The variance of prediction at each corresponding design point in ξ is respectively

7.2500

7.2500

7.2500

7.2500

5.0000

5.0000

5.0000

5.0000

$$M^{-1}=\begin{pmatrix} 5.0000 & 0 & 0 & 0 & -3.0000 & -3.0000 \\ 0 & 1.5000 & 0 & 0 & 0 & 0 \\ 0 & 0 & 1.5000 & 0 & 0 & 0 \\ 0 & 0 & 0 & 2.2500 & 0 & 0 \\ -3.0000 & 0 & 0 & 0 & 4.5000 & 0 \\ -3.0000 & 0 & 0 & 0 & 0 & 4.5000 \end{pmatrix}$$

The Eigen values of M^{-1} are

0.5000

1.5000

1.5000

2.2500

4.5000

9.0000

The determinant value and the trace of M^{-1} are respectively 102.5156 and 19.2500.

Case 2: K=2, N=9; α=1.414

The 9-point central composite design is

$$\xi = \begin{pmatrix} -1 & -1 \\ 1 & -1 \\ -1 & 1 \\ 1 & 1 \\ 1.414 & 0 \\ -1.414 & 0 \\ 0 & 1.414 \\ 0 & -1.414 \\ 0 & 0 \end{pmatrix}$$

Using the design and the model in (3) yields the normalized information matrix

$$\mathbf{M} = \begin{pmatrix} 1.0000 & 0 & 0 & 0 & 0.8888 & 0.8888 \\ 0 & 0.8888 & 0 & 0 & 0 & 0 \\ 0 & 0 & 0.8888 & 0 & 0 & 0 \\ 0 & 0 & 0 & 0.4444 & 0 & 0 \\ 0.8888 & 0 & 0 & 0 & 1.3328 & 0.4444 \\ 0.8888 & 0 & 0 & 0 & 0.4444 & 1.3328 \end{pmatrix}$$

The Eigen values of M are

0.0730

0.4444

0.8884

0.8888

0.8888

2.7042

The determinant value and the trace of M are respectively 0.0616 and 5.8875.

The variance of prediction at each corresponding design point in ξ is respectively

5.6257

5.6257

5.6257

5.6257

5.6243

5.6243

5.6243

5.6243

9.0000

$$\mathbf{M}^{-1} = \begin{pmatrix} 9.0000 & 0 & 0 & 0 & -4.5007 & -4.5007 \\ 0 & 1.1252 & 0 & 0 & 0 & 0 \\ 0 & 0 & 1.1252 & 0 & 0 & 0 \\ 0 & 0 & 0 & 2.2500 & 0 & 0 \\ -4.5007 & 0 & 0 & 0 & 3.0949 & 1.9692 \\ -4.5007 & 0 & 0 & 0 & 1.9692 & 3.0949 \end{pmatrix}$$

The Eigen values of M^{-1} are

13.6942

0.3698

1.1257

1.1252

1.1252

2.2500

The determinant value and the trace of M^{-1} are respectively 16.2379 and 19.6902.

Case 3: K=3, N=15; α=1

The 15-point central composite design is

$$\xi = \begin{pmatrix} -1 & -1 & -1 \\ 1 & -1 & -1 \\ -1 & 1 & -1 \\ 1 & 1 & -1 \\ -1 & -1 & 1 \\ 1 & -1 & 1 \\ -1 & 1 & 1 \\ 1 & 1 & 1 \\ 1 & 0 & 0 \\ -1 & 0 & 0 \\ 0 & 1 & 0 \\ 0 & -1 & 0 \\ 0 & 0 & 1 \\ 0 & 0 & -1 \\ 0 & 0 & 0 \end{pmatrix}$$

Using the design and the model in (3) yields the normalized information matrix

$$M = \begin{pmatrix} 1.0000 & 0 & 0 & 0 & 0 & 0 & 0 & 0.6667 & 0.6667 & 0.6667 \\ 0 & 0.6667 & 0 & 0 & 0 & 0 & 0 & 0 & 0 & 0 \\ 0 & 0 & 0.6667 & 0 & 0 & 0 & 0 & 0 & 0 & 0 \\ 0 & 0 & 0 & 0.6667 & 0 & 0 & 0 & 0 & 0 & 0 \\ 0 & 0 & 0 & 0 & 0.5333 & 0 & 0 & 0 & 0 & 0 \\ 0 & 0 & 0 & 0 & 0 & 0.5333 & 0 & 0 & 0 & 0 \\ 0 & 0 & 0 & 0 & 0 & 0 & 0.5333 & 0 & 0 & 0 \\ 0.6667 & 0 & 0 & 0 & 0 & 0 & 0 & 0.6667 & 0.5333 & 0.5333 \\ 0.6667 & 0 & 0 & 0 & 0 & 0 & 0 & 0.5333 & 0.6667 & 0.5333 \\ 0.6667 & 0 & 0 & 0 & 0 & 0 & 0 & 0.5333 & 0.5333 & 0.6667 \end{pmatrix}$$

The Eigen values of M are

0.1333

0.1333

0.1551

0.5333

0.5333

0.5333

0.6667

0.6667

0.6667

2.5782

The determinant value and the trace of M are respectively 3.1964e-004 and 6.6000.

The variance of prediction at each corresponding design point in ξ is respectively

11.9583

11.9583

11.9583

11.9583

11.9583

11.9583

11.9583

11.9583

8.3333

8.3333

8.3333

8.3333

8.3333

8.3333

4.3333

M^{-1} =

4.3333	0	0	0	0	0	0	-1.6667	-1.6667	-1.6667
0	1.5000	0	0	0	0	0	0	0	0
0	0	1.5000	0	0	0	0	0	0	0
0	0	0	1.5000	0	0	0	0	0	0
0	0	0	0	1.8750	0	0	0	0	0
0	0	0	0	0	1.8750	0	0	0	0
0	0	0	0	0	0	1.8750	0	0	0
-1.6667	0	0	0	0	0	0	5.8333	-1.6667	-1.6667
-1.6667	0	0	0	0	0	0	-1.6667	5.8333	-1.6667
-1.6667	0	0	0	0	0	0	-1.6667	-1.6667	5.8333

The Eigen values of M^{-1} are

6.4455

0.3879

7.5000

7.5000

1.5000

1.5000

1.5000

1.8750

1.8750

1.8750

The determinant value and the trace of M^{-1} are respectively 3.1285e+003 and 31.9583.

Case 4: N=15 P=10, 1.682=α, K=3

The 15-point central composite design is

$$\xi = \begin{pmatrix} -1 & -1 & -1 \\ 1 & -1 & -1 \\ -1 & 1 & -1 \\ 1 & 1 & -1 \\ -1 & -1 & 1 \\ 1 & -1 & 1 \\ -1 & 1 & 1 \\ 1 & 1 & 1 \\ 1.682 & 0 & 0 \\ -1.682 & 0 & 0 \\ 0 & 1.682 & 0 \\ 0 & -1.682 & 0 \\ 0 & 0 & 1.682 \\ 0 & 0 & -1.682 \\ 0 & 0 & 0 \end{pmatrix}$$

Using the design and the model in (3) yields the normalized information matrix

$$\mathbf{M} = \begin{pmatrix} 1.0000 & 0 & 0 & 0 & 0 & 0 & 0 & 0.9105 & 0.9105 & 0.9105 \\ 0 & 0.9105 & 0 & 0 & 0 & 0 & 0 & 0 & 0 & 0 \\ 0 & 0 & 0.9105 & 0 & 0 & 0 & 0 & 0 & 0 & 0 \\ 0 & 0 & 0 & 0.9105 & 0 & 0 & 0 & 0 & 0 & 0.0000 \\ 0 & 0 & 0 & 0 & 0.5333 & 0 & 0 & 0 & 0 & 0 \\ 0 & 0 & 0 & 0 & 0 & 0.5333 & 0 & 0 & 0 & 0 \\ 0 & 0 & 0 & 0 & 0 & 0 & 0.5333 & 0 & 0 & 0 \\ 0.9105 & 0 & 0 & 0 & 0 & 0 & 0 & 1.6005 & 0.5333 & 0.5333 \\ 0.9105 & 0 & 0 & 0 & 0 & 0 & 0 & 0.5333 & 1.6005 & 0.5333 \\ 0.9105 & 0 & 0 & 0 & 0 & 0 & 0 & 0.5333 & 0.5333 & 1.6005 \end{pmatrix}$$

The Eigen values of M are

0.0497

0.5333

0.5333

0.5333

0.9105

0.9105

0.9105

1.0672

1.0672

3.6175

The determinant value and the trace of M are respectively 0.0235 and 10.1332.

The variance of prediction at each corresponding design point in ξ is respectively

10.0532

10.0532

10.0532

10.0532

10.0532

10.0532

10.0532

10.0532

9.1247

9.1247

9.1247

9.1247

9.1247

9.1247

14.8269

$$M^{-1} = \begin{pmatrix} 14.8268 & 0 & 0 & 0 & 0 & 0 & 0 & -5.0617 & -5.0617 & -5.0617 \\ 0 & 1.0982 & 0 & 0 & 0 & 0 & 0 & 0 & 0 & 0 \\ 0 & 0 & 1.0982 & 0 & 0 & 0 & 0 & 0 & 0 & 0 \\ 0 & 0 & 0 & 1.0982 & 0 & 0 & 0 & 0 & 0 & 0 \\ 0 & 0 & 0 & 0 & 1.8750 & 0 & 0 & 0 & 0 & 0 \\ 0 & 0 & 0 & 0 & 0 & 1.8750 & 0 & 0 & 0 & 0 \\ 0 & 0 & 0 & 0 & 0 & 0 & 1.8750 & 0 & 0 & 0 \\ -5.0617 & 0 & 0 & 0 & 0 & 0 & 0 & 2.4777 & 1.5406 & 1.5406 \\ -5.0617 & 0 & 0 & 0 & 0 & 0 & 0 & 1.5406 & 1.5406 & 1.5406 \\ -5.0617 & 0 & 0 & 0 & 0 & 0 & 0 & 1.5406 & 1.5406 & 2.4777 \end{pmatrix}$$

The Eigen values of M^{-1} are

20.1094

0.2764

1.0982

0.9370

0.9370

1.0982

1.0982

1.8750

1.8750

1.8750

The determinant value and the trace of M^{-1} are respectively 42.6197 and 31.1796.

Case 5 :N=15, P=10, 1.7321=α, K=3

The 15-point central composite design is

$$\xi = \begin{pmatrix} -1 & -1 & -1 \\ 1 & -1 & -1 \\ -1 & 1 & -1 \\ 1 & 1 & -1 \\ -1 & -1 & 1 \\ 1 & -1 & 1 \\ -1 & 1 & 1 \\ 1 & 1 & 1 \\ 1.7321 & 0 & 0 \\ -1.7321 & 0 & 0 \\ 0 & 1.7321 & 0 \\ 0 & -1.7321 & 0 \\ 0 & 0 & 1.7321 \\ 0 & 0 & -1.7321 \\ 0 & 0 & 0 \end{pmatrix}$$

Using the design and the model in (3) yields the normalized information matrix

$$M = \begin{pmatrix} 1.0000 & 0 & 0 & 0 & 0 & 0 & 0 & 0.9334 & 0.9334 & 0.9334 \\ 0 & 0.9334 & 0 & 0 & 0 & 0 & 0 & 0 & 0 & 0 \\ 0 & 0 & 0.9334 & 0 & 0 & 0 & 0 & 0 & 0 & 0 \\ 0 & 0 & 0 & 0.9334 & 0 & 0 & 0 & 0 & 0 & -0.0000 \\ 0 & 0 & 0 & 0 & 0.5333 & 0 & 0 & 0 & 0 & 0 \\ 0 & 0 & 0 & 0 & 0 & 0.5333 & 0 & 0 & 0 & 0 \\ 0 & 0 & 0 & 0 & 0 & 0 & 0.5333 & 0 & 0 & 0 \\ 0.9334 & 0 & 0 & 0 & 0 & 0 & 0 & 1.7335 & 0.5333 & 0.5333 \\ 0.9334 & 0 & 0 & 0 & 0 & 0 & 0 & 0.5333 & 1.7335 & 0.5333 \\ 0.9334 & 0 & 0 & 0 & 0 & 0 & 0 & 0.5333 & 0.5333 & 1.7335 \end{pmatrix}$$

The Eigen values of M are

0.0498

0.5333

0.5333

0.5333

0.9334

0.9334

0.9334

1.2001

1.2001

3.7504

The determinant value and the trace of M are respectively 0.0332 and 10.6005.

The variance of prediction at each corresponding design point in ξ is respectively

9.9106

9.9106

9.9106

9.9106

9.9106

9.9106

9.9106

9.9106

9.2859

9.2859

9.2859

9.2859

9.2859

9.2859

15.000

$$M^{-1} = \begin{pmatrix} 15.0000 & 0 & 0 & 0 & 0 & 0 & 0 & -4.9999 & -4.9999 & -4.9999 \\ 0 & 1.0714 & 0 & 0 & 0 & 0 & 0 & 0 & 0 & 0 \\ 0 & 0 & 1.0714 & 0 & 0 & 0 & 0 & 0 & 0 & 0 \\ 0 & 0 & 0 & 1.0714 & 0 & 0 & 0 & 0 & 0 & 0 \\ 0 & 0 & 0 & 0 & 1.8750 & 0 & 0 & 0 & 0 & 0 \\ 0 & 0 & 0 & 0 & 0 & 1.8750 & 0 & 0 & 0 & 0 \\ 0 & 0 & 0 & 0 & 0 & 0 & 1.8750 & 0 & 0 & 0 \\ -4.9999 & 0 & 0 & 0 & 0 & 0 & 0 & 2.3411 & 1.5079 & 1.5079 \\ -4.9999 & 0 & 0 & 0 & 0 & 0 & 0 & 1.5079 & 2.3411 & 1.5079 \\ -4.9999 & 0 & 0 & 0 & 0 & 0 & 0 & 1.5079 & 1.5079 & 2.3411 \end{pmatrix}$$

The Eigen values of M^{-1} are

20.0902

0.2666

1.0714

0.8332

0.8332

1.0714

1.8750

1.8750

1.8750

The determinant value and the trace of M^{-1} are respectively 30.1510 and 30.8625.

Case 6:K=4 , I.0 P=15, N=25

The 15-point central composite design is

$$\xi = \begin{pmatrix} -1 & -1 & -1 & -1 \\ 1 & -1 & -1 & -1 \\ -1 & 1 & -1 & -1 \\ 1 & 1 & -1 & -1 \\ -1 & -1 & 1 & -1 \\ 1 & -1 & 1 & -1 \\ -1 & 1 & 1 & -1 \\ 1 & 1 & 1 & -1 \\ -1 & -1 & -1 & 1 \\ 1 & -1 & -1 & 1 \\ -1 & 1 & -1 & 1 \\ 1 & 1 & -1 & 1 \\ -1 & -1 & 1 & 1 \\ 1 & -1 & 1 & 1 \\ -1 & 1 & 1 & 1 \\ 1 & 1 & 1 & 1 \\ 1 & 0 & 0 & 0 \\ -1 & 0 & 0 & 0 \\ 0 & 1 & 0 & 0 \\ 0 & -1 & 0 & 0 \\ 0 & 0 & 1 & 0 \\ 0 & 0 & -1 & 0 \\ 0 & 0 & 0 & 1 \\ 0 & 0 & 0 & -1 \\ 0 & 0 & 0 & 0 \end{pmatrix}$$

Using the design and the model in (3) yields the normalized information matrix

$$M = \begin{pmatrix}
1.0000 & 0 & 0 & 0 & 0 & 0 & 0 & 0 & 0 & 0 & 0 & 0.7200 & 0.7200 & 0.7200 & 0.7200 \\
0 & 0.7200 & 0 & 0 & 0 & 0 & 0 & 0 & 0 & 0 & 0 & 0 & 0 & 0 & 0 \\
0 & 0 & 0.7200 & 0 & 0 & 0 & 0 & 0 & 0 & 0 & 0 & 0 & 0 & 0 & 0 \\
0 & 0 & 0 & 0.7200 & 0 & 0 & 0 & 0 & 0 & 0 & 0 & 0 & 0 & 0 & 0 \\
0 & 0 & 0 & 0 & 0.7200 & 0 & 0 & 0 & 0 & 0 & 0 & 0 & 0 & 0 & 0 \\
0 & 0 & 0 & 0 & 0 & 0.6400 & 0 & 0 & 0 & 0 & 0 & 0 & 0 & 0 & 0 \\
0 & 0 & 0 & 0 & 0 & 0 & 0.6400 & 0 & 0 & 0 & 0 & 0 & 0 & 0 & 0 \\
0 & 0 & 0 & 0 & 0 & 0 & 0 & 0.6400 & 0 & 0 & 0 & 0 & 0 & 0 & 0 \\
0 & 0 & 0 & 0 & 0 & 0 & 0 & 0 & 0.6400 & 0 & 0 & 0 & 0 & 0 & 0 \\
0 & 0 & 0 & 0 & 0 & 0 & 0 & 0 & 0 & 0.6400 & 0 & 0 & 0 & 0 & 0 \\
0 & 0 & 0 & 0 & 0 & 0 & 0 & 0 & 0 & 0 & 0.6400 & 0 & 0 & 0 & 0 \\
0.7200 & 0 & 0 & 0 & 0 & 0 & 0 & 0 & 0 & 0 & 0 & 0.7200 & 0.6400 & 0.6400 & 0.6400 \\
0.7200 & 0 & 0 & 0 & 0 & 0 & 0 & 0 & 0 & 0 & 0 & 0.6400 & 0.7200 & 0.6400 & 0.6400 \\
0.7200 & 0 & 0 & 0 & 0 & 0 & 0 & 0 & 0 & 0 & 0 & 0.6400 & 0.6400 & 0.7200 & 0.6400 \\
0.7200 & 0 & 0 & 0 & 0 & 0 & 0 & 0 & 0 & 0 & 0 & 0.6400 & 0.6400 & 0.6400 & 0.7200
\end{pmatrix}$$

The Eigen values of M are

0.0800

0.0800

0.0800

0.1629

0.6400

0.6400

0.6400

0.6400

0.6400

0.6400

0.7200

0.7200

0.7200

0.7200

3.4771

The determinant value and the trace of M are respectively 5.3555×10^{-6} and 10.6000.

The variance of prediction at each corresponding design point in ξ is respectively

16.4842

16.4842

16.4842

16.4842

16.4842

16.4842

16.4842

16.4842

16.4842

16.4842

16.4842

16.4842

16.4842

16.4842

16.4842

16.4842

17.4905

17.4905

17.4905

17.4905

17.4905

17.4905

17.4905

17.4905

4.6610

$M^{-1} =$

4.6610	0	0	0	0	0	0	0	0	0	0	-1.2712	-1.2712	-1.2712	-1.2712
0	1.3889	0	0	0	0	0	0	0	0	0	0	0	0	0
0	0	1.3889	0	0	0	0	0	0	0	0	0	0	0	0
0	0	0	1.3889	0	0	0	0	0	0	0	0	0	0	0
0	0	0	0	1.3889	0	0	0	0	0	0	0	0	0	0
0	0	0	0	0	1.5625	0	0	0	0	0	0	0	0	0
0	0	0	0	0	0	1.5625	0	0	0	0	0	0	0	0
0	0	0	0	0	0	0	1.5625	0	0	0	0	0	0	0
0	0	0	0	0	0	0	0	1.5625	0	0	0	0	0	0
0	0	0	0	0	0	0	0	0	1.5625	0	0	0	0	0
0	0	0	0	0	0	0	0	0	0	1.5625	0	0	0	0
-1.2712	0	0	0	0	0	0	0	0	0	0	9.8164	-2.6836	-2.6836	-2.6836
-1.2712	0	0	0	0	0	0	0	0	0	0	-2.6836	9.8164	-2.6836	-2.6836
-1.2712	0	0	0	0	0	0	0	0	0	0	-2.6836	-2.6836	9.8164	-2.6836
-1.2712	0	0	0	0	0	0	0	0	0	0	-2.6836	-2.6836	-2.6836	9.8164

The Eigen values of M^{-1} are

0.2876

6.1390

12.500

12.500

12.500

1.3889

1.3889

1.3889

1.3889

1.5625

1.5625

1.5625

1.5625

1.5625

1.5625

The determinant value and the trace of M^{-1} are respectively 1.8672e+005 and 58.8571

Case 7:K=4, 2, P=15, N=25

The 9-point central composite design is

$\xi =$

-1	-1	-1	-1
1	-1	-1	-1
-1	1	-1	-1
1	1	-1	-1
-1	-1	1	-1
1	-1	1	-1
-1	1	1	-1
1	1	1	-1
-1	-1	-1	1
1	-1	-1	1
-1	1	-1	1
1	1	-1	1
-1	-1	1	1
1	-1	1	1
-1	1	1	1
1	1	1	1
2	0	0	0
-2	0	0	0
0	2	0	0
0	-2	0	0
0	0	2	0
0	0	-2	0
0	0	0	2
0	0	0	-2
0	0	0	0

Using the design and the model in (3) yields the normalized information matrix

M =

1.0000	0	0	0	0	0	0	0	0	0	0	0.9600	0.9600	0.9600	0.9600
0	0.9600	0	0	0	0	0	0	0	0	0	0	0	0	0
0	0	0.9600	0	0	0	0	0	0	0	0	0	0	0	0
0	0	0	0.9600	0	0	0	0	0	0	0	0	0	0	0
0	0	0	0	0.9600	0	0	0	0	0	0	0	0	0	0
0	0	0	0	0	0.6400	0	0	0	0	0	0	0	0	0
0	0	0	0	0	0	0.6400	0	0	0	0	0	0	0	0
0	0	0	0	0	0	0	0.6400	0	0	0	0	0	0	0
0	0	0	0	0	0	0	0	0.6400	0	0	0	0	0	0
0	0	0	0	0	0	0	0	0	0.6400	0	0	0	0	0
0	0	0	0	0	0	0	0	0	0	0.6400	0	0	0	0
0.9600	0	0	0	0	0	0	0	0	0	0	1.9200	0.6400	0.6400	0.6400
0.9600	0	0	0	0	0	0	0	0	0	0	0.6400	1.9200	0.6400	0.6400
0.9600	0	0	0	0	0	0	0	0	0	0	0.6400	0.6400	1.9200	0.6400
0.9600	0	0	0	0	0	0	0	0	0	0	0.6400	0.6400	0.6400	1.9200

The Eigen values of M are

0.0319

0.6400

0.6400

0.6400

0.6400

0.6400
0.6400
0.9600
0.9600
0.9600
0.9600
1.2800
1.2800
1.2800
4.8081

The determinant value and the trace of M are respectively 0.0188 and 16.3600.

The variance of prediction at each corresponding design point in ξ is respectively

14.5833
14.5833
14.5833
14.5833
14.5833
14.5833
14.5833
14.5833
14.5833
14.5833
14.5833
14.5833
14.5833
14.5833
14.5833
14.5833
18.8802
18.8802
18.8802
18.8802
18.8802
18.8802
18.8802
18.8802
25.0000

$$M^{-1} = \begin{pmatrix}
25.0000 & 0 & 0 & 0 & 0 & 0 & 0 & 0 & 0 & 0 & 0 & -6.2500 & -6.2500 & -6.2500 & -6.2500 \\
0 & 1.0417 & 0 & 0 & 0 & 0 & 0 & 0 & 0 & 0 & 0 & 0 & 0 & 0 & 0 \\
0 & 0 & 1.0417 & 0 & 0 & 0 & 0 & 0 & 0 & 0 & 0 & 0 & 0 & 0 & 0 \\
0 & 0 & 0 & 1.0417 & 0 & 0 & 0 & 0 & 0 & 0 & 0 & 0 & 0 & 0 & 0 \\
0 & 0 & 0 & 0 & 1.0417 & 0 & 0 & 0 & 0 & 0 & 0 & 0 & 0 & 0 & 0 \\
0 & 0 & 0 & 0 & 0 & 1.5625 & 0 & 0 & 0 & 0 & 0 & 0 & 0 & 0 & 0 \\
0 & 0 & 0 & 0 & 0 & 0 & 1.5625 & 0 & 0 & 0 & 0 & 0 & 0 & 0 & 0 \\
0 & 0 & 0 & 0 & 0 & 0 & 0 & 1.5625 & 0 & 0 & 0 & 0 & 0 & 0 & 0 \\
0 & 0 & 0 & 0 & 0 & 0 & 0 & 0 & 1.5625 & 0 & 0 & 0 & 0 & 0 & 0 \\
0 & 0 & 0 & 0 & 0 & 0 & 0 & 0 & 0 & 1.5625 & 0 & 0 & 0 & 0 & 0 \\
0 & 0 & 0 & 0 & 0 & 0 & 0 & 0 & 0 & 0 & 1.5625 & 0 & 0 & 0 & 0 \\
-6.2500 & 0 & 0 & 0 & 0 & 0 & 0 & 0 & 0 & 0 & 0 & 2.2135 & 1.4323 & 1.4323 & 1.4323 \\
-6.2500 & 0 & 0 & 0 & 0 & 0 & 0 & 0 & 0 & 0 & 0 & 1.4323 & 2.2135 & 1.4323 & 1.4323 \\
-6.2500 & 0 & 0 & 0 & 0 & 0 & 0 & 0 & 0 & 0 & 0 & 1.4323 & 1.4323 & 2.2135 & 1.4323 \\
-6.2500 & 0 & 0 & 0 & 0 & 0 & 0 & 0 & 0 & 0 & 0 & 1.4323 & 1.4323 & 1.4323 & 2.2135
\end{pmatrix}$$

The Eigen values of M^{-1} are

31.3024

0.2080

0.7813

0.7813 + 0.0000i

0.7813 - 0.0000i

1.0417

1.0417

1.0417

1.0417

1.5625

1.5625

1.5625

1.5625

1.5625

1.5625

The determinant value and the trace of M^{-1} are respectively 53.1881and 47.3958.

We summarize in Table 1 some optimality constants for the three categories of designs considered.

Table 1. Optimality Constants for Response Surface Methodology Designs

Model Type	Experimental Design	Control Variables k, Axial distance α	Design Size N	A-Optimality constant (Trace of M^{-1})	D-Optimality constant (Determinant of M)	D-Optimality constant (Determinant of M^{-1})	E-Optimality constant (minimum Eigen value of M)	G-Optimality constant (maximum scaled predictive variance)	T-Optimality constant (Trace of M)
Main effects model	Plackett-Burman design	k=3	4	4.0	1.0	1.0	1.0	4.0	4.0
		k=7	8	8.0	1.0	1.0	1.0	8.0	8.0
		k=11	12	12.0	1.0	1.0	1.0	12.0	12.0
		k=15	16	16.0	1.0	1.0	1.0	16.0	16.0
First-order Complete model	2^k Factorial design	k=2	4	4.0	1.0	1.0	1.0	4.0	4.0
		k=3	8	8.0	1.0	1.0	1.0	8.0	80
		k=4	16	16.0	1.0	1.0	1.0	16.0	16.0
Second-order complete model	Central Composite Design	k=2 (α=1.0)	9	19.2500	0.0098	102.5156	0.1111	7.2500	4.1111
		k=2 (α=1.414)	9	19.6900	0.0616	16.2379	0.0730	9.0000	5.8875
		k=3 (α=1.0)	15	31.1796	3.1954×10^{-4}	3.1285e+003	0.1333	11.9583	6.6000
		k=3 (α=1.682)	15	31.1796	0.0235	42.6197	0.0497	14.8269	10.1332
		k=3 (α=1.7321)	15	30.8625	0.0332	30.1510	0.0498	15.0000	10.6005
		k=4 (α=1.0)	25	58.8571	5.3555×10^{-6}	1.8672e+005	0.0800	17.4906	10.6000
		k=4 (α=2.0)	15	53.1881	0.0188	53.1881	0.0319	25.0000	16.3600

4. Discussion

Many areas of Mathematical Science deal with approximation properties and asymptotic behaviours as could be seen in Mishra (2007), Mishra and Mishra (2012), Mishra *et al.* (2013) and Deepmala (2014)). With regards to estimation problems and asymptotic behaviours of optimality criteria, several literatures on optimal design of experiments have theoretically shown that relationships exist among the optimality criteria. However, in reviewing the literatures, there arose the need for concise numerical illustrations appreciably relating many of the criteria. In this paper, illustrative examples have been considered and presented with helpful evaluations thus offering first-hand information regarding response surface methodology designs and their optimality properties. As mentioned in the introduction, Plackett-Burman designs and the 2^k factorial designs are suitable for modeling first order effects. These categories of designs share similar optimality properties as seen in this paper. For all N-point k variable Plackett-Burman designs, the information matrices were diagonal and hence orthogonal with uniform diagonal elements. The Plackett-Burman designs exhibit uniform precision as the variance of prediction at each design point remained constant. Moreover, the maximum variance of prediction in each case was exactly equal to the number of model parameters. These two properties of the Plackett-Burman designs indicate D-optimality and G-optimality of the designs respectively. In fact, the design matrices being orthogonal satisfy rotatability requirements for first order designs. For each normalized information matrix, the determinant value and the minimum Eigen value were unity. These were also true for the determinant values of the variance-covariance matrices. The maximum scaled predictive variance, the trace of information matrix and the trace of the variance-covariance matrix shared the property of being equal to the number of design size. Each of these properties was also exhibited by the 2^k factorial designs.

Unlike the first-order designs, the second-order central composite designs do not show orthogonality of associated information matrices. This is true from existing literatures. However, the central composite designs have nice variance property of having the same variance of prediction at any two design points that are equidistant from the center point. For k=2, the 9-point design with α=1.0 which was A-optimal when compared with the 9-point design with α=1.414 was also E- and G-optimal. Hence the design with α=1.0 which minimized the trace of the variance-covariance matrix, M^{-1}, also maximized the minimum Eigen value of the information matrix as well as minimizing the maximum scaled predictive variance. On the other hand, the 9-point design with α=1.414 showed preference over the 9-point design with α=1.0 under the criteria of D- and T-optimality as the design with α=1.414 which maximized the determinant of information matrix also maximized the trace of information matrix.

For k=3. The design with α=1.7321, which minimized the trace of the variance-covariance matrix also maximized the determinant of the information matrix and correspondingly maximized the trace of the information matrix. By these, the design with α=1.7321 showed preference in terms of A-, D- and T-optimality criteria when compared with the designs with α=1.0 and α=1.682. However, design with α=1.0 was preferred in terms of E- and G-optimality criteria. For k=4, design with α=1.0 was again preferred in terms of E- and G-optimality criteria when compared with the design with α=2.0. The design with α=2.0 was preferred in terms of A-, D- and T-optimality criteria when compared with design with α=1.0. It was observed as mentioned in existing literatures that designs that maximized the determinant of information matrices equivalently minimized the determinant of the variance-covariance matrices. One interesting finding is that central composite designs defined on cubes and hypercubes (i.e. with unit axial distance) were uniformly preferred in terms of E-optimality and G-optimality criteria.

References

Box, G. E. P., & Wilson, K. B. (1951). On the Experimental Attainment of Optimum conditions. *Journal of the Royal statistical society, Series B*(1), 1-45.

Chigbu, P. E., & Nduka, U. C. (2006). On the optimal choice of the cube and star replications in restricted second-order designs. United Nations Educational, Scientific and Cultural Organization and International Atomic Energy Agency: The Abdus Salam International Centre for Theoretical Physics, Trieste, Italy. Available at: http://www.ictp.it/~pub−off

Chigbu, P. E., Ukaegbu, E. C., & Nwanya, J. C. (2009). On comparing the prediction variances of some Central Composite Designs in Spherical regions: A Review. *Statistica*, anno LXIX, n, 4.

Deepmala. (2014). *A study on fixed-point theorems for nonlinear contractions and its applications*. PhD Thesis, Pt. Ravishankar Shukla University, Raipur 492010, Chhatisgarh, India.

Dette, H., & Grigoriev, Y. (2014). Construction of efficient and optimal experimentl design for response surface models. *The annals of statistics, 42*(4), pg. 1635-1656. http://dx.doi.org/10.1214/14-AOS1241

Giovannitti-Jensen, A., & Myers, R. H. (1989). Graphical Assessment of the Prediction Capability of Response Surface

Designs. *Technometrics, 31*(2), 159-171. http://dx.doi.org/10.1080/00401706.1989.10488510

Iwundu, M. P. (2015). Optimal Partially Replicated Cube, Star and Center Runs in Face-centered Central Composite Designs. *International Journal of Statistics and Probability, 4*(4). http://dx.doi.org/10.5539/ijsp.v4n4p1

Khuri, A. I., & Mukhopadhyay, S. (2010). Response Surface Methodology. *Wiley Interdisciplinary Reviews: Computational Statistics, 2*(2), 128-149. http://dx.doi.org/10.1002/wics.73

Lucas, J. M. (1976). Response surface design is the best: A performance comparison of several types of quadratic response surface designs in symmetric regions. *Technometrics, 18*, 411-417. http://dx.doi.org/10.1080/00401706.1976.10489472

Mishra, V. N., & Mishra, L. N. (2012). Trigonometric approximation of signals (functions) in L_p $(p \geq 1)$-norm. *International Jourmal Of Contemporary Mathematical Sciences, 7*(19), 909-918.

Mishra, V. N. (2007). *Some problems on approximations of functions in Banach spaces*. PhD Thesis, Indian Institute of Technology, Roorkee, 247 667, Uttarakhand, India.

Mishra, V. N. , Khatri, K., Mishra, L. N., & Deepmala. (2013). Inverse result in Simultaneous approximation. *Journal of Inequalities and Approximation.* http://dx.doi.org/10.1186/1029-242X-2013-586

Myers, R. H., Vinning, G. G., Giovannitti-Jensen, A., & Myers, S. L. (1992). Variance Dispersion properties of Second-order response surface designs. *Journal of Quality technology, 24*, 1-11.

Myers, R. H., Montgomery, D. C., & Anderson-Cook, C. M. (2009). *Response Surface Methodology: Process and Product Optimization using designed experiments.* 3rd Edition. John Wiley & Sons, Inc. New Jersey.

Plackett, R. L., & Burman, J. P. (1946). The design of optimum multifactorial experiments. *Biometrika, 33*, 305-325. http://dx.doi.org/10.1093/biomet/33.4.305

Rady, E. A., Abd El-Monsef, M. M. E., & Seyam, M. M. (2009). Relationship among several optimality criteria, interstat. statjournals. net>YEAR>articles

Ukaegbu, E.C., & Chigbu, P.E. (2015). Graphical Evaluation of the Prediction Capabilities of Partially Replicated Orthogonal Central Composite Designs. *Quality and Reliability Engineering International, 31*, 707-717 http://dx.doi.org/10.1002/qre.1630.

Zahran, A., Anderson-Cook, C.M., & Myers, R.H. (2003). Fraction of Design Space to Assess the Prediction Capability of Response Surface Designs. *Journal of Quality Technology, 35*, 377-386.

Dimension Formulae for the Polynomial Algebra as a Module over the Steenrod Algebra in Degrees Less than or Equal to 12

Mbakiso Fix Mothebe[1], Professor Kaelo[2] & Orebonye Ramatebele[3]

[1,2,3] Department of Mathematics, University of Botswana, Pvt Bag 00704, Gaborone, Botswana

Correspondence: Mbakiso Fix Mothebe, Department of Mathematics, University of Botswana, Pvt Bag 00704, Gaborone, Botswana. E-mail: mothebemf@mopipi.ub.bw

Abstract

Let $\mathbf{P}(n) = \mathbb{F}_2[x_1, \ldots, x_n]$ be the polynomial algebra in n variables x_i, of degree one, over the field $\mathbb{F}_2$ of two elements. The mod-2 Steenrod algebra $\mathcal{A}$ acts on $\mathbf{P}(n)$ according to well known rules. A major problem in algebraic topology is that of determining $\mathcal{A}^+\mathbf{P}(n)$, the image of the action of the positively graded part of $\mathcal{A}$. We are interested in the related problem of determining a basis for the quotient vector space $\mathbf{Q}(n) = \mathbf{P}(n)/\mathcal{A}^+\mathbf{P}(\mathbf{n})$. Both $\mathbf{P}(n) = \bigoplus_{d\geq 0} \mathbf{P}^d(n)$ and $\mathbf{Q}(n)$ are graded, where $\mathbf{P}^d(n)$ denotes the set of homogeneous polynomials of degree d. In this paper we give explicit formulae for the dimension of $\mathbf{Q}(n)$ in degrees less than or equal to 12.

Keywords: Steenrod squares, polynomial algebra, hit problem.

1. Introduction

For $n \geq 1$, let $\mathbf{P}(n)$ be the mod-2 cohomology group of the n-fold product of $\mathbb{R}P^\infty$ with itself. Then $\mathbf{P}(n)$ is the polynomial algebra

$$\mathbf{P}(n) = \mathbb{F}_2[x_1, \ldots, x_n]$$

in n variables x_i, each of degree 1, over the field $\mathbb{F}_2$ of two elements. The mod-2 Steenrod algebra $\mathcal{A}$ is the graded associative algebra generated over $\mathbb{F}_2$ by symbols Sq^i for $i \geq 0$, called Steenrod squares subject to the Adem relations (Adem 1957) and $Sq^0 = 1$. Let $\mathbf{P}^d(n)$ denote the homogeneous polynomials of degree d. The action of the Steenrod squares $Sq^i : \mathbf{P}^d(n) \to \mathbf{P}^{d+i}(n)$ is determined by the formula:

$$Sq^i(u) = \begin{cases} u, & i = 0 \\ u^2, & \deg(u) = i \\ 0, & \deg(u) < i, \end{cases}$$

and the Cartan formula

$$Sq^i(uv) = \sum_{r=0}^{i} Sq^r(u)Sq^{i-r}(v).$$

A polynomial $u \in \mathbf{P}^d(n)$ is said to be hit if it is in the image of the action of $\mathcal{A}$ on $\mathbf{P}(n)$, that is, if

$$u = \sum_{i>0} Sq^i(u_i),$$

for some $u_i \in \mathbf{P}(n)$ of degree $d - i$. Let $\mathcal{A}^+\mathbf{P}(n)$ denote the subspace of all hit polynomials. The problem of determining $\mathcal{A}^+\mathbf{P}(n)$ is called the hit problem and has been studied by several authors, (Silverman, 1998; Singer, 1991) and (Wood, 1989). We are interested in the related problem of determining a basis for the quotient vector space

$$\mathbf{Q}(n) = \mathbf{P}(n)/\mathcal{A}^+\mathbf{P}(n)$$

which has also been studied by several authors, (Kameko, 1990; 2003; Peterson, 1987) and (Sum, 2007). Some of the motivation for studying these problems is mentioned in (Nam, 2004). It stems from the Peterson conjecture proved in (Wood, 1989) and various other sources (Peterson, 1989; Singer, 1989).

The following result is useful for determining $\mathcal{A}$-generators for $\mathbf{P}(n)$. Let $\alpha(m)$ denote the number of digits 1 in the binary expansion of m.

In [(Wood, 1989) Theorem 1], R.M.W. Wood proved that:

Theorem 1 (Wood, 1989). *Let $u \in \mathbf{P}(n)$ be a monomial of degree d. If $\alpha(n+d) > n$, then u is hit.*

Thus $\mathbf{Q}^d(n)$ is zero unless $\alpha(n+d) \leq n$ or, equivalently, unless d can be written in the form, $d = \sum_{i=1}^{n}(2^{\lambda_i} - 1)$ where $\lambda_i \geq 0$. Thus $\mathbf{Q}^d(n) \neq 0$ only if $\mathbf{P}^d(n)$ contains monomials $v = x_1^{2^{\lambda_1}-1} \cdots x_n^{2^{\lambda_n}-1}$ called spikes. For convenience we shall assume that $\lambda_1 \geq \lambda_2 \geq \ldots \geq \lambda_n \geq 0$. We, in addition, shall consider a special one when $\lambda_1 \geq \lambda_2 \geq \ldots \geq \lambda_s \geq 0$ and $\lambda_{j-1} = \lambda_j$ only if $j = s$ or $\lambda_{j+1} = 0$. In this case v is called a **minimal spike**.

$\mathbf{Q}(n)$ has been explicitly calculated by Peterson in (Peterson, 1987) for $n = 1, 2$, by (Kameko, 1990) in his thesis, for $n = 3$ and independently by (Kameko, 2003) and (Sum, 2007) for $n = 4$. In this work we shall, unless otherwise stated, be concerned with a basis for $\mathbf{Q}(n)$ consisting of 'admissible monomials', as defined below. Thus when we write $u \in \mathbf{Q}^d(n)$ we mean that u is an admissible monomial of degree d.

We define what it means for a monomial $b = x_1^{e_1} \cdots x_n^{e_n} \in \mathbf{P}(n)$ to be admissible. Write $e_i = \sum_{j\geq 0} \alpha_j(e_i)2^j$ for the binary expansion of each exponent e_i. The expansions are then assembled into a matrix $\beta(b) = (\alpha_j(e_i))$ of digits 0 or 1 with $\alpha_j(e_i)$ in the (i, j)-th position of the matrix.

We then associate with b, two sequences,

$$w(b) = (w_0(b), w_1(b), \ldots, w_j(b), \ldots),$$
$$e(b) = (e_1, e_2, \ldots, e_n),$$

where $w_j(b) = \sum_{i=1}^{n} \alpha_j(e_i)$ for each $j \geq 0$. $w(b)$ is called the **weight vector** of the monomial b and $e(b)$ is called the **exponent vector** of the monomial b.

Given two sequences $p = (u_0, u_1, \ldots, u_l, 0, \ldots)$, $q = (v_0, v_1, \ldots, v_l, 0, \ldots)$, we say $p < q$ if there is a positive integer k such that $u_i = v_i$ for all $i < k$ and $u_k < v_k$. We are now in a position to define an order relation on monomials.

Definition 1. Let a, b be monomials in $\mathbf{P}(n)$. We say that $a < b$ if one of the following holds:

1. $w(a) < w(b)$,
2. $w(a) = w(b)$ and $e(a) < e(b)$.

Note that the order relation on the set of sequences is the lexicographical one.

Following (Kameko, 1990) we define:

Definition 2. A monomial $b \in \mathbf{P}(n)$ is said to be **inadmissible** if there exist monomials $b_1, b_2, \ldots, b_r \in \mathbf{P}(n)$ with $b_j < b$ for each j, $1 \leq j \leq r$, such that

$$b \equiv \left(\sum_{j=1}^{r} b_j\right) \mod \mathcal{A}^+\mathbf{P}(n).$$

b is said to be **admissible** if it is not inadmissible.

Clearly the set of all admissible monomials in $\mathbf{P}(n)$ form a basis for $\mathbf{Q}(n)$.

To put our work in context we require some preliminary observations. For each r, $1 \leq r \leq n$, let

$$\mathbf{X}(r) = \mathrm{Span}\{x_1^{m_1} \cdots x_r^{m_r} \in \mathbf{P}(r) \mid m_1 m_2 \cdots m_r \neq 0\}.$$

Then $\mathbf{X}(r)$ is an $\mathcal{A}$-submodule of $\mathbf{P}(r)$. Let

$$\mathbf{W}(r) = \mathbf{X}(r)/\mathcal{A}^+\mathbf{X}(r).$$

Then for each $n \geq 1$ we have a direct sum decomposition:

$$\mathbf{Q}(n) \cong \bigoplus_{r=1}^{n} \bigoplus_{k=1}^{\binom{n}{r}} \mathbf{W}(r).$$

Thus for any integer $d > 0$ we have the following inexplicit formula for the dimension of $\mathbf{Q}^d(n)$:

$$\dim(\mathbf{Q}^d(n)) = \sum_{r=1}^{n} \binom{n}{r} \dim(\mathbf{W}^d(r)). \tag{1}$$

But $\mathbf{Q}(n)$ is known for , $1 \leq n \leq 4$, and in some cases when $n = 5$. This gives rise to the following explicit formula for $\mathbf{Q}^d(n)$ for $d \leq 5$.

$$\begin{aligned}
\dim(\mathbf{Q}^1(n)) &= \tbinom{n}{1}.\\
\dim(\mathbf{Q}^2(n)) &= \tbinom{n}{2}.\\
\dim(\mathbf{Q}^3(n)) &= \tbinom{n}{1} + \tbinom{n}{2} + \tbinom{n}{3}.\\
\dim(\mathbf{Q}^4(n)) &= \tbinom{n}{2}\cdot 2 + \tbinom{n}{3}\cdot 2 + \tbinom{n}{4}.\\
\dim(\mathbf{Q}^5(n)) &= \tbinom{n}{3}\cdot 3 + \tbinom{n}{4}\cdot 3 + \tbinom{n}{5}.
\end{aligned}$$

We follow the convention that $\binom{n}{i} = 0$ if $i > n$.

Our main result is Theorem 2 below which consists of explicit formulae for $\dim(\mathbf{Q}^d(n))$, $6 \leq d \leq 12$.

Theorem 2. *For all* $n \geq 1$:

$$\begin{aligned}
\dim(\mathbf{Q}^6(n)) &= \tbinom{n}{2} + \tbinom{n}{3}\cdot 3 + \tbinom{n}{4}\cdot 6 + \tbinom{n}{5}\cdot 4 + \tbinom{n}{6}.\\
\dim(\mathbf{Q}^7(n)) &= \tbinom{n}{1} + \tbinom{n}{2} + \tbinom{n}{3}\cdot 4 + \tbinom{n}{4}\cdot 9 + \tbinom{n}{5}\cdot 10 + \tbinom{n}{6}\cdot 5 + \tbinom{n}{7}.\\
\dim(\mathbf{Q}^8(n)) &= \tbinom{n}{2}\cdot 3 + \tbinom{n}{3}\cdot 6 + \tbinom{n}{4}\cdot 13 + \tbinom{n}{5}\cdot 19 + \tbinom{n}{6}\cdot 15 + \tbinom{n}{7}\cdot 6 + \tbinom{n}{8}.\\
\dim(\mathbf{Q}^9(n)) &= \tbinom{n}{3}\cdot 7 + \tbinom{n}{4}\cdot 18 + \tbinom{n}{5}\cdot 31 + \tbinom{n}{6}\cdot 34 + \tbinom{n}{7}\cdot 21 + \tbinom{n}{8}\cdot 7 + \tbinom{n}{9}.\\
\dim(\mathbf{Q}^{10}(n)) &= \tbinom{n}{2}\cdot 2 + \tbinom{n}{3}\cdot 8 + \tbinom{n}{4}\cdot 26 + \tbinom{n}{5}\cdot 50 + \tbinom{n}{6}\cdot 65 + \tbinom{n}{7}\cdot 55\\
&\quad + \tbinom{n}{8}\cdot 28 + \tbinom{n}{9}\cdot 8 + \tbinom{n}{10}.\\
\dim(\mathbf{Q}^{11}(n)) &= \tbinom{n}{3}\cdot 8 + \tbinom{n}{4}\cdot 32 + \tbinom{n}{5}\cdot 75 + \tbinom{n}{6}\cdot 115 + \tbinom{n}{7}\cdot 120 + \tbinom{n}{8}\cdot 83\\
&\quad + \tbinom{n}{9}\cdot 36 + \tbinom{n}{10}\cdot 9 + \tbinom{n}{11}.\\
\dim(\mathbf{Q}^{12}(n)) &= \tbinom{n}{4}\cdot 21 + \tbinom{n}{5}\cdot 85 + \tbinom{n}{6}\cdot 176 + \tbinom{n}{7}\cdot 231 + \tbinom{n}{8}\cdot 203\\
&\quad + \tbinom{n}{9}\cdot 109 + \tbinom{n}{10}\cdot 45 + \tbinom{n}{11}\cdot 10 + \tbinom{n}{12}.
\end{aligned}$$

Our proof of Theorem 2 is deferred until Section 3.

Ignoring the terms for which $\mathbf{W}^d(r)$ is trivial, the above formulae yield Table 1 below for the dimension of $\mathbf{Q}^d(n)$, $1 \leq d \leq 12$. The numbers on the top row represent n, the number of variables while the serial numbers in the first column represent d, the degree of $\mathbf{Q}(n)$.

Our work is organized as follows. In Section 2, we recall some results on admissible monomials and hit monomials in $\mathbf{P}(n)$. In Section 3 we prove Theorem 2.

2. Preliminaries

In this section we recall some results in (Kameko, 1990; Silverman, 1998; Singer, 1991) and (Mothebe, 2016) on admissible monomials and hit monomials in $\mathbf{P}(n)$.

The following theorem has been used to great effect by Kameko and Sum in computing a basis for $\mathbf{Q}(3)$ and $\mathbf{Q}(4)$ respectively.

Theorem 3 (Kameko, 1990; Sum, 2010)**.** *Let a, b be monomials in $\mathbf{P}(n)$ such that $w_j(a) = 0$ for $j > r > 0$. If b is inadmissible, then ab^{2^r} is also inadmissible.*

Up to permutation of representatives weight order provides a total order relation amongst spikes in a given degree.

Table 1. Results obtained using Theorem 2

d \ n	1	2	3	4	5	6	7	8	9	10	11	12
1	1											
2	0	1										
3	1	3	7									
4	0	2	8	21								
5	0	0	3	15	46							
6	0	1	6	24	74	190						
7	1	3	10	35	110	301	729					
8	0	3	15	55	174	489	1238	2863				
9	0	0	7	46	191	630	1785	4515	10438			
10	0	2	14	70	280	945	2792	7412	18020	40701		
11	0	0	8	64	315	1205	3900	11151	28917	69234	155035	
12	0	0	0	21	190	1001	3983	13209	38402	100880	243737	550847

It is easy to show that a spike $v = x_1^{2^{\lambda_1}-1}\cdots x_n^{2^{\lambda_n}-1} \in \mathbf{P}^d(n)$ is a minimal spike if its weight order is minimal with respect to other spikes of degree d. In [(Kameko 1990), Theorem 4.2] Masaki Kameko proved that:

Theorem 4 (Kameko, 1990). *Let d be a positive integer and let v be the minimal spike of degree $2d + n$. Define a linear mapping, $f : \mathbf{P}^d(n) \to \mathbf{P}^{2d+n}(n)$, by,*

$$f(x_1^{m_1}\cdots x_n^{m_n}) = x_1^{2m_1+1}\cdots x_n^{2m_n+1}.$$

If $w_0(v) = n$, then f induces an isomorphism $f_* : \mathbf{Q}^d(n) \to \mathbf{Q}^{2d+n}(n)$.

From Wood's theorem and the above result of Kameko the problem of determining $\mathcal{A}$-generators for $\mathbf{P}(n)$ is reduced to the cases for which $w_0(v) \le n-1$ whenever v is a minimal spike of a given degree d.

We recall the following result of Singer on hit polynomials in $\mathbf{P}(n)$. In [(Singer, 1991), Theorem 1.2] W. M. Singer proved that:

Theorem 5 (Singer, 1991). *Let $b \in \mathbf{P}(n)$ be a monomial of degree d, where $\alpha(n + d) \le n$. Let v be the minimal spike of degree d. If $w(b) < w(v)$, then b is hit.*

We note the following stronger version of Theorem 5. Let b be a monomial of degree d. For $l > 0$ define $d_l(b)$ to be the integer $d_l(b) = \sum_{j\ge l} w_j(b)2^{j-l}$.

In [(Silverman, 1998), Theorem 1.2], J. H. Silverman proved that:

Theorem 6 (Silverman, 1998). *Let $b \in \mathbf{P}(n)$ be a monomial of degree d, where $\alpha(n + d) \le n$. Let v be the minimal spike of degree d. If $d_l(b) > d_l(v)$ for some $l \ge 1$, then b is hit.*

We shall require the following result of (Mothebe, 2016):

Theorem 7 (Mothebe, 2016). *If $u = x_1^{m_1}\cdots x_k^{m_k} \in \mathbf{P}^d(k)$ and $v = x_1^{e_1}\cdots x_r^{e_r} \in \mathbf{P}^{d'}(r)$ are admissible monomials, then for each permutation $\sigma \in S_{k+r}$ for which $\sigma(i) < \sigma(j)$, $i < j \le k$ and $\sigma(s) < \sigma(t)$, $k < s < t \le k + r$, the monomial*

$$x_{\sigma(1)}^{m_1}\cdots x_{\sigma(k)}^{m_k}x_{\sigma(k+1)}^{e_1}\cdots x_{\sigma(k+r)}^{e_r} \in \mathbf{P}^{d+d'}(k+r)$$

is admissible.

Theorem 7 is a generalization of the following result of the (Mothebe & Uys, 2015).

Let $u = x_1^{m_1}\cdots x_{n-1}^{m_{n-1}} \in \mathbf{P}(n-1)$ be a monomial of degree d'. Given any pair of integers (j, λ), $1 \le j \le n$, $\lambda \ge 0$, let $h_j^\lambda(u)$ denote the monomial $x_1^{m_1}\cdots x_{j-1}^{m_{j-1}}x_j^{2^\lambda-1}x_{j+1}^{m_j}\cdots x_n^{m_{n-1}} \in \mathbf{P}^{d'+(2^\lambda-1)}(n)$.

Theorem 8 (Mothebe & Uys, 2015). *Let $u \in \mathbf{P}(n-1)$ be a monomial of degree d', where $\alpha(d' + n - 1) \le n - 1$. If u is admissible, then for each pair of integers (j, λ), $1 \le j \le n$, $\lambda \ge 0$, $h_j^\lambda(u)$ is admissible.*

3. Proof of Theorem 2

The result of Theorem 2 is a consequence of Lemma 9 and Lemma 10 which we prove below.

Lemma 9. *If $a = x_1^{m_1}...x_n^{m_n} \in \mathbf{X}(n)$ is an admissible monomial then $m_1 = 2^\lambda - 1$ for some $\lambda \geq 1$.*

Proof. The lemma is clearly true if $n = 1$. Suppose that $m_1 = 2^\lambda - 2$. Let $b = x_1^{e_1}...x_n^{e_n}$ be the monomial obtained from a by replacing m_1 by $2^\lambda - 3$. Then $a = Sq^1(b) + x_1^{2^\lambda-3}Sq^1(x_2^{m_2}...x_n^{m_n})$ and the fact that all terms in $x_1^{2^\lambda-3}Sq^1(x_2^{m_2}...x_n^{m_n})$ are of lower order than a shows that a is inadmissible. But every monomial with $m_1 \neq 2^\lambda - 1$ is of the form cd^{2^r} for some monomial $d = x_1^{t_1}...x_n^{t_n}$ with $t_1 = 2^\lambda - 2$ so the general result follows from Theorem 3.

Further to Lemma 9, suppose that $a = x_1^{2^\lambda-1}...x_n^{m_n}$ with $\lambda \geq 2$. If $m_2 = 2^k - 2$ for some $k \geq 2$, then a is inadmissible.

Recall that for each r, $1 \leq r \leq n$,

$$\mathbf{X}(r) = \text{Span}\{x_1^{m_1} \cdots x_r^{m_r} \in \mathbf{P}(r) \mid m_1 m_2 \cdots m_r \neq 0\}.$$

Then $\mathbf{X}(r)$ is an $\mathcal{A}$-submodule of $\mathbf{P}(r)$. Let

$$\mathbf{W}(r) = \mathbf{X}(r)/\mathcal{A}^+\mathbf{X}(r).$$

Then for each $n \geq 1$ we have a direct sum decomposition:

$$\mathbf{Q}(n) \cong \bigoplus_{r=1}^{n} \bigoplus_{k=1}^{\binom{n}{r}} \mathbf{W}(r).$$

Thus for any integer $d > 0$ we have the following inexplicit formula for the dimension of $\mathbf{Q}^d(n)$

$$\dim(\mathbf{Q}^d(n)) = \sum_{r=1}^{n} \binom{n}{r} \dim(\mathbf{W}^d(r)). \quad (2)$$

The following lemma evaluates Formula (2) explicitly in some cases. It gives explicit formulae for $\dim(\mathbf{W}^d(n))$ for those cases that enables us to obtain $\dim(\mathbf{Q}^d(n))$ for all values of n in the range , $1 \leq n \leq 12$.

Lemma 10.

$$\dim(\mathbf{W}^n(n)) = 1 \quad \text{for all } n \geq 1.$$

$$\dim(\mathbf{W}^n(n-1)) = n-2 \quad \text{for all } n \geq 3.$$

$$\dim(\mathbf{W}^n(n-2)) = \binom{n-2}{2} \quad \text{for all } n \geq 6.$$

$$\dim(\mathbf{W}^n(n-3)) = \binom{n-4}{3} + (n-3)(n-5) \quad \text{for all } n \geq 7.$$

$$\dim(\mathbf{W}^n(n-4)) = \left(\binom{n-5}{4} - 1\right) + \binom{n-4}{2} + (n-4).\binom{n-6}{2} + \binom{n-5}{2} \quad \text{for all } n \geq 10.$$

$$\begin{aligned}\dim(\mathbf{W}^n(n-5)) = {} & n - 6 + \tfrac{(n-5)!}{2\cdot(n-7)(n-9)!} + \binom{n-5}{2} + \binom{n-6}{2} + (n-5)\cdot\binom{n-7}{3} \\ & +2\cdot\left(\textstyle\sum_{i=2}^{n-8}\binom{i}{2}\right) + \binom{n-8}{2} + \left(\binom{n-7}{2} - 1\right) \quad 10 \leq n \leq 12.\end{aligned}$$

$$\begin{aligned}\dim(\mathbf{W}^n(n-6)) = {} & n - 6 + ((n-7)(n-8) - 2) + \binom{n-6}{2} + \binom{n-6}{3} \\ & +(n-6)\cdot\binom{n-8}{2} + \tfrac{(n-7)!}{4\cdot(n-11)!} + (n-8)\cdot\binom{n-9}{2} + \binom{n-9}{2} \\ & +\left(\binom{n-6}{3} - 1\right) + 2\cdot\binom{n-7}{3} - (n-9) \quad 11 \leq n \leq 12.\end{aligned}$$

Proof. For $n \geq 1$ the basis of $\mathbf{X}^n(n)$ consists of the monomial $x_1 x_2 \cdots x_{n-1} x_n$. For $n \geq 3$ the basis of $\mathbf{X}^n(n-1)$ consists of the monomial $x_1 x_2 \cdots x_{n-2} x_{n-1}^2$ and its permutation representatives. For $n \geq 6$ the basis of $\mathbf{X}^n(n-2)$ consists of the monomials:

$$\{a_{n-2} = x_1 x_2 \cdots x_{n-4} x_{n-3}^2 x_{n-2}^2,\ b_{n-2} = x_1 x_2 \cdots x_{n-4} x_{n-3} x_{n-2}^3\}$$

and their permutation representatives. For convenience we shall ignore those monomials in $\mathbf{X}^n(n-i)$ which, by Theorem 6, are hit. For $n \geq 7$ the basis of $\mathbf{X}^n(n-3)$ consists of the monomials:

$$\{a_{n-3} = x_1 x_2 \cdots x_{n-6} x_{n-5}^2 x_{n-4}^2 x_{n-3}^2,\ b_{n-3} = x_1 x_2 \cdots x_{n-5} x_{n-4}^2 x_{n-3}^3,\ c_{n-3} = x_1 x_2 \cdots x_{n-4} x_{n-3}^4\}$$

and their permutation representatives. If $n \geq 8$, the basis of $\mathbf{X}^n(n-4)$ consists of the monomials:

$$\{ a_{n-4} = x_1 \cdots x_{n-7}x_{n-6}^2x_{n-5}^2x_{n-4}^3,\ b_{n-4} = x_1 \cdots x_{n-6}x_{n-5}^2x_{n-4}^4,\ c_{n-4} = x_1 \cdots x_{n-6}x_{n-5}^3x_{n-4}^3,\ e_{n-4} = x_1 \cdots x_{n-6}x_{n-5}x_{n-4}^5\}$$

and their permutation representatives, while

$$d_{n-4} = x_1x_2 \cdots x_{n-8}x_{n-7}^2x_{n-6}^2x_{n-5}^2x_{n-4}^2$$

and its permutation representatives have to be included in the list when $n \geq 10$.

If $n \geq 9$ the basis of $\mathbf{X}^n(n-5)$ consists of the monomials:

$$\{a_{n-5} = x_1 \cdots x_{n-8}x_{n-7}^2x_{n-6}^3x_{n-5}^3,\ b_{n-5} = x_1x_2 \cdots x_{n-7}x_{n-6}^3x_{n-5}^4,\ c_{n-5} = x_1x_2 \cdots x_{n-7}x_{n-6}^2x_{n-5}^5,\ d_{n-5} = x_1x_2 \cdots x_{n-6}x_{n-5}^6\}$$

and their permutation representatives, while

$$e_{n-5} = x_1x_2 \cdots x_{n-9}x_{n-8}^2x_{n-7}^2x_{n-6}^2x_{n-5}^3,\ f_{n-5} = x_1x_2 \cdots x_{n-8}x_{n-7}^2x_{n-6}^2x_{n-5}^4$$

and their permutation representatives have to be added to the list when $n \geq 10$ and

$$g_{n-5} = x_1x_2 \cdots x_{n-11}x_{n-10}x_{n-9}^2x_{n-8}^2x_{n-7}^2x_{n-6}^2x_{n-5}^2$$

and its permutation representatives have to be added to the list when $n \geq 13$.

If $n \geq 11$ the basis of $\mathbf{X}^n(n-6)$ consists of the monomials:

$$\{a_{n-6} = x_1 \cdots x_{n-9}x_{n-8}^3x_{n-7}^3x_{n-6}^3,\ b_{n-6} = x_1 \cdots x_{n-7}x_{n-6}^7,\ c_{n-6} = x_1 \cdots x_{n-8}x_{n-7}^3x_{n-6}^5,\ d_{n-6} = x_1 \cdots x_{n-10}x_{n-9}^2x_{n-8}^2x_{n-7}^3x_{n-6}^3,$$

$$e_{n-6} = x_1x_2 \cdots x_{n-9}x_{n-8}^2x_{n-7}^3x_{n-6}^4,\ f_{n-6} = x_1x_2 \cdots x_{n-8}x_{n-7}^2x_{n-6}^6,\ g_{n-6} = x_1x_2 \cdots x_{n-9}x_{n-8}^2x_{n-7}^2x_{n-6}^5\}$$

and their permutation representatives, while

$$h_{n-6} = x_1x_2 \cdots x_{n-11}x_{n-10}^2x_{n-9}^2x_{n-8}^2x_{n-7}^2x_{n-6}^3,\ k_{n-6} = x_1x_2 \cdots x_{n-10}x_{n-9}^2x_{n-8}^2x_{n-7}^2x_{n-6}^4$$

and their permutation representatives have to be added to the list when $n \geq 13$ and

$$l_{n-6} - x_1x_2 \cdots x_{n-12}x_{n-11}^2x_{n-10}^2x_{n-9}^2x_{n-8}^2x_{n-7}^2x_{n-6}^2$$

and its permutation representatives has to be added to the list when $n \geq 14$.

We claim that for all pairs (n, i), $1 \leq i \leq 4$, and all n as specified in Lemma 10, $\mathbf{W}^{n+1}(n-i+1)$ is the vector space sum

$$\mathbf{W}^{n+1}(n-i+1) = \sum_{j=1}^{n-i+1} h_j^1(\mathbf{W}^n(n-i)), \tag{3}$$

while this is true for $n = 10, 11$ when $i = 5$ and for $n = 11$ when $i = 6$. This shall suffice for a proof of Lemma 10 since the formulae in the lemma give the number of elements $\mathbf{W}^{n+1}(n-i+1)$ obtained from $\mathbf{W}^n(n-i)$ in this inductive manner.

For each i, $0 \leq i \leq 6$, we identify, for the initial value n_0 of n, all elements in $\mathbf{X}^{n_0}(n_0-i)$ which are admissible, that is, we determine the admissible monomial basis for $\mathbf{W}^{n_0}(n_0-i)$. We then show, for a given integer i, and all $n \geq n_0$, that each element of $\mathbf{X}^{n+1}(n-i+1)$ which does not belong to the set

$$\{h_j^1(x_1^{m_1} \cdots x_{n-i}^{m_{n-i}}) \mid x_1^{m_1} \cdots x_{n-i}^{m_{n-i}} \in \mathbf{W}^n(n-i),\ 1 \leq j \leq n-i+1\},$$

is inadmissible.

Clearly $\dim(\mathbf{W}^n(n)) = 1$ for all $n \geq 1$ since $x_1x_2 \cdots x_{n-1}x_n$ is the only admissible element in $\mathbf{X}^n(n)$, while $\dim(\mathbf{W}^n(n-1)) = n-2$ for all $n \geq 3$ since there are $n-2$ permutation representatives of $x_1x_2 \cdots x_{n-2}x_{n-1}^2$ which are admissible.

If $i = 2$, then $n_0 = 6$ and it is known that $\mathbf{W}^6(4)$ is generated by $x_1x_2x_3x_4^3$ and all its permutation representatives together with the monomials $x_1x_2^2x_3x_4^2$ and $x_1x_2x_3^2x_4^2$. If $n > 6$, then all permutation representatives of b_{n-2} are admissible and the only permutation representatives of a_{n-2} in $\mathbf{X}^n(n-2)$ that may not be obtained from the basis of $\mathbf{W}^6(4)$ by inductively applying Equation (3) are those of the form $x_1^2 \cdots x_{n-2}^{m_{n-2}}$ as well as the monomial $x_1x_2^2x_3^2x_4 \cdots x_{n-2}$, all of which are clearly inadmissible. Thus $\dim(\mathbf{W}^n(n-2)) = \binom{n-2}{2}$, for all $n \geq 6$, since a_{n-2}, b_{n-2} have, respectively, $\binom{n-3}{2} - 1$, $n-2$ permutation representatives which are admissible and $\binom{n-3}{2} - 1 + n - 2 = \binom{n-2}{2}$.

If $i = 3$, then $n_0 = 7$ and it is known that $\mathbf{W}^7(4)$ is generated by the monomials

$$x_1x_2^2x_3^2x_4^2 \text{ and } x_1x_2x_3^2x_4^3,\ x_1x_2x_3^3x_4^2,\ x_1x_2^3x_3x_4^2,\ x_1^3x_2x_3x_4^2,\ x_1x_2^2x_3x_4^3\ x_1x_2^2x_3^3x_4,\ x_1x_2^3x_3^2x_4,\ x_1^3x_2x_3^2x_4.$$

It is easy to show that the monomial c_{n-3} and all its permutation representatives are inadmissible for all $n \geq 7$.

If $n > 7$, then the only permutation representatives of a_{n-3} and b_{n-3} in $\mathbf{X}^n(n-3)$ that may not be obtained from the basis of $\mathbf{W}^7(4)$ by inductively applying Equation (3) are those of the form $x_1^2 \cdots x_{n-3}^{m_{n-3}}$ as well as the monomial $x_1^3x_2^2x_3x_4 \cdots x_{n-3}$, all of which are clearly inadmissible. Thus $\dim(\mathbf{W}^n(n-3)) = \binom{n-4}{3} + (n-3)(n-5)$, for all $n \geq 7$, since a_{n-3}, b_{n-3} have, respectively, $\binom{n-4}{3}$, $(n-3)(n-5)$ permutation representatives which are admissible.

If $i = 4$, then $n_0 = 10$. We first note that $\mathbf{W}^8(4)$ is generated by $x_1x_2x_3^3x_4^3$ and all its permutation representatives, as well as the monomials in the following sets:

$A = \{x_1x_2x_3^2x_4^4,\ x_1x_2^2x_3x_4^4,\ x_1x_2^2x_3^4x_4\}$ and $B = \{x_1^3x_2x_3^2x_4^2,\ x_1x_2^3x_3^2x_4^2,\ x_1x_2^2x_3^3x_4^2,\ x_1x_2^2x_3^2x_4^3\}$.

It is easy to show that the monomial e_{n-4} and all its permutation representatives are inadmissible for all $n \geq 8$.

If $n > 8$, then we know that all the permutation representatives of c_{n-4} are admissible. The only permutation representatives of a_{n-4} and b_{n-4} in $\mathbf{X}^n(n-4)$ that may not be obtained from the basis of $\mathbf{W}^8(4)$ by inductively applying Equation (3) are those of the form $x_1^2 \cdots x_{n-4}^{m_{n-4}}$ and of the form $x_1^3x_2^2x_3^{m_3}x_4^{m_4} \cdots x_{n-4}^{m_{n-4}}$ as well as those with a factor of the form $x_i^4x_j^2$, $i < j$, all of which are clearly inadmissible. If $n = 10$, then by Theorem 7, the following permutation representatives of d_6 in $\mathbf{X}^{10}(6)$ are admissible, namely those in the set:

$C = \{x_1x_2x_3^2x_4^2x_5^2x_6^2, x_1x_2^2x_3x_4^2x_5^2x_6^2, x_1x_2^2x_3^2x_4x_5^2x_6^2, x_1x_2^2x_3^2x_4^2x_5x_6^2\}$,

The monomial $x_1x_2^2x_3^2x_4^2x_5^2x_6$ is inadmissible since

$$x_1x_2^2x_3^2x_4^2x_5^2x_6 = Sq^4(x_1x_2x_3x_4x_5x_6) + Sq^3(x_1^2x_2x_3x_4x_5x_6) + Sq^1(x_1^4x_2x_3x_4x_5x_6) + x_1x_2^2x_3^2x_4^2x_5x_6^2 + x_1x_2^2x_3^2x_4x_5^2x_6^2$$
$$+ x_1x_2^2x_3x_4^2x_5^2x_6^2 + x_1x_2x_3^2x_4^2x_5^2x_6^2.$$

The other permutation representatives of d_6 are inadmissible since they are of the form $x_1^2 \cdots x_6^{m_6}$.

If $n > 10$, then the only permutation representatives of d_{n-4} in $\mathbf{X}^n(n-4)$ that may not be obtained from d_6 by inductively applying Equation (3) are those of the form $x_1^2 \cdots x_{n-4}^{m_{n-4}}$ as well the monomial $x_1x_2^2x_3^2x_4^2x_5^2x_6 \cdots x_{n-4}$, all of which are clearly is inadmissible. Thus

$$\dim(\mathbf{W}^n(n-4)) \;=\; \left(\binom{n-5}{4} - 1\right) + \binom{n-4}{2} + (n-4).\binom{n-6}{2} + \binom{n-5}{2}, \quad \text{for all } n \geq 10,$$

since a_{n-4}, b_{n-4}, c_{n-4}, d_{n-4}, have, respectively,

$$(n-4).\binom{n-6}{2}, \binom{n-5}{2}, \binom{n-4}{2}, \left(\binom{n-5}{4} - 1\right)$$

permutation representatives which are admissible. If , $8 \leq n < 10$, then we omit the term $\left(\binom{n-5}{4} - 1\right)$ in the expression for the value of $\dim(\mathbf{W}^n(n-4))$.

If $i = 5$, then $n_0 = 10$. We first note that $\mathbf{W}^9(4)$ is generated by monomials in the following sets:

$A = \{x_1x_2x_3x_4^6,\ x_1x_2x_3^6x_4,\ x_1x_2^6x_3x_4\}$, $\quad B = \{x_1x_2x_3^2x_4^5,\ x_1x_2^2x_3x_4^5,\ x_1x_2^2x_3^5x_4\}$,

$C = \{x_1x_2^2x_3^3x_4^3,\ x_1x_2^3x_3^2x_4^3,\ x_1x_2^3x_3^3x_4^2,\ x_1^3x_2x_3^2x_4^3, x_1^3x_2x_3^3x_4^2,\ x_1^3x_2^3x_3x_4^2\}$, and

$D = \{x_1x_2x_3^3x_4^4,\ x_1x_2^3x_3x_4^4,\ x_1x_2^3x_3^4x_4,\ x_1^3x_2x_3x_4^4, x_1^3x_2x_3^4x_4,\ x_1^3x_2^4x_3x_4\}$.

If $n > 9$, then the only permutation representative of a_{n-5} in $\mathbf{X}^n(n-5)$ that may not be obtained from the basis of $\mathbf{W}^9(4)$ by inductively applying Equation (3) is that of the form $x_1^6x_2 \cdots x_{n-5}$ which is clearly inadmissible while the only permutation representatives of b_{n-5}, c_{n-5} and d_{n-5} in $\mathbf{X}^n(n-5)$ that may not be obtained from the basis of $\mathbf{W}^9(4)$ by inductively applying Equation (3) are, respectively, those with a factor of the form $x_i^5x_j^2$, $i < j$, $x_i^2x_j$, $i < j$, and $x_i^4x_j^3$, $i < j$, all of which are clearly inadmissible.

If $i = 5$ and $n = 10$, then by Theorem 7, the following permutation representatives of e_5 and f_5 are admissible in $\mathbf{X}^{10}(5)$, namely those in the sets:

$E = \{x_1^3x_2x_3^2x_4^2x_5^2,\ x_1x_2^3x_3^2x_4^2x_5^2,\ x_1x_2^2x_3^3x_4^2x_5^2,\ x_1x_2^2x_3^2x_4^3x_5^2,\ x_1x_2^2x_3^2x_4^2x_5^3\}$,
$F = \{x_1x_2x_3^2x_4^2x_5^4,\ x_1x_2x_3^2x_4^4x_5^2,\ x_1x_2^2x_3x_4^2x_5^4,\ x_1x_2^2x_3^4x_4x_5^2,\ x_1x_2^2x_3x_4^4x_5^2\}$.

Permutation representatives of e_5 which are not in E are those of the form $x_1^2\cdots x_5^{m_5}$ as well as those of the form $x_1^3x_2^2x_3^{m_3}\cdots x_5^{m_5}$, all of which are clearly inadmissible. Permutation representatives of f_5 which are not in F are those of the form $x_1^2\cdots x_5^{m_5}$ as well as those with a factor of the form $x_i^4x_j^2x_k^2,\ i<j,k$, and the three monomials in the set $G = \{x_1x_2^2x_3^4x_4^2x_5,\ x_1x_2^2x_3^2x_4^4x_5,\ x_1x_2^2x_3^2x_4x_5^4\}$, all of which are clearly inadmissible. For example

$$x_1x_2^2x_3^4x_4^2x_5 = Sq^2(x_1x_2x_3^4x_4x_5) + Sq^1(x_1^2x_2x_3^4x_4x_5) + x_1x_2^2x_3^4x_4x_5^2 + x_1x_2x_3^4x_4^2x_5^2.$$

If $n > 10$, then the only permutation representatives of e_{n-5} in $\mathbf{X}^n(n-5)$ that may not be obtained from the basis of $\mathbf{W}^{10}(5)$ by inductively applying Equation (3) are those of the form $x_1^2\cdots x_{n-5}^{m_{n-5}}$ as well as those of the form $x_1^3x_2^2x_3^{m_3}\cdots x_{n-5}^{m_{n-5}}$, all of which are clearly inadmissible. On the other hand the permutation representatives of f_{n-5} that may not be obtained from the basis of $\mathbf{W}^{10}(5)$ by inductively applying Equation (3) are those of the form $x_1^2\cdots x_{n-5}^{m_{n-5}}$ as well as those with a factor of the form $x_i^4x_j^2x_k^2,\ i<j,k$, and the three monomials in the set:

$$G = \{x_1x_2^2x_3^4x_4^2x_5\cdots x_{n-5},\ x_1x_2^2x_3^2x_4^4x_5\cdots x_{n-5},\ x_1x_2^2x_3^2x_4x_5^4\cdots x_{n-5}\},$$

all of which are clearly inadmissible. Thus

$$\dim(\mathbf{W}^n(n-5)) \geq n-6+\tfrac{(n-5)!}{2\cdot(n-7)(n-9)!}+\binom{n-5}{2}+\binom{n-6}{2}+(n-5)\cdot\binom{n-7}{3}+2\cdot\left(\sum_{i=2}^{n-8}\binom{i}{2}\right)+\binom{n-8}{2}+\left(\binom{n-7}{2}-1\right),$$

since $a_{n-5}, b_{n-5}, c_{n-5}, d_{n-5}, e_{n-5}, f_{n-5}$, have, respectively,

$$\frac{(n-5)!}{2\cdot(n-7)(n-9)!},\ \binom{n-5}{2},\ \binom{n-6}{2},\ n-6,\ (n-5)\cdot\binom{n-7}{3},\ 2\cdot\left[\sum_{i=2}^{n-8}\binom{i}{2}\right]+\binom{n-8}{2}+\left(\binom{n-7}{2}-1\right),$$

permutation representatives which are admissible. Equality holds when, $10 \leq n \leq 12$.

If $i = 6$ then $n_0 = 11$. We know that all permutation representatives of a_{n-6} and b_{n-6} are admissible for all $n \geq 11$. By Theorem 7, the following permutation representatives of $c_5,\ d_5,\ e_5,\ f_5$ and g_5 are admissible in $\mathbf{X}^{11}(5)$, namely those in the sets:

$A = \{x_1x_2^2x_3^2x_4^3x_5^3,\ x_1x_2^2x_3^3x_4^2x_5^3,\ x_1x_2^3x_3^2x_4^2x_5^3,\ x_1^3x_2x_3^2x_4^2x_5^3,\ x_1x_2^2x_3^3x_4^3x_5^2,\ x_1x_2^3x_3^2x_4^3x_5^2,\ x_1^3x_2x_3^2x_4^3x_5^2,\ x_1x_2^3x_3^3x_4^2x_5^2,\ x_1^3x_2x_3^3x_4^2x_5^2,$ $x_1^3x_2^3x_3x_4^2x_5^2\}$,

$B = \{x_1x_2^2x_3x_4^6x_5,\ x_1x_2^2x_3x_4x_5^6,\ x_1x_2x_3^2x_4x_5^6,\ x_1x_2x_3x_4^2x_5^6,\ x_1x_2x_3x_4^6x_5^2,\ x_1x_2^6x_3x_4^2x_5,\ x_1x_2^6x_3x_4x_5^2,\ x_1x_2x_3^6x_4^2x_5,\ x_1x_2x_3^6x_4x_5^2,$ $x_1x_2x_3^2x_4^6x_5\}$,

$C = \{x_1x_2^3x_3^4x_4x_5^2,\ x_1x_2^3x_3^4x_4^2x_5,\ x_1x_2^3x_3x_4^4x_5^2,\ x_1x_2x_3^3x_4^4x_5^2,\ x_1^3x_2x_3^4x_4x_5^2,\ x_1^3x_2x_3x_4^4x_5^2,\ x_1^3x_2^4x_3x_4^2x_5,\ x_1^3x_2^4x_3x_4x_5^2,\ x_1^3x_2x_3^4x_4^2x_5,$ $x_1^3x_2x_3x_4^2x_5^4,\ x_1x_2^3x_3x_4^2x_5^4,\ x_1x_2x_3^3x_4^2x_5^4,\ x_1x_2x_3^2x_4^3x_5^4,\ x_1x_2x_3^2x_4^4x_5^3,\ x_1^3x_2x_3^2x_4x_5^4,\ x_1x_2^3x_3^2x_4x_5^4,\ x_1x_2^2x_3^3x_4x_5^4,\ x_1x_2^2x_3x_4^3x_5^4,$ $x_1x_2^2x_3x_4^4x_5^3,\ x_1^3x_2x_3^2x_4^4x_5,\ x_1x_2^3x_3^2x_4^4x_5,\ x_1x_2^2x_3^3x_4^4x_5,\ x_1x_2^2x_3^4x_4^3x_5,\ x_1x_2^2x_3^4x_4x_5^3\}$,

$D = \{x_1x_2x_3x_4^3x_5^5,\ x_1x_2x_3^3x_4x_5^5,\ x_1x_2^3x_3x_4x_5^5,\ x_1^3x_2x_3x_4x_5^5,\ x_1x_2x_3^3x_4^5x_5,\ x_1x_2^3x_3x_4^5x_5,\ x_1^3x_2x_3x_4^5x_5,\ x_1x_2^3x_3^5x_4x_5,\ x_1^3x_2x_3^5x_4x_5,$ $x_1^3x_2^5x_3x_4x_5\}$,

$E = \{x_1x_2x_3^2x_4^5x_5^2,\ x_1x_2^2x_3x_4^5x_5^2,\ x_1x_2^2x_3^5x_4x_5^2,\ x_1x_2^2x_3^5x_4^2x_5,\ x_1x_2^2x_3x_4^2x_5^5,\ x_1x_2x_3^2x_4^2x_5^5\}$.

It is known, (Mothebe 2009), that $\dim(\mathbf{Q}^{11}(5)) = 315$, so the above permutation representatives of the monomials of $a_5,\ b_5\ c_5,\ d_5,\ e_5,\ f_5$ and g_5 form a complete list of admissible monomials in $\mathbf{X}^{11}(5)$. We claim that

$$\mathbf{W}^{12}(6) = \sum_{j=1}^{6} h_j^1(\mathbf{W}^{11}(5)). \tag{4}$$

If $n > 11$, then the only permutation representatives of $c_{n-6}, d_{n-6}, e_{n-6}, f_{n-6}$ and g_{n-6} in $\mathbf{X}^n(n-6)$ that may not be obtained from the basis of $\mathbf{W}^{11}(5)$ by inductively applying Equation (3) are those of the form $x_1^2\cdots x_{n-6}^{m_{n-6}},\ x_1^3x_2^2\cdots x_{n-6}^{m_{n-6}}$ or, respectively, those of the form $x_1^3x_2^3x_3^2\cdots x_{n-6}^{m_{n-6}},\ x_1^6\cdots x_{n-6}^{m_{n-6}}$ or $x_1x_2^2x_3^6x_4\cdots x_{n-6}$ or $x_1x_2^6x_3^2x_4\cdots x_{n-6},\ x_1^4\cdots x_{n-6}^{m_{n-6}}$ or $x_1^3x_2^4x_3^2\cdots x_{n-6}$ or $x_1x_2^4\cdots x_{n-6}^{m_{n-6}}$ or $x_1x_2x_3^4\cdots x_{n-6}^{m_{n-6}}$, those with a factors of the form $x_i^5x_j^3,\ i<j,\ x_1x_2^2x_3^2\cdots x_{n-6}^{m_{n-6}}$ or those with a factors of the form $x_i^5x_j^2x_k^2,\ i<j,k$, all of which are clearly inadmissible. Thus

$$\begin{aligned}\dim(\mathbf{W}^n(n-6)) \geq\ & n-6+((n-7)(n-8)-2)+\binom{n-6}{2}+\binom{n-6}{3}+2\cdot\binom{n-7}{3}-(n-9)\\ & +(n-6).\binom{n-8}{2}+\binom{n-6}{3}-1+\tfrac{(n-7)!}{4\cdot(n-11)!}+(n-8).\binom{n-9}{2}+\binom{n-9}{2},\end{aligned}$$

since $a_{n-6}, b_{n-6}, c_{n-6}, d_{n-6}, e_{n-6}, f_{n-6}, g_{n-6}$, have, respectively,

$$\binom{n-6}{3},\ n-6,\ \binom{n-6}{2},\ \frac{(n-7)!}{4\cdot(n-11)!}+(n-8)\cdot\binom{n-9}{2}+\binom{n-9}{2},\ (n-6)\cdot\binom{n-8}{2}+\binom{n-6}{3}-1,\ (n-7)(n-8)-2,\ 2\cdot\binom{n-7}{3}-(n-9)$$

permutation representatives which are admissible. Equality holds when $n = 11, 12$.

It is known, (Sum & Phuc, 2013), that $\dim(\mathbf{Q}^{12}(5)) = 190$. Since $\dim(\mathbf{Q}^{12}(4)) = 21$, we must have $\dim(\mathbf{W}^{12}(5)) = 85$. This completes the proof of the lemma hence that of Theorem 2.

References

Adem, J. (1957). *The relations on Steenrod powers of cohomology classes.* Algebraic Geometry and Topology, a symposium in honour of S. Lefschetz, Princeton Univ. Press, Princeton NJ, 191-238.

Kameko, M. (1990). *Products of projective spaces as Steenrod modules.* Ph.D. Thesis, John Hopkins University, USA.

Kameko, M. (2003). *Generators of the cohomology of BV_4* Preprint: Toyama Univ.

Mothebe, M. F. (2016). Products of admissible monomials in the polynomial algebra as a module over the Steenrod algebra. *Journal of Mathematics Research, 8*(3), 112-116. http://dx.doi.org/10.5539/jmr.v8n3p112

Mothebe, M. F., & Uys, L. (2015). Some Relations between admissible monomials for the polynomial algebr. *International Journal of Mathematics and Mathematical Sciences.* http://dx.doi.org/10.1155/2015/235806

Mothebe, M. F. (2009). Dimensions of the polynomial algebra $\mathbb{F}_2[x_1, \ldots, x_n]$ as a module over the Steenrod algebra. *JP Journal of Algebra, Number Theory and its Applications, 13*(2), 161-170.

Nam, T. N. (2004). $\mathcal{A}$-générateurs génériques pour l' algèbre polynomiale. *Advances in Mathematics, 186*, 334-362. http://dx.doi.org/10.1016/j.aim.2003.08.004

Peterson, F. P. (1987). Generators of $H^*(RP^\infty \wedge RP^\infty)$ as a module over the Steenrod algebra. *Abstracts American Mathematical Society, 833*, 55-89.

Peterson, F. P. (1989). $\mathcal{A}$-generators for certain polynomial algebras. *Mathematical Proceedings of the Cambridge Philosophical Society, 105*, 311-312. http://dx.doi.org/10.1017/S0305004100067803

Silverman, J. H. (1998). Hit polynomials and conjugation in the dual Steenrod algebra *Mathematical Proceedings of the Cambridge Philosophical Societys, 123*, 531-547. http://dx.doi.org/10.1017/S0305004197002302

Singer, W. M. (1989). The transfer in homological algebra. *Mathematische Zeitschrift, 202*, 493-523. http://dx.doi.org/10.1007/BF01221587

Singer, W. M. (1991). On the action of Steenrod squares on polynomials. *Proceedings of the American Mathematical Society, 111*, 577-583. http://dx.doi.org/ 10.2307/2048351

Sum, N. (2007). *The hit problem for the polynomial algebra of four variables.* Preprint, University of Quy Nhon, Viet Nam.

Sum, N. (2010). The negative answer to Kameko's conjecture on the hit problem. *Advances in Mathematics, 225*, 2365-2390.

Sum, N., & Phuc, D. V. (2013). *On a minimal set of generators for the polynomial algebra of five variables as a module over the Steenrod algebra.* Preprint, University of Quy Nhon, Viet Nam.

Wood, R. M. W. (1989). Steenrod squares of polynomials and the Peterson conjecture. *Mathematical Proceedings of the Cambridge Philosophical Societys, 105*, 307-309. http://dx.doi.org/10.1017/S0305004100067797

16

Asymptotic Behavior of Higher Order Quasilinear Neutral Difference Equations

V. Sadhasivam[1], Pon. Sundar[2] & A. Santhi[1]

[1] PG and Research Department of Mathematics, Thiruvalluvar Government Arts College, Rasipuram, Namakkal - 637 401, Tamil Nadu, India

[2] Om Muruga College of Arts and Science, Salem - 636 303, Tamil Nadu, India

Correspondence: V. Sadhasivam, PG and Research Department of Mathematics, Thiruvalluvar Government Arts College, Rasipuram, Namakkal - 637 401, Tamil Nadu, India. E-mail: ovsadha@gmail.com

Abstract

We study, the asymptotic behavior of solutions to a class of higher order quasilinear neutral difference equations under the assumptions that allow applications to even and odd-order difference equations with delayed and advanced arguments, as well as to functional difference equations with more complex arguments that may for instance, alternate infinitely between delayed and advanced types. New theorems extend a number of results reported in the literature. Illustrative examples are presented.

Keywords: asymptotic behavior, higher order, quasilinear, even and odd order, delay and advanced arguments, functional difference equation

1. Introduction

In this paper, we study the asymptotic behavior of solutions to a class of higher-order quasilinear neutral functional difference equations

$$\Delta\left(r(n)\left(\Delta^{m-1}z(n)\right)^{\alpha}\right) + q(n)x^{\beta}(\sigma(n)) = 0 \tag{1}$$

where $n \geq n_0 \in N_0 = \{n_0, n_0+1, n_0+2, \cdots\}$, $z(n) = x(n) + p(n)x(\tau(n))$ and Δ is the forward difference operator defined by

$$\Delta x(n) = x(n+1) - x(n) \text{ and } \Delta^i x(n) = \Delta\left(\Delta^{i-1}x(n)\right), \quad i = 1, 2, \cdots, m-1.$$

We assume the following conditions on equation (1), without further mention

(c_1) $\{r(n)\}, \{\sigma(n)\}$ and $\{\tau(n)\}$ are positive sequences of real numbers such that $\Delta r(n) \geq 0$, $r(n) > 0$, $\lim_{n\to\infty} \sigma(n) = \infty$.

(c_2) $\{p(n)\}$ and $\{q(n)\}$ are sequences of positive real numbers such that $q(n) \geq 0$ and $q(n)$ does not vanish eventually.

(c_3) α and $\beta \in \Re$ where $\Re$ stands for the set consisting of all quotients of odd positive integers.

Analysis of qualitative properties of equation (1) is important not only for the sake of further development of the oscillation theory, but for practical reasons too. In fact, an Emden-Fowler type differential equation

$$\left(r(t)\left(x^{(n-1)}(t)\right)^{\alpha}\right)' + q(t)x^{\beta}(\sigma(t)) = 0 \tag{2}$$

has numerous applications in physics and engineering: that is; for instance, the papers by Ou and Wong [2004].

By a solution of equation (1), we mean a real sequence $\{x(n)\}$ which is defined for $n \geq -\mu = \max\{\sup\{\tau(n), \sigma(n)\}$ and which satisfies the equation (1) for $n \geq n_0$. We deal only with proper solution $x(n)$ of (1) that satisfy the condition $\sup\{|x(n)| : n \geq N\} > 0$ for all $N \geq N_\mu$ and tacitly assume that (1) possesses such solutions. A solution of (1) is said to be oscillatory if it has arbitrarily large zeros on $n \geq N_\mu$. Otherwise, it is termed nonoscillatory.

For several decades, an increasing interest in obtaining sufficient conditions for oscillatory and nonoscillatory behavior of different classes of difference equations has been observed; see, for instance, the monographs [Agarwal, 1992; Agarwal and Wong, 1997; Elaydi, 1995; Gyori and Ladas, 1991; Lakshmikanthan and Trigiante, 1988], the papers [Graef et al, 1996; Guan and Yang, 1999; He, 1993; Jurang and Bin, 1997; Karpuz, 2009; Sundar and Kishorkumar, 2014; Wong,

1975; Yu and Wang, 1994; Zafer, 1995] and the references cited therein. Let us briefly comment on a number of related results which motivated our study.

S.S. Cheng and W.T. Patula [1993] studied the difference equation

$$\Delta(\Delta y_{k-1})^{p-1} + s_k y_k^{p-1} = 0 \tag{3}$$

where $p > 1$ and proved an existence theorem for the equation (3).

X.Zhou and J. Yan [1998] studied the difference equation

$$\Delta\left[p_{n-1}(\Delta y_{n-1})^{\delta}\right] + s_n y_n^{\delta} = 0 \tag{4}$$

and they obtained some comparison results and necessary and sufficient conditions for the oscillation of equation (4).

I. Kvbiaczyk and S.H. Sekar [2002] studied the second order sublinear delay difference equation

$$\Delta(p_n(\Delta x_n)) + q_n x_{n-\tau}^{\gamma} = 0, \quad 0 \le \gamma \le 1. \tag{5}$$

By using Riccati transformation techniques the authors obtained oscillation criteria for the equation (5) under the conditions

$$\sum_{n=n_0}^{\infty} \frac{1}{p_n} = \infty \quad \text{and} \quad \sum_{n=n_0}^{\infty} \frac{1}{p_n} < \infty.$$

Y. Bolat and O. Alzabot [2012] considered the half-linear delay difference equation

$$\Delta\left[p_n\left(\Delta^{m-1}\left(x_n + q_n x_{\tau(n)}\right)\right)^{\alpha}\right] + \gamma_n x_{\sigma_n}^{\beta} = 0, \quad n \ge n_0 \tag{6}$$

under the condition $\sum\limits_{n_0}^{\infty} \frac{1}{(p_n)^{1/\alpha}} < \infty$ and with using that $\Delta p_n \ge 0$ and derived some oscillation and asymptotic criteria for the equation (6).

M.K. Yildiz and O. Ocalan [2007] studied the neutral difference equation

$$\Delta^m(y_n - p_n y_{n-k}) + q_n y_{n-l}^{\alpha} = 0, \quad n \ge n_1 \tag{7}$$

where $1 > \alpha > 0$ is a quotient of odd positive integers and $\{p_n\}$ satisfies $-1 < p_n < 1$.

J. Luo in 2002 [2002] considered the second order quasilinear neutral delay difference equation

$$\Delta\left[a_{n-1}\left|\Delta(x_{n-1} + p_{n-1}x_{n-1-\sigma})\right|^{\alpha-1}\Delta(x_{n-1} + p_{n-1}x_{n-1-\sigma})\right] + q_n f(x_{n-\tau}) = 0 \tag{8}$$

under the condition $\sum^{\infty} \frac{1}{a_n^{1/\alpha}} = \infty$ and obtained several oscillation results.

Pon. Sundar and E. Thandapani [2000], considered the second order quasilinear functional difference equation

$$\Delta\left(|\Delta y(n)|^{\alpha-1}\Delta y(n)\right) + f(n, y(\sigma(n))) = 0, \quad n \ge n_0 \tag{9}$$

and established some necessary and sufficient conditions for the equation (9) to have various types of nonoscillatory solutions.

Our principal goal is to analyze the asymptotic behavior of solutions to (1) in the case where the condition

$$\sum_{n_0}^{\infty} \frac{1}{(r(n))^{1/\alpha}} < \infty \tag{10}$$

holds. We provide sufficient conditions which ensure that solution to (1) are either oscillatory or approach zero at infinity. In some cases, we reveal oscillatory nature of (1).

As usual, all functional inequalities are supposed to hold for all t large enough. Without loss of generality, we deal only with positive solutions of (1) since, under our assumptions, if $x(n)$ is a solution, then $-x(n)$ is a solution of this equation too.

In the sequel, we denote by τ^{-1} the function which is inverse to τ. We also adopt the following notations for the compact presentation of our results.

$$\begin{aligned}
A(n) &= \sum_{n}^{\infty} \frac{1}{(r(n))^{1/\alpha}} \\
Q(n) &= \min\{q(n), q(\tau(n))\} \\
R(n) &= \max\{r(n), r(\tau(n))\} \\
Q_\gamma(n) &= Q(n)\left(\frac{(\eta_1(n))^{m-1}}{\gamma^{1/\alpha}(\eta_1(n))}\right)^{\gamma} \\
Q_\beta(n) &= Q(n)((\sigma(n))^{(m-2)}\gamma)^{\beta}, \quad \tilde{Q}_\beta(n) = Q(n)\left(\frac{(\eta_1(n))^{(m-1)}}{\gamma^{1/\beta}(\eta_1(n))}\right)^{\beta} \\
Q_\theta(n) &= Q(n)\left(\sum_{\eta_3(n)}^{\infty} (\eta - \eta_3(n))^{(m-3)} A(\eta(n))\right)^{\theta} \\
\bar{Q}_\beta(n) &= Q(n)\left(\sum_{\eta_3(n)}^{\infty} (\eta - \eta_3(n))^{(m-3)} A(\eta(n))\right)^{\beta} \\
\widehat{Q}_\beta(n) &= Q(n)\left(\frac{\sigma_n^{(m-1)}}{\gamma^{1/\beta}(\sigma(n))}\right)^{\beta}, \quad \breve{Q}_\gamma(n) = Q(n)\left(\frac{\sigma^{(m-1)}(n)}{\gamma^{1/\alpha}(\sigma(n))}\right)^{\gamma}
\end{aligned}$$

where the meaning of γ, θ, η_i and η_3 will be explained later.

2. Asymptotic Behavior of Solutions to Even-order Equations

In what follows, $\tau(n)$ can be both a delayed or an advanced argument. Throughout this section, in addition to the basic assumptions listed in the introduction, it is also supposed that (10) holds along with

(c_4) $0 \le p(n) \le p_0 < \infty$ for some constant p_0;

(c_5) $\Delta\tau(n) \ge \tau_* > 0$ and $\tau \circ \sigma = \sigma \circ \tau$

We shall need the following lemmas which are useful is the sequel.

Lemma 1. [Gyori and Ladas, 1991] *Assume that* $q(n) \ge 0$ *for all* $n \in N$ *and*

$$\liminf_{n\to\infty} \left[\sum_{n-l}^{n-1} q(s)\right] > \left(\frac{l}{l+1}\right)^{l+1}. \tag{11}$$

Then,

(i) $v(n+1) - v(n) + q(n)\, v(n-l) \le 0,\ n \in N$ *has no eventually positive solution.*

(ii) $v(n+1) - v(n) + q(n)\, v(n-l) \ge 0, n \in N$ *has no eventually negative solution*

(iii) $v(n+1) - v(n) + q(n)\, v(n-l) = 0$ *is oscillatory.*

Lemma 2. [Agarwal, 1992](Discrete Kingser's Theorem) *Let* $z(n)$ *be defined for* $n \ge a$, *and* $z(n) > 0$ *with* $\Delta^m z(n)$ *of constant sign for* $n \ge a$ *and not identically zero. Then, there exists an integer* j, $0 \le j \le m$ *with* $(m+j)$ *odd for* $\Delta^m z(n) \le 0$ *and* $(m+j)$ *even for* $\Delta^m z(n) \ge 0$, *such that*

$$j \le m-1 \ \text{implies}\ (-1)^{j+i}\Delta^i z(n) > 0, \ \text{for all}\ n \ge a, \ \ j \le i \le m-1, \tag{12}$$

and

$$j \ge 1 \ \text{implies}\ \Delta^i z(n) > 0 \ \text{for all large}\ n \ge a,\ 1 \le i \le j-1. \tag{13}$$

Lemma 3. [Agarwal, 1992] *Let $z(n)$ be defined for $n \geq a$ and $z(n) > 0$ with $\Delta^m z(n) \leq 0$ for $n \geq a$ and not identically zero. Then, there exists a large $n_1 \geq a$ such that*

$$z(n) \geq \frac{(n - n_2)^{(m-1)}}{(m-1)!} \Delta^{(m-1)} z(2^{m-j-1})(n), \quad n \geq n_1, \tag{14}$$

where $\lambda = \left(\frac{1}{2^{m-1}}\right)^{(m-1)}$ *and j is defined us in Lemma 2. Further $z(n)$ is increasing, then*

$$z(n) \geq \frac{\lambda}{(m-1)!} n^{(m-1)} \Delta^{m-1} z(n), \quad n \geq 2^{m-1} n. \tag{15}$$

Lemma 4. *Assume that $A \geq 0, B \geq 0, \alpha \geq 1$. Then*

$$(A + B)^\alpha \geq A^\alpha + B^\alpha.$$

Proof. If $A = 0$ or $B = 0$, then the inequality holds trivially. For $A \neq 0$, setting $x = \frac{B}{A}$. The inequality takes the form $(1 + x)^\alpha \geq 1 + x^\alpha$ which is for $x > 0$ evidently true.

Lemma 5. *Assume that $A \geq 0, B \geq 0, 0 < \alpha \leq 1$. Then*

$$(A + B)^\alpha \geq \frac{A^\alpha + B^\alpha}{2^{1-\alpha}}.$$

Proof. We may assume that $0 < A < B$. Consider the function $f(n) = u^\alpha$. Since $\Delta^\alpha f(n) < 0$ for $n > 0$, the sequence $f(n)$ is concave down; that is

$$f\left(\frac{A + B}{2}\right) \geq \frac{f(A) + f(B)}{2}$$

which implies the desired inequality.

Theorem 6. *Let $m \geq 2$ be even and $0 < \beta \leq 1$. Assume that conditions (c_4) and (c_5) are satisfied, and there exist two real numbers $\gamma, \lambda \in R$ such that $\gamma \leq \beta \leq \lambda$ and $\gamma < \alpha < \lambda$. Suppose further that there exist two sequence $\eta_1^{(n)}, \eta_2^{(n)}$ such that*

$$\eta_1(n) \leq \sigma(n) \leq \eta_2(n), \quad \eta_1(n) < n < \tau(n) \leq \eta_2(n) \tag{16}$$

$$\lim_{n \to \infty} \eta_1(n) = \infty.$$

If

$$\sum^{\infty} Q_\gamma(n) = \infty \tag{17}$$

and

$$\sum^{\infty} Q_\beta(n) \Delta^\lambda(\eta_2(n)) = \infty. \tag{18}$$

Then every solution $x(n)$ of (1) is either oscillatory or satisfies

$$\lim_{n \to \infty} x(n) = 0. \tag{19}$$

Proof. Assume that equation (1) has a nonoscillatory solutions $x(n)$ which is eventually positive and such that

$$\lim_{n \to \infty} x(n) \neq 0. \tag{20}$$

Then $z(n)$ satisfies

$$\begin{aligned} z(\sigma(n)) &= x(\sigma(n)) + p(\sigma(n)) x(\tau(\sigma(n))) \\ &\leq x(\sigma(n)) + p_0 x(\tau(\sigma(n))). \end{aligned} \tag{21}$$

In view of (1), we have

$$0 = p_0^\beta \Delta\left[r(\tau(n))\left(\Delta^{m-1}z(\tau(n))\right)^\alpha\right] + p_0^\beta q(\tau(n))x^\beta(\sigma(\tau(n))) \tag{22}$$

Using (22), we obtain

$$\begin{aligned} &q(n)x^\beta(\sigma(n)) + p_0^\beta q(\tau(n))x^\beta(\sigma(\tau(n))) \\ &\quad = q(n)x^\beta(\sigma(n)) + p_0^\beta q(\tau(n))x^\beta(\tau(\sigma(n))) \\ &\quad \geq Q(n)z^\beta(\sigma(n)). \end{aligned} \tag{23}$$

It follows from equation (1), (22) and (23) that

$$\Delta\left[r(n)(\Delta^{m-1}z(n))^\alpha + p_0^\beta r(\tau(n))(\Delta^{m-1}z(\tau(n)))^\alpha\right] + Q(n)z^\beta(\sigma(n)) \leq 0. \tag{24}$$

Then there exist two possible cases:

(i)

$$z(n) > 0,\ \Delta^{m-1}z(n) > 0,\ \Delta^m z(n) \leq 0,\ \Delta\left[r(n)(\Delta^{m-1}z(n))^\alpha\right] \leq 0 \tag{25}$$

(ii)

$$z(n) > 0,\ \Delta^{m-2}z(n) > 0,\ \Delta^{m-1}z(n) < 0,\ \Delta\left[r(n)(\Delta^{m-1}z(n))^\alpha\right] \leq 0 \tag{26}$$

for $n \geq n_1$ where $n_1 \geq n_0$ is large enough.

Case 1: Suppose first that conditions (25) hold. Using inequality (24) and assumption $\eta_1(n) \leq \sigma(n)$, we conclude that

$$\Delta\left[r(n(\Delta^{m-1}z(n))^\alpha) + p_0^\beta r(\tau(n))(\Delta^{m-1}z(\tau(n)))^\alpha\right] + Q(n)z^\beta(\eta_1(n)) \leq 0. \tag{27}$$

Furthermore, by the monotonicity of $z(n)$, there exists a constants $M > 0$ such that

$$z^\beta(\eta_1(n)) = z^{\beta-\gamma}(\eta_1(n))z^\gamma(\eta_1(n)) \geq M^{\beta-\gamma}z^\gamma(\eta_1(n)). \tag{28}$$

Combining (27) and (28), we have

$$\Delta\left[r(n)(\Delta^{m-1}z(n))^\alpha + p_0^\beta r(\tau(n))(\Delta^{m-1}z(\tau(n)))^\alpha\right] + M_1 Q(n)z^\gamma(\eta_1(n)) \leq 0, \tag{29}$$

where $M_1 = M^{\beta-\gamma}$. An application of conditions (25) allows us to deduce that the sequence

$$w(n) = r(n)(\Delta^{m-1}z(n))^\alpha \tag{30}$$

is positive and nonincreasing. By Lemma 3, we have

$$\begin{aligned} z(n) &\geq \frac{\lambda n^{(m-1)}}{(m-1)!r^{1/\alpha}(n)} r^{1/\alpha}\Delta^{m-1}z(n) \\ &= \frac{\lambda n^{(m-1)}}{(m-1)!r^{1/\alpha}} w^{1/\alpha}(n) \end{aligned} \tag{31}$$

for every $\lambda \in (0,1)$ and for sufficiently large n. Using (31) in (29), we conclude that $w(n)$ is a positive solution of a delay difference inequality

$$\Delta\left(w(n) + p_0 w(\tau(n))\right) + M_1\left(\frac{\lambda}{(m-1)!}\right)^\gamma Q_r(n)w^{\gamma/\alpha}(\eta_1(n)) \leq 0. \tag{32}$$

Define now a function $y(n)$ by

$$y(n) = w(n) + p_0 w(\tau(n)). \tag{33}$$

Then by the monotonicity of $w(n)$,

$$y(n) \le w(n)(1 + p_0^\beta). \tag{34}$$

Substituting (34) into (32), we observe that $y(n)$ is a positive solution of a delay difference inequality

$$\Delta y(n) + M_1\left(\frac{\lambda}{(m-1)!}\right)^\gamma \left(\frac{1}{1+p_0^\beta}\right)^{\gamma/\alpha} Q_\gamma(n) y^{\gamma/\alpha}(\eta_1(n)) \le 0. \tag{35}$$

Then, by virtue of [Gyori and Ladas, 1991], the associated delay difference equation

$$\Delta y(n) + M_1\left(\frac{\lambda}{(m-1)!}\right)\left(\frac{1}{1+p_0^\beta}\right) Q_\gamma(n) y^{\gamma/\alpha}(\eta_1(n)) = 0 \tag{36}$$

also has a positive solution. However, by Lemma 1 implies that, under assumption (17), equation (36) is oscillatory. Therefore, (1) cannot have positive solutions.

Case 2: Assume now that conditions (26) hold. By virtue of (20), we have that

$$\lim_{n\to\infty} z(n) \ne 0. \tag{37}$$

An application of Lemma 3 yields

$$z(n) \ge \frac{\lambda}{(m-2)!} n^{(m-2)} \Delta^{m-2} z(n) \tag{38}$$

for any $\lambda \in (0, 1)$ and for sufficiently large n. Hence, by (24) and (38), we obtain

$$\Delta\left[r(n)\left(\Delta^{m-1}z(n)\right)^\alpha + p_0^\beta r(\tau(n))\left(\Delta^{m-1}z(\tau(n))\right)^\alpha\right] + \left(\frac{\lambda}{(m-2)!}\right)^\beta Q_\beta(n)\left(\Delta^{m-2}z(\sigma(n))\right)^\beta \le 0. \tag{39}$$

Using conditions $\Delta^{m-1}z(n) < 0$, $\sigma(n) \le \eta_2(n)$, and inequality (39), we have

$$\Delta\left[r(n)\left(\Delta^{m-1}z(n)\right)^\alpha + p_0^\beta r(\tau(n))\left(\Delta^{m-1}z(\tau(n))\right)^\alpha\right] + \left(\frac{\lambda}{(m-2)!}\right)^\beta Q_\beta(n)\left(\Delta^{m-2}z(\eta_1(n))\right)^\beta \le 0. \tag{40}$$

Furthermore, by the monotonicity of $\Delta^{m-2}z(n)$, there exists a constant $N > 0$ such that

$$\begin{aligned}\left(\Delta^{m-2}z(\eta_1(n))\right)^\beta &= \left(\Delta^{m-2}z(\eta_1(n))\right)^{\beta-\lambda}\left(\Delta^{m-2}z(\eta_1(n))\right)^\lambda \\ &\ge N^{\beta-\lambda}\left(\Delta^{m-2}z(\eta_1(n))\right)^\lambda\end{aligned} \tag{41}$$

Combining (40) and (41), we arrive at

$$\Delta\left[r(n)\left(\Delta^{m-1}z(n)\right)^\alpha + p_0^\beta r(\tau(n))\left(\Delta^{m-1}z(\tau(n))\right)^\alpha\right] + N_1\left(\frac{\lambda}{(m-2)!}\right)^\beta Q_\beta(n)\left(\Delta^{m-2}z(\eta_2(n))\right)^\lambda \le 0 \tag{42}$$

where $N_1 = N^{\beta-\lambda}$. Using the monotonicity of $w(n)$, for $s \ge n \ge n_1$ we conclude that

$$r^{1/\alpha}(s)\Delta^{m-1}z(s) \le r^{1/\alpha}\Delta^{m-1}z(n). \tag{43}$$

Dividing (43) by $r^{1/\alpha}(s)$ and summing the resulting inequality from n to $l-1$, we obtain

$$\Delta^{m-2}z(l) \le \Delta^{m-2}z(n) + r^{1/\alpha}(n)\Delta^{m-1}z(n)\sum_{n}^{l-1}\frac{1}{r^{1/\alpha}(s)}. \tag{44}$$

Passing to the limit as $l \to \infty$, we deduce that

$$0 \le \Delta^{m-1}z(n) + r^{1/\alpha}\Delta^{m-1}z(n)A(n) \tag{45}$$

which yields

$$\Delta^{m-2}z(n) \geq -A(n)r^{1/\alpha}(n)\Delta^{m-1}z(n) = -A(n)w^{1/\alpha}(n) \tag{46}$$

Combining (42) and (46), we have

$$\Delta\left[r(n)\left(\Delta^{m-1}z(n)\right)^{\alpha} + p_0^{\beta}r(\tau(n))\left(\Delta^{m-1}z(\tau(n))\right)^{\alpha}\right] - N_1\left(\frac{\lambda}{(m-2)!}\right)^{\beta} Q_{\beta}(n)A^{\lambda}(\eta_2(n))w^{\lambda/\alpha}(\eta_2(n)) \leq 0 \tag{47}$$

Using the monotonicity of $w(n)$, we conclude that

$$y(n) \geq w(\tau(n))[1 + p_0^{\beta}] \tag{48}$$

Using (48) into (47), we observe that $y(n)$ is a negative solution of an advanced difference inequality

$$\Delta y(n) - N_1\left(\frac{\lambda}{(m-2)!}\right)^{\beta}\left(\frac{1}{1+p_0^{\beta}}\right)^{\lambda/\alpha} Q_{\beta}(n)A^{\lambda}(\eta_2(n))A^{\lambda}(\eta_2(n))y^{\lambda/\alpha}(\tau^{-1}(\eta_2(n))) \leq 0 \tag{49}$$

which implies that $u(n) = -y(n)$ is a positive solution of an advanced difference inequality

$$\Delta u(n) - N_1\left(\frac{\lambda}{(m-2)!}\right)^{\beta}\left(\frac{1}{1+p_0^{\beta}}\right)^{\lambda/\alpha} Q_{\beta}(n)A^{\lambda}(\eta_2(n))A^{\lambda}(\eta_2(n))u^{\lambda/\alpha}(\tau^{-1}(\eta_2(n))) = 0 \tag{50}$$

consequently, by virtue of [Gyori and Ladas, 1991] the associated advanced difference equation

$$\Delta u(n) - N_1\left(\frac{\lambda}{(m-2)!}\right)^{\beta}\left(\frac{1}{1+p_0^{\beta}}\right)^{\lambda/\alpha} Q_{\beta}(n)A^{\lambda}(\eta_2(n))A^{\lambda}(\eta_2(n))u^{\lambda/\alpha}(\tau^{-1}(\eta_2(n))) = 0 \tag{51}$$

also has a positive solution. However, by [Sundar and Murugesan, 2010] implies that, under assumption (18) equation (51) is oscillatory. Therefore, (1) cannot have a positive solution this contradiction with initial assumption completes the proof.

Theorem 7. *Let $m \geq 2$ be even, and let $0 < \alpha = \beta \leq 1$. Assume that conditions (c_4) and (c_5) hold, and there exist two sequences $\eta_1(n)$, $\eta_2(n)$ satisfying (16). Suppose also that conditions*

$$\frac{1}{((m-1)!)^{\beta}}\frac{1}{(1+p_0^{\beta})}\liminf_{n\to\infty}\sum_{\eta_1(n)}^{n-1}\tilde{Q}_{\beta}(n) > \left(\frac{k}{k+1}\right)^{k+1} \tag{52}$$

where k is the delay argument and

$$\frac{1}{((m-2)!)^{\beta}}\frac{1}{(1+p_0^{\beta})}\liminf_{n\to\infty}\sum_{n}^{\tau^{-1}(\eta_2(n))-1} Q_{\beta}(s)A^{\beta}(\eta_2(s)) > \left(\frac{l-1}{l}\right)^{l} \tag{53}$$

where l is the advance argument are satisfied. Then conclusion of Theorem 6 remains in fact.

Proof. Assume that $x(n)$ is an eventually positive solution of equation (1) that satisfies (20). Proceeding as in the proof of Theorem 6, one comes to the conclusion that, for every $\lambda \in (0, 1)$, a delay difference equation

$$\Delta y(n) + \left(\frac{\lambda}{(m-1)!}\right)^{\beta}\frac{1}{1+p_0^{\beta}}\tilde{Q}_{\beta}(n)y(\eta_1(n)) = 0 \tag{54}$$

and an advanced difference equation

$$\Delta y(n) - \left(\frac{\lambda}{(m-2)!}\right)^{\beta}\frac{1}{1+p_0^{\beta}}Q_{\beta}(n)A^{\beta}(\eta_2(n))u(\tau^{-1}(\eta_1(n))) = 0 \tag{55}$$

both have positive solutions. On the other hand, condition (52) and [Gyori and Ladas, 1991] imply that equation (54) is oscillatory, a contradiction. Likewise, by virtue of [Sundar and Murugesan, 2010, Lemma 2.3.2] condition (53) yields that equation (55) is oscillatory. This contradiction completes the proof.

Theorem 8. *Let $m \geq 2$ be even and $0 < \beta \leq 1$. Assume that conditions (c_4) and (c_5) are satisfied, and there exist two numbers $\lambda, \gamma \in R$ as in Theorem 6 and two real sequences $\eta_1(n)$, $\eta_2(n)$ such that*

$$\eta_1(n) \leq \sigma(n) \leq \eta_2(n), \quad \eta_1(n) < \tau(n) \leq n < \eta_2(n), \quad \lim_{n\to\infty} \eta_1(n) = \infty. \tag{56}$$

If conditions (17) and (18) hold, the conclusion of Theorem 6 remains infact.

Proof. As above let $x(n)$ be an eventually positive solution of equation (1) that satisfies (20). As in the proof of Theorem 6, we split the arguments into two parts.

Case 1: Assume first that (25) is satisfied. If has been established in the proof of Theorem 6 that the sequence $w(n)$ defined by (30) is positive, nonincreasing and satisfies inequality (32). Introducing again $y(n)$ by (33) and using the monotonicity of $w(n)$, we conclude that

$$y(n) \leq w(\tau(n))(1 + p_0^\beta) \tag{57}$$

substituting of (57) into (32) implies that, for sufficiently large n, $y(n)$ is a positive solution of a delay difference inequality

$$\Delta y(n) + M_1 \left(\frac{\lambda}{(m-1)!}\right)^\gamma \left(\frac{1}{1+p_0^\beta}\right)^{\gamma/\alpha} Q_\gamma(n) y^{\gamma/\alpha}(\tau^{-1}\eta_1(n)) \leq 0 \tag{58}$$

Then, the associated difference equation

$$\Delta y(n) + M_1 \left(\frac{\lambda}{(m-1)!}\right)^\gamma \left(\frac{1}{1+p_0^\beta}\right)^{\gamma/\alpha} Q_\gamma(n) y^{\gamma/\alpha}(\tau^{-1}\eta_1(n)) = 0 \tag{59}$$

also has a positive solution. However, by [Sundar and Murugesan, 2010, Lemma 2.3.2] implies that , under assumption (17) equation (59) is oscillatory. Therefore equation (1) cannot have positive solutions.

Case 2: Assume that (26) is satisfied. It has been established in the proof of Theorem 6 that the sequence $w(n)$ defined by (30) is negative, nonincreasing and satisfies the inequality (47). Introducing again $y(n)$ by (33) and using the monotonicity of $w(n)$, we conclude that

$$y(n) \geq w(n)(1 + p_0^\beta) \tag{60}$$

substituting (60) into (47), we observe that $y(n)$ is a negative solution of an advanced difference inequality

$$\Delta y(n) - N_1 \left(\frac{\lambda}{(m-2)!}\right)^\beta \left(\frac{1}{1+p_0^\beta}\right)^{\lambda/\alpha} Q_\beta(n) A^\lambda(\eta_2(n)) y^{\lambda/\alpha}(\eta_2(n)) \leq 0. \tag{61}$$

That is, $u(n) = -y(n)$ is a positive solution of an advanced difference equality

$$\Delta u(n) - N_1 \left(\frac{\lambda}{(m-2)!}\right)^\beta \left(\frac{1}{1+p_0^\beta}\right)^{\lambda/\alpha} Q_\beta(n) A^\lambda(\eta_2(n)) u^{\lambda/\alpha}(\eta_2(n)) \geq 0. \tag{62}$$

Then, by virtue of [Gyori and Ladas, 1991] the associated advanced difference equation

$$\Delta u(n) - N_1 \left(\frac{\lambda}{(m-2)!}\right)^\beta \left(\frac{1}{1+p_0^\beta}\right)^{\lambda/\alpha} Q_\beta(n) A^\lambda(\eta_2(n)) u^{\lambda/\alpha}(\eta_2(n)) = 0. \tag{63}$$

also has a positive solution. However, by [Sundar and Murugesan, 2010, Lemma 2.3.2] implies that; under assumption (18), equation (63) is oscillatory. Therefore, equation (1) cannot have positive solutions. This contradiction with our initial assumption complete the proof.

Theorem 9. *Let $m \geq 2$ be even and $0 < \alpha = \beta \leq 1$. Assume that conditions (c_4) and (c_5) are satisfied, and there exist two real sequences $\eta_1(n), \eta_2(n)$ satisfying (56). Suppose also that*

$$\frac{1}{((m-1)!)^\beta (1 + p_0^\beta)} \liminf_{n\to\infty} \sum_{\tau^{-1}(\eta_1(n))}^{n-1} \tilde{Q}_\beta(s) > \left(\frac{k}{k+1}\right)^{k+1} \tag{64}$$

where k denotes the delay arguments and

$$\frac{1}{((m-2)!)^{\beta}(1+p_0^{\beta})} \liminf_{n\to\infty} \sum_{n}^{\eta_1(n)-1} Q_{\beta}(s)A^{\beta}(\eta_2(s)) > \left(\frac{l-1}{l}\right)^{l} \tag{65}$$

where l denotes the advance arguments, are satisfied. Then conclusion of Theorem 6 remains infact.

Proof. Assume that $x(n)$ is an eventually positive solution of equation (1) that satisfies (20) and proceeding as in the proof of Theorem 8, one concludes that; for every $\lambda \in (0, 1)$ a delay difference equation

$$\Delta y(n) + \left(\frac{\lambda}{((m-1)!}\right)^{\beta} \frac{1}{1+p_0^{\beta}} \tilde{Q}_{\beta}(n) y(\tau^{-1}(\eta_1(n))) = 0 \tag{66}$$

and an advanced difference equation

$$\Delta u(n) - \left(\frac{\lambda}{(m-1)!}\right)^{\beta} \frac{1}{1+p_0^{\beta}} Q_{\beta}(n) A^{\beta}(\eta_2(n)) u(\eta_1(n)) = 0 \tag{67}$$

have positive solutions. On the other hand, application of condition (64) along with [Gyori and Ladas, 1991] imply that equation (66) is oscillatory, a contradiction. Likewise, by virtue of [Sundar and Murugesan, 2010, Lemma 2.3.2] condition (65) yields that equation (67) is oscillatory. This contradiction completes the proof.

Note that Theorems 6 to 9 ensure that every solution $x(n)$ of equation (1) is either oscillatory or tends to zero as $n \to \infty$ and unfortunately cannot distinguish solutions with different behaviors. In the remaining part of this section, we establish several results which guarantee that all solutions of equation (1) are oscillatory.

Theorem 10. *Let $m \geq 4$ be even and $0 < \beta \leq 1$. Assume that conditions (c_4) and (c_5) are satisfied, and there exist there numbers $\gamma, \lambda, \theta \in R$ such that $\gamma \leq \beta \leq \lambda$, $\gamma < \alpha < \lambda$, $\theta \geq \beta$ and $\theta > \alpha$. Suppose further that there exist three real sequences $\eta_1(n), \eta_2(n), \eta_3(n)$*

$$\eta_3(n) \geq \sigma(n), \qquad \eta_3(n) > c(n) \tag{68}$$

and such that (16) holds. Assume that conditions (17), (18) and

$$\sum^{\infty} Q_{\theta}(n) = \infty \tag{69}$$

hold. Then equation (1) is oscillatory.

Proof. Without loss of generality, suppose that $x(n)$ is a nonoscillatory solution of equation (1) which is eventually positive. As in the proof of Theorem 6, we obtain (24). In view of equation (1) and Lemma 2, in addition to the case (25), there are two more possible types of behavior of solutions for $n \geq n_1$ where $n_1 \geq n_0$ is large enough in the proof of Theorem 6. Namely, one can also have

$$z(n) > 0,\ \Delta z(n) > 0,\ \Delta^{m-2}z(n) > 0,\ \Delta^{m-1}z(n) < 0,\ \Delta\left[r(n)(\Delta^{m-1}z(n))^{\alpha}\right] \leq 0 \tag{70}$$

or

$$z(n) > 0,\ \Delta^{j}z(n) < 0,\ \Delta^{j+1}z(n) > 0,\ \Delta^{m-1}z(n) < 0,\ \Delta\left[r(n)(\Delta^{m-1}z(n))^{\alpha}\right] \leq 0 \tag{71}$$

for all odd integers $j \in \{1, 2, \cdots, m-3\}$. However conditions (17) and (18) yields that neither (24) nor (70) is possible. Therefore, we have to analyze the only remaining case, and we assume now that all the conditions in (71) are satisfied. Then, inequality (46) holds. Summing (46) from n to ∞ $(n-2)$ times, we obtain

$$\begin{aligned} z(n) &\geq -\sum_{n}^{\infty} \frac{(\eta-n)^{(m-3)}}{(m-3)!} A(\eta) r^{1/\alpha}(n) \Delta^{m-1} z(n) \\ &= -\sum_{n}^{\infty} \frac{(\eta-n)^{(m-3)}}{(m-3)!} A(\eta) w^{1/\alpha}(n) \end{aligned} \tag{72}$$

where $w(n)$ is defined by (30). Taking into account that $\Delta z(n) < 0$, $\sigma(n) \leq \eta_3(n)$ and using (24), we have

$$\Delta\left[r(n)(\Delta^{m-1}z(n))^{\alpha} + p_0^{\beta} r(\tau(n)) \left(\Delta^{m-1}z(\tau(n))\right)^{\alpha}\right] + Q(n) z^{\beta}(\eta_3(n)) \leq 0. \tag{73}$$

By virtue of monotonicity of $z(n)$, there exists a constant $M_2 > 0$ such that

$$z^{\beta}(\eta_3(n)) = z^{\beta-\theta}(\eta_3(n))z^{\theta}(\eta_3(n)) \geq M_2 z^{\theta}(\eta_3(n)) \tag{74}$$

combining (73) and (74), we obtain

$$\Delta\left[r(n)(\Delta^{m-1}z(n))^{\alpha} + p_0^{\beta} r(\tau(n))\left(\Delta^{m-1}z(\tau(n))\right)^{\alpha}\right] + M_2 Q(n) z^{\theta}(\eta_3(n)) \leq 0. \tag{75}$$

Using (72) in (75), we conclude that in this case, the sequence $w(n)$ defined by (30) is negative, nonincreasing, and satisfies the inequality

$$\Delta\left[r(n)(\Delta^{m-1}z(n))^{\alpha} + p_0^{\beta} r(\tau(n))\left(\Delta^{m-1}z(\tau(n))\right)^{\alpha}\right] + \frac{M_2}{((m-3)!)^{\theta}} Q_{\theta}(n) w^{\theta/\alpha}(\eta_3(n)) \leq 0. \tag{76}$$

Introduction again $y(n)$ by (33) and using the monotonicity of $w(n)$, we arrive at (48). Substitution of (48) into (76) leads to the conclusion that $y(n)$ is a negative solution of an advanced difference equation

$$\Delta y(n) - \frac{M_2}{((m-3)!)^{\theta}}\left(\frac{1}{1+p_0^{\beta}}\right)^{\theta/\alpha} Q_{\theta}(n) y^{\theta/\alpha}(\eta_3(n)) \leq 0. \tag{77}$$

In which case the function $u(n) = -y(n)$ is a positive solution of an advanced difference inequality

$$\Delta u(n) - \frac{M_2}{((m-3)!)^{\theta}}\left(\frac{1}{1+p_0^{\beta}}\right)^{\theta/\alpha} Q_{\theta}(n) u^{\theta/\alpha}(\tau^{-1}(\eta_3(n))) \geq 0. \tag{78}$$

Then, by [Gyori and Ladas, 1991], the associated advanced difference equation

$$u(n) - \frac{M_2}{((m-3)!)^{\theta}}\left(\frac{1}{1+p_0^{\beta}}\right)^{\theta/\alpha} Q_{\theta}(n) u^{\theta/\alpha}(\tau^{-1}(\eta_3(n))) = 0, \tag{79}$$

also has a positive solution. However, by [Sundar and Murugesan, 2010, Lemma 2.3.2] implies that (79) is oscillatory under assumption (69). Therefore, equation (1) cannot have positive solutions. This contradiction with our initial assumption completes the proof.

Theorem 11. *Let $m \geq 4$ be even and $0 < \alpha = \beta \leq 1$. Assume that conditions (c_4) and (c_5) are satisfied, and that there exist three real sequences $\eta_1(n), \eta_2(n)$ and $\eta_3(n)$ as in Theorem 10. Suppose also that conditions (52) and (55) hold. If*

$$\frac{1}{((m-3)!)^{\beta}(1+p_0^{\beta})} \liminf_{n\to\infty} \sum_{n}^{\tau^{-1}(\eta_3(n))-1} \tilde{Q}_{\beta}(s) > \left(\frac{l-1}{l}\right)^{l} \tag{80}$$

where l denotes the advanced argument, then equation (1) is oscillatory.

Proof. Let $x(n)$ be a nonoscillatory solution of (1) which is eventually positive. As in the proof of Theorem 10, one can have either (25) or (70) or (71). However, conditions (51) and (53) exclude cases (25) and (70). Then all the inequalities in (71) should be satisfied. Along the same lines as in the proof of Theorem 10, one comes to the conclusion that an advanced difference equation

$$\Delta u(n) - \frac{1}{((m-3)!)^{\beta}(1+p_0^{\beta})}\tilde{Q}_{\beta}(n) u(\tau^{-1}(\eta_3(n))) = 0$$

has positive solutions. On the other hand, if condition (80) holds, then by virtue of [Sundar and Murugesan, 2010] implies that equation (80) is oscillatory. This contradiction completes the proof.

Theorem 12. *Let $m \geq 4$ be even, $0 < \beta \leq 1$, and assume that conditions (c_4) and (c_5) are satisfied. Suppose further that there exist three numbers $\lambda, \gamma, \theta \in R$ as in Theorem 11 and three real sequences $\eta_1(n), \eta_2(n), \eta_3(n)$ such that (56) is satisfied, $\eta_3(n) \geq \sigma(n)$ and $\eta_3(n) \geq n$. If (17), (18) and (69) hold, then equation (1) is oscillatory.*

Proof. Let $x(n)$ be an eventually positive nonoscillatory solution of equation (1). The same argument as in the proof of Theorem 10 yields that (71) holds. Define the sequence $w(n)$ by (30). From the proof of Theorem 10, we already know that $w(n)$ is negative, nonincreasing and satisfies the inequality (76). Introducing the sequence $y(n)$ by (32) and using the monotonicity of $w(n)$, we arrive at (60). Substituting (60) into (76) we observe that $y(n)$ is a negative solution of an advanced difference inequality

$$\Delta y(n) - \frac{M_2}{((m-3)!)^\theta}\left(\frac{1}{1+p_0^\beta}\right)^{\theta/\alpha} Q_\theta(n) y^{\theta/\alpha}(\eta_3(n)) \le 0, \tag{81}$$

while $u(n) = -y(n)$ is a positive solution of an advanced difference inequality

$$\Delta u(n) - \frac{M_2}{((m-3)!)^\theta}\left(\frac{1}{1+p_0^\beta}\right)^{\theta/\alpha} Q_\theta(n) u^{\theta/\alpha}(\eta_3(n)) \ge 0. \tag{82}$$

In this case, the result due to Gyori [1991] allows one to deduce that the associated advanced difference equation

$$\Delta u(n) - \frac{M_2}{((m-3)!)^\theta}\left(\frac{1}{1+p_0^\beta}\right)^{\theta/\alpha} Q_\theta(n) u^{\theta/\alpha}(\eta_3(n)) = 0, \tag{83}$$

also has a positive solution. However, it has been established [Sundar and Murugesan, 2010, Lemma 2.3.2] that if condition (69) is satisfied, then equation (83) is oscillatory. Therefore equation (1) cannot have positive solutions, and this contradiction with the assumptions of the theorem completes the proof.

Theorem 13. *Let $m \ge 4$ be even and $0 < \alpha = \beta \le 1$. Assume that conditions (c_4) and (c_5) are satisfied, and there exist three real sequences $\eta_1(n), \eta_2(n)$ and $\eta_3(n)$ as in Theorem 12. Suppose further that* (64), (65) *hold, and*

$$\frac{1}{((m-3)!)^\theta(1+p_0^\beta)} \liminf_{n\to\infty} \sum_{n}^{\eta_3(n)-1} \bar{Q}_\beta(s) > \left(\frac{l-1}{l}\right)^l \tag{84}$$

where l denotes the advanced argument. Then equation (1) *is oscillatory.*

Proof. Assume that $x(n)$ is an eventually positive nonoscillatory solution of equation (1) and increasing as in the proof of Theorem 10 one concludes that (71) holds. As in the proof of Theorem 12 we observed that an advance difference equation

$$\Delta u(n) - \frac{1}{((m-3)!)^\theta(1+p_0^\beta)} \bar{Q}_\beta(n) u(\eta_3(n)) = 0, \tag{85}$$

has positive solution. On the other hand, if condition (84) is satisfied, a result reported by [Sundar and Murugesan, 2010, Lemma 2.3.2] yields that equation (85) is oscillatory. This contradiction completes the proof.

3. Asymptotic Behavior of Solutions to Odd-Order Equations

In this section, in addition to conditions $(c_4), (c_5)$ and (10), we also assume that

(c_6) $\sigma(n) < n$.

The validity of the following four propositions can be established in the same manner as it has been done for Theorems 6, 9. Therefore, to avoid unnecessary repetition, we only formulate counterparts of Theorems 9 and following for the case of odd-order equations.

Theorem 14. *Let $m \ge 3$ be odd and $0 < \beta \le 1$. Assume that conditions $(c_4) - (c_6)$ are satisfied, and there exist two real numbers $\gamma, \lambda \in R$ as in Theorem 6 and a real sequence $\eta_4(n)$ such that $n \le \tau(n) < \eta_4(n)$. Suppose further that*

$$\sum^{\infty} \tilde{Q}_\gamma(n) = \infty \tag{86}$$

and

$$\sum^{\infty} Q_\beta(n) A^\lambda(\eta_4(n)) = \infty. \tag{87}$$

Then the conclusion of Theorem 6 remains intact.

Theorem 15. *Let $m \geq 3$ be odd, and let $0 < \alpha = \beta \leq 1$. Assume that conditions $(c_4) - (c_6)$ are satisfied, and there exists a real sequence $\eta_4(n)$ as in Theorem 14. Suppose also that*

$$\frac{1}{((m-1)!)^{\beta}} \frac{1}{1+p_0^{\beta}} \liminf_{n\to\infty} \sum_{\sigma(n)}^{n-1} \widehat{Q}_{\beta}(s) > \frac{1}{e} \tag{88}$$

and

$$\frac{1}{((m-2)!)^{\beta}} \frac{1}{1+p_0^{\beta}} \liminf_{n\to\infty} \sum_{n}^{\tau^{-1}(\eta_4(n))-1} Q_{\beta}(s) A^{\beta}(\eta_4(s)) > \frac{1}{e}. \tag{89}$$

Then the conclusion of Theorem 6 remains intact.

Theorem 16. *Let $m \geq 3$ be odd and let $0 < \beta \leq 1$. Assume that conditions $(c_4) - (c_6)$ are satisfied, and there exist two real numbers $\gamma, \lambda \in R$ as in Theorem 6 and a sequence $\eta_4(n)$ such that $\sigma(n) < \tau(n) \leq n \leq \eta_4(n)$. Suppose further that conditions (86) and (87) are satisfied. Then the conclusion of Theorem 6 remains intact.*

Theorem 17. *Let $m \geq 3$ be odd, and $0 < \alpha = \beta \leq 1$. Assume that conditions $(c_4) - (c_6)$ are satisfied, and there exists a real sequence $\eta_4(n)$ as in Theorem 16. If*

$$\frac{1}{((m-1)!)^{\beta}} \frac{1}{1+p_0^{\beta}} \liminf_{n\to\infty} \sum_{\tau^{-1}(\sigma(n))}^{n-1} \widehat{Q}_{\beta}(s) > \frac{1}{e} \tag{90}$$

and

$$\frac{1}{((m-2)!)^{\beta}} \frac{1}{1+p_0^{\beta}} \liminf_{n\to\infty} \sum_{n}^{\eta_4(n)-1} Q_{\beta}(s) A^{\beta} \eta_4(s) > \frac{1}{e}. \tag{91}$$

Then the conclusions of Theorem 6 remain intact.

Note that Theorems 14-17 apply only if σ is a delayed argument, $\sigma(n) < n$. Hence it is important to complement such results with the following theorems that can be applied in the case where σ is an advanced argument, $\sigma(n) \geq n$.

Theorem 18. *Let $m \geq 3$ be odd and let $0 < \beta \leq 1$. Assume that conditions (c_4) and (c_5) are satisfied, and there exist two real numbers $\gamma, \lambda \in R$ as in Theorem 6 and two real sequences $\eta_1(n), \eta_2(n)$ satisfying (16). Suppose also that*

$$\sum_{n_0}^{\infty} \xi^{(m-2)} \left[\frac{1}{R(\xi)} \sum_{\xi}^{\infty} Q(s) \right]^{1/\alpha} = \infty. \tag{92}$$

If (17) and (18) are satisfied, then the conclusions of Theorem 6 remains intact.

Proof. Assume that equation (1) has an eventually positive solution $x(n)$ satisfying (20). Proceeding as in the proof of Theorem 6, we arrive at (23) and observe that equation (1) yields that either (24) or (25) holds. Indeed, it follows from the condition $\Delta\left[r(n)(\Delta^{m-1}z(n))^{\alpha}\right] \leq 0$ that either $\Delta^{m-1}z(n) > 0$ or $\Delta^{m-1}z(n) < 0$. Assume first that $\Delta^{m-1}z(n) < 0$; this immediately leads us to conditions (25).

On the other hand, if $\Delta^{m-1}z(n) > 0$, then $\Delta^{m}z(n) \leq 0$ to the fact that $\Delta r(n) > 0$. We claim that $\Delta z(n) > 0$ eventually. Infact, if this is not the case, then $\Delta z(n) < 0$ eventually. Since $z(n) > 0$, $\Delta z(n) < 0$, and (20) holds, there should exist a positive constant a such that

$$\lim_{n\to\infty} z(n) = a. \tag{93}$$

On the other hand, if $\Delta^{m-1}z(n) > 0$ and $\Delta^{m}z(n) \leq 0$, there exists a constant $b \geq 0$ such that

$$\lim_{n\to\infty} \Delta^{m-1}z(n) = b \geq 0. \tag{94}$$

Hence,

$$\lim_{n\to\infty} \Delta^{(i)}z(n) = 0, \tag{95}$$

for $i = 1, 2, \cdots, m-1$. Summing (23) from n to ∞ and using the fact that the limit

$$\lim_{n\to\infty} r(n)\left(\Delta^{m-1}z(n)\right)^{\alpha} \geq 0 \tag{96}$$

is finite, we have

$$-r(n)\left(\Delta^{m-1}z(n)\right)^{\alpha} - p_0^{\beta}r(\tau(n))\left(\Delta^{m-1}z(\tau(n))\right)^{\alpha} + \sum_{n}^{\infty} Q(s)z^{\beta}(\sigma(s)) \leq 0. \tag{97}$$

Consequently,

$$-R(n)\left[\left(\Delta^{m-1}z(n)\right)^{\alpha} + ((p_0^{\beta})^{1/\alpha})^{\alpha}\left(\Delta^{m-1}z(\tau(n))\right)^{\alpha}\right] + \sum_{n}^{\infty} Q(s)z^{\beta}(\sigma(s)) \leq 0. \tag{98}$$

Assume first that $\alpha \leq 1$. Using Lemma 5, we obtain

$$\begin{aligned}&\left(\Delta^{m-1}z(n)\right)^{\alpha} + ((p_0^{\beta})^{1/\alpha})^{\alpha}\left(\Delta^{m-1}z(\tau(n))\right)^{\alpha}\\ &\quad\leq \frac{1}{2^{\alpha-1}}\left[\Delta^{m-1}z(n) + (p_0^{\beta})^{1/\alpha}\Delta^{m-1}z(\tau(n))\right]^{\alpha}.\end{aligned} \tag{99}$$

Substituting (99) into (98), we have

$$-2^{1-\alpha}R(n)\left[\Delta^{m-1}z(n) + (p_0^{\beta})^{1/\alpha}\Delta^{m-1}z(\tau(n))\right]^{\alpha} + \sum_{n}^{\infty} Q(s)z^{\beta}(\sigma(s)) \leq 0 \tag{100}$$

which implies

$$-\left[\Delta^{m-1}z(n) + (p_0^{\beta})^{1/\alpha}\Delta^{m-1}z(\tau(n))\right]^{\alpha} \leq -\frac{1}{2^{1-\alpha}R(n)}\sum_{n}^{\infty} Q(s)z^{\beta}(\sigma(s)). \tag{101}$$

Therefore,

$$-\left[\Delta^{m-1}z(n) + (p_0^{\beta})^{1/\alpha}\Delta^{m-1}z(\tau(n))\right] + \left[\frac{1}{2^{1-\alpha}R(n)}\sum_{n}^{\infty} Q(s)z^{\beta}(\sigma(s))\right]^{1/\alpha} \leq 0. \tag{102}$$

Summing (102) $(m-2)$ times from n to ∞ and then one more time from n_1 to ∞. Using (95) and changing the order of summation, we obtain

$$\sum_{n_1}^{\infty}\frac{(\xi-\eta_1)^{(m-2)}}{(m-2)!}\left[\frac{1}{2^{1-\alpha}R(\xi)}\sum_{\xi}^{\infty} Q(s)z^{\beta}(\sigma(s))\right]^{1/\alpha} < \infty. \tag{103}$$

Inequality (103) yields

$$\sum_{n_1}^{\infty}\xi^{(m-2)}\left[\frac{1}{R(\xi)}\sum_{\xi}^{\infty} Q(s)\right]^{1/\alpha} < \infty, \tag{104}$$

which contradicts (92).

For the case $\alpha > 1$, one arises at the contradiction with the assumptions of the theorem by using Lemma 4. Then we conclude that $\Delta z(\alpha) > 0$ eventually. The rest of the proof follows the same lines as in Theorem 6 and is omitted.

Combining the ideas exploited in the proofs of Theorems 7-9 and 18, one can derive the following results.

Theorem 19. *Let $m \geq 3$ be odd, and let $0 < \alpha = \beta \leq 1$. Assume that conditions (c_4) and (c_5) are satisfied, and there exist two real sequence $\eta_1(n), \eta_2(n)$ satisfying (16). If (52), (53) and (92) hold, then the conclusion of Theorem 6 remains intact.*

Theorem 20. *Let $m \geq 3$ be odd, and $0 < \beta \leq 1$. Assume that conditions (c_4) and (c_5) are satisfied, and there exist two real numbers $\gamma, \lambda \in R$ as in Theorem 6 and two real sequences $\eta_1(n), \eta_2(n)$ satisfying (56). If conditions (17) and (18) and (92) are satisfied, the conclusion of Theorem 6 remains intact.*

Theorem 21. *Let $m \geq 3$ be odd, and $0 < \alpha = \beta \leq 1$. Assume that conditions (c_4) and (c_5) are satisfied, and there exist two real sequences $\eta_1(n), \eta_2(n)$ satisfying* (56). *If conditions* (64) *and* (65) *and* (92) *are satisfied, then the conclusion of Theorem 6 remains intact.*

4. Examples and Discussions

The following examples illustrate applications of some of theoretical results presented in the previous sections. In all the examples, p_0 is a constant such that $0 \leq p_0 < \infty$.

Example 1. *For $n \geq 1$, consider the fourth order neutral difference equation*

$$\Delta\left(e^n\left(\Delta^3\left(x(n) + p_0 x(n-2)\right)\right)\right) + (1 + p_0 e)(1 - \sqrt{e})^3\left(\frac{1}{\sqrt{e}} - 1\right)e^{n-\frac{3}{2}}x(n-1) = 0. \tag{105}$$

Let $\eta_1(n) = n-3$ and $\eta_2(n) = n+1$. An application of Theorem 9 yields that every solution $x(n)$ of equation (105) *is either oscillatory or satisfies $\lim_{n\to\infty} x(n) = 0$. As a matter of that $x(n) = e^{-\frac{n}{2}}$ is an exact solution to* (105) *satisfying $\lim_{n\to\infty} x(n) = 0$.*

Example 2. *For $n \geq 1$, consider a fourth order neutral difference equation*

$$\Delta\left(e^n\left(\Delta^3\left(x(n) + p_0 x(n+2)\right)\right)\right) + (e^2 + 1)(1 + e^3)(1 + p_0 e)^2 e^{n+3}x(n-3) = 0. \tag{106}$$

Let $\eta_1(n) = n-3$ and $\eta_2 = \eta_3 = n+3$. Using Theorem 11, we deduce that equation (106) *is oscillatory. It is easy to verify that one oscillatory solution of the equation $x(n) = (-1)^n e^n$.*

Example 3. *For $n \geq 1$, consider a third order neutral difference equation*

$$\Delta\left(e^n\left(\Delta^2\left(x(n) + p_0 x(n-2)\right)\right)\right) + (1 + p_0 e)(\frac{1}{e} - 1)^2(1 - \sqrt{e})e^{n-\frac{1}{2}}x(n-1) = 0. \tag{107}$$

Let $\eta(n) = n+1$. It follows from Theorem 17 that every solution $x(n)$ of equation (107) *is either oscillatory or satisfies $\lim_{n\to\infty} x(n) = 0$ is $x(n) = e^{-\frac{n}{2}}$.*

Remark 1. *By using inequality*

$$x_1^\beta + x_2^\beta \geq 2^{1-\beta}(x_1 + x_2)^\beta \tag{108}$$

which holds for any $\beta \geq 1$ and for all $x_1, x_2 \in [0, \infty)$, results reported in this paper can be extended to equation (1) *for all $\beta \in R$ which satisfy $\beta > 1$. In this case one has to replace $Q(n) = \min\{q(n), q(\tau(n))\}$ with a function $Q(n) = 2^{1-\beta}\min\{q(n), q(\tau(n))\}$ and proceed in above.*

Remark 2. *Our main assumptions on functional arguments do not specify whether $\tau(n)$ is delayed or an advanced argument. Remarkably, $\sigma(n)$ can even switch its nature between an advanced and delayed argument. However, such flexibility is achieved at the cost of requiring that the function τ is monotonic and satisfies $\tau \circ \sigma = \sigma \circ \tau$. The question regarding the analysis of the asymptotic behavior of solutions to* (1) *with other methods that do not require these assumptions remains open at the moment.*

Acknowledgements

The authors wish to express their sincere thanks to the referee for valuable comments and suggestions.

References

Agarwal. R. P. (1992). Difference Equations and Inequalities Theory, Method and Applications. *Marcel Dekker, New York.*

Agarwal. R. P. & Wong. P. J. Y. (1997). Advanced Topics in Difference Equations. *Kluwer Academic Publishers Grovi, Dordrecht.*

Bolat, Y., & Alzabut, J. O. (2012). On the oscillation of higher-order half-linear delay difference equations. *Appl. Math. Int. Sci., 6*, 423-427.

Cheng, S. S., & Patula, W. T. (1993). An existence theorem for a nonlinear difference equations, *Nonlinear Analysis, Theory Method and Applications, 20*(3), 193-203.

Elaydi, S. N. (1995). *An Introduction to Difference Equations.* Springer Verlag, New York.

Graef, J. R, Mician, A. S, Spokes, P. W, Sundar, P., & Thandapani, E. (1996). Oscillatory and asymptotic behavior of neutral type difference equation. *J. Austra. Math. Soc. Ser. B., 38*, 163-171. https://doi.org/10.1017/S0334270000000552

Guan, X., & Yang, J. (1999). Asymptotic and oscillatory behavior of higher-order nonlinear neutal difference equations, *Kyun. Math. I., 39*, 251-259.

Gyori, I., & Ladas, G. (1991). *Oscillation Theory of Delay Differential Equations with Applications*. Clarendon Press, Oxford.

He, X. Z. (1993). Oscillatory and Asymptotic behavior of second order nonlinear difference equations. *J. Math. Anal. Appl., 175*, 482-498. https://doi.org/10.1006/jmaa.1993.1186

Jurang, Y., & Bin, L. (1997). Oscillatory and asymptotic behavior of fourth order nonlinear difference equations. *Acta. Math. Sinica. New Series, 13*, 105-115. https://doi.org/10.1007/BF02560530

Karpuz, B. (2009). On Oscillation and asymptotic behavior of a higher order functional difference equation of neutral type, *Inter. J. Differen. Eqn., 4*(1), 69-96.

Kubiaczyk, I., & Sekar, S. H. (2002). Oscillation theorems for second order sublinear delay difference equations. *Math. Slovaca, 52*, 343-359.

Lakshmikanthan, V., & Trigiante, O. (1988). Theory of Difference Equations: Numerical Method and Application. Academic Press, New York.

Luo, J. (2002). Oscillation criteria for second order quasilinear neutral difference equations. *Computers Math. Appl., 43*, 1549-1557. https://doi.org/10.1016/S0898-1221(02)00118-9

Ou, C. H., & Wong, J. S. W. (2004). Oscillation and nonoscillation theorems for superlinear Emden-Fowler equations of the fourth order. *Annali di Mathematica Pura e Applicata, Series IV, 183*(1), 25-43. https://doi.org/10.1007/s10231-003-0079-z

Sundar, P., & Kishorkumar. (2014). Nonoscillatory solution of even-order nonlinear neutral difference equations. *Journal of Indian Academy of Mathematics, 36*(1).

Sundar, P., & Kishorkumar. (2014). Oscillation results for even-order quasilinear neutral functional difference equation, *IOSR Journal of Mathematics, 10*(4), 8-18.

Sundar, P., & Murugesan, A. (2010). Asymptotic Behavior and Oscillation of Solutions of Neutral/Non-Neutral Advanced Difference Equations, Ph.D., Dissertation, Periyar University, Tamil Nadu, India, October.

Sundar, P., & Thandapani, E. (2000). Oscillation and non-oscillation theorems for second order quasilinear functional difference equations. *Indian J. Pure Appl. Math., 31*(1), 37-47.

Wong, J. S. W. (1975). On the generalized Emden-Fowler equation. *SIAM Review, 17*, 339-360. https://doi.org/10.1137/1017036

Yu, J. S., & Wang, Z. C. (1994). Asymptotic behavior and oscillation in neutral delay difference equations. *Funk. Ekvac., 37*, 241-248.

Yildiz, M. K., & Ocalan, O. (2007). Oscillation results for higher order nonlinear neutral delay difference equation. *Appl. Math. Lett., 20*(3), 243-247. https://doi.org/10.1016/j.aml.2006.05.001

Zafer, A. (1995). Oscillation and asymptotic behavior of higher order difference equations. *Math. Comput. Modelling, 21*(4), 43-50. https://doi.org/10.1016/0895-7177(95)00005-M

Zhou, X., & Yan, J. (1998). Oscillatory and asymptotic behavior of higher order nonlinear difference equations. *Nonlinear Analysis, Theory Methods and Applications, 31*(3-4), 493-502. https://doi.org/10.1016/S0362-546X(97)00417-3

17

Fermat's Theorem – a Geometrical View

Luis Teia[1]

[1] Department of Energy Sciences, Lund University Ole Römers Väg 1, M-Building SE-22100 Lund, Sweden

Correspondence: Dr. Luis Teia, Department of Energy Sciences, Lund University Ole Römers Väg 1, M-Building SE-22100 Lund, Sweden. E-mail: luistheya@gmail.com

Abstract

Fermat's Last Theorem questions not only what is a triple, but more importantly, what is an integer in the context of equations of the type $x^n + y^n = z^n$. This paper explores these questions in one, two and three dimensions. It was found that two conditions are required for an integer element to exist in the context of the Pythagoras' theorem in 1D, 2D and 3D. An integer must satisfy the Pythagoras' theorem of the respective dimension – condition 1. And, it must be completely successfully split into multiple unit scalars – condition 2. In 1D, the fundamental unit scalar is the line length 1. All integers in 1D satisfy $x + y = z$, and can be decomposed into multiples of the unit line, hence integers exist and can form 1D triples (x, y, z). In 2D, the fundamental unit scalar is the square side 1. Only some groups of integers (called triples) satisfy $x^2 + y^2 = z^2$, and can be decomposed into multiples of the unit square, forming 2D triples. In 3D, the fundamental unit scalar is the octahedron side 1. The geometry of the 3D Pythagoras' theorem dictates that $x^3 + y^3 = z^3$ is governed by octahedrons, validating condition 1. However, octahedrons with side length integer cannot be completely divided into unit octahedrons (as tetrahedrons appear), invalidating condition 2. Hence, if integers do not exist in the context of the 3D Pythagoras' theorem, then neither do triples. This confirms Fermat's Last Theorem for three dimensions ($n = 3$). The geometrical interdependency between integers in 1D and 2D suggests that all integers of higher dimensions are built, and hence are dependent, on the integers of lower dimensions. This interdependency coupled with the absence of integers in 3D suggests that there are no integers above $n > 2$, and therefore there are also no triples that satisfy $x^n + y^n = z^n$ for $n > 2$.

Keywords: geometry, Fermat, Pythagoras, theorem

1. Introduction

Fermat's conjecture, also known as Fermat's Last Theorem (Dickson, 1919), is more than just about triples, it is about the fundamental nature of an integer number, and it's mathematical and geometrical meaning. It raises the philosophical question: What is a unit? In the language of mathematics, a unit is defined by the number 1. In the language of geometry, a unit is defined by an element of side length one. A perspective of a problem depends on the language we use to observe it. And a change in perspective is often all it takes to see the solution. In mathematics, the manipulation of the number 1 (i.e., summation, subtraction, multiplication, division, etc) gives rise to the whole universe of numbers. Whether it is a complex, real or integer number, they all can be expressed as a manipulation of one or a group of ones (Figure 1). Hence, the number 1 is the fundamental building block of the mathematical universe of numbers.

$$\textbf{Complex}: i=\sqrt{-1} \quad ; \quad \textbf{Real}: 1{,}2 = 1+\frac{1+1}{1+1+1+1+1+1+1+1+1+1} \quad ; \quad \textbf{Integer}: 2 = 1+1$$

Figure 1. Examples of numbers built mathematically with 1.

Geometry, the field of knowledge dealing with spatial relationships, predates modern mathematics (Cooke, 2005), being later revolutionized by the book *Elements* (Euclid, 300BC) into the foundation of contemporary mathematics. Hence, every mathematical operation has a geometrical meaning. For example, mathematical operations with powers 1, 2 and 3 translate into geometric operations in one, two, and three dimensions (Figure 2). Visualization of such geometrical operations can be overwhelmingly complex, and hence mathematics is used to bypass such complexity, allowing the manipulation of certain aspects of the underlying geometrical operation.

At the end, a mathematical answer that has some geometrical meaning is obtained. Completeness is sacrificed when changing from a geometrical perspective to mathematical, resulting in an easier and speedier answer. In geometry, the manipulation of the unit element gives rise to the universe of elements, equivalent to the mathematical universe of numbers. A number has different meanings in different dimensions. In the 1D Pythagoras' theorem, i.e. the fundamental process of

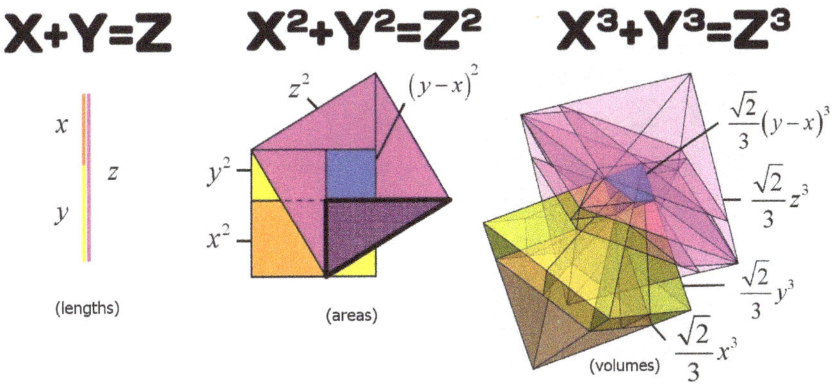

Figure 2. The 1D, 2D and 3D Pythagoras' theorem (Teia, 2016)

summation $x + y = z$, numbers are governed by lines with given lengths. In 2D, numbers are governed by squares with given areas. And in 3D, numbers are governed by octahedrons with given volumes (Teia, 2016). These geometrical units (i.e., line, square and octahedron) establish the foundation for the geometric interpretation of integers. Since triples are aggregates of three integer numbers, this raises the question – how does the mathematical and geometrical meaning of triples evolves from 1D to 2D, and finally to 3D?

2. Theory

2.1 The 1D and 2D Pythagoras' Theorem

Let us start with the simple one dimensional version of the Pythagoras' theorem. It is governed by line segments, where the geometrical addition of two line segments gives a third, or $x+y = z$ (Figure 3a). This is by definition the mathematical process of summation. One process that transforms the 1D Pythagoras' theorem to 2D is defined by the following two steps:

Step 1. Rotation by 90 degrees between two lines about the middle point. This gives a right-angled triangle (Figure 3b).

Step 2. Lines are extruded perpendicularly to their length and form squares. The result is the two dimensional version of the Pythagoras' theorem, where the geometric addition of two squares gives a third, or $x^2 + y^2 = z^2$ (Figure 3c).

The necessity for an integer within a triple to satisfy the geometric part of the Pythagoras' theorem is, from now on, termed Condition 1.

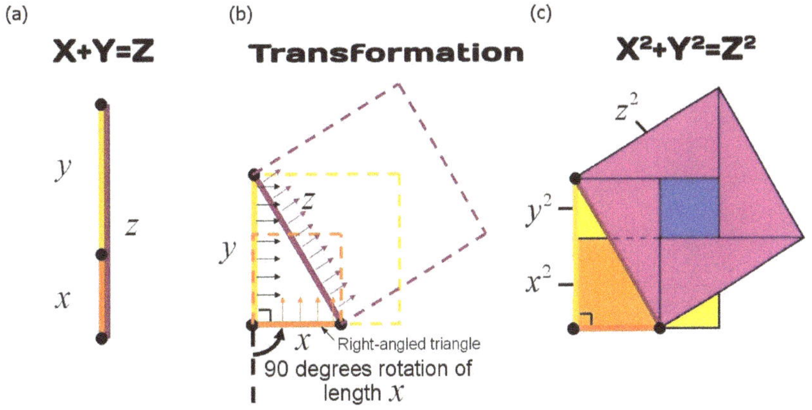

Figure 3. The geometry of (a) $x + y = z$, (b) the transformation, and (c) $x^2 + y^2 = z^2$

Mathematical numbers are grouped into different categories, like real and integer. Real numbers ($\mathbb{R}$) compose the entire universe of numbers and can be thought of as points on an infinite long number line. They include all the rational numbers, such as fractions and integers (Clapham & Nicholson, 2014). Integers ($\mathbb{Z}$) are a subgroup of real numbers (Figure 4). An integer is a number that can be written without a fractional component (i.e., 1,2,3). Triples are a set of three integer numbers that satisfy the Pythagoras' theorem, hence a subgroup of integer numbers. Figure 4 shows examples of mathematical numbers that belong to these three divisions, and satisfy the 1D and 2D Pythagoras' theorem. Mathematical summations of these numbers and their geometric equivalent are presented.

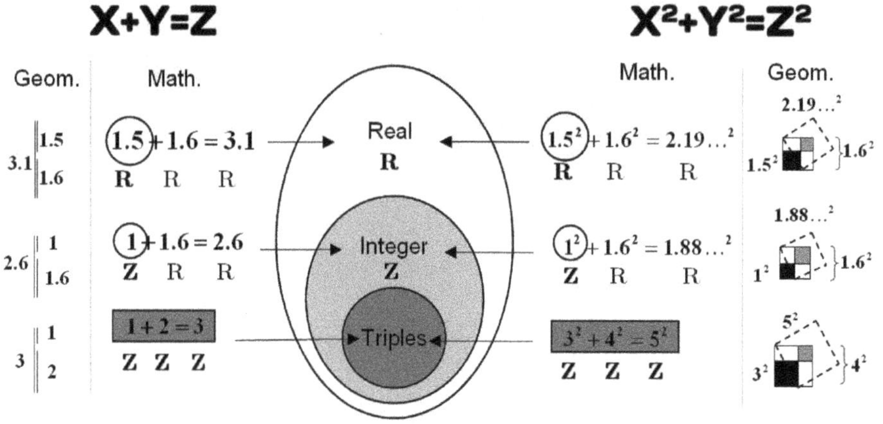

Figure 4. Groups of numbers in the Real, Integer, and Triple domains

Example 1 (Figure 4 left) shows two lines of length real numbers 1.5 and 1.6 give another line of length real number 3.1. Example 2 shows an integer 1 plus a real number 1.6 give another real number 2.6. Finally, example 3 shows a group of integer numbers 1, 2 and 3 that satisfies the 1D Pythagoras' theorem, and hence forms the 1D triple $(1, 2, 3)$. Similar summations are observed for 2D. The area of two squares 1.5^2 and 1.6^2 gives the area of a third square $2.19\ldots^2$ (Figure 4 right). And, a group of squares 3^2, 4^2 and 5^2 satisfies the 2D Pythagoras' theorem, and hence forms the 2D triple $(3, 4, 5)$.

2.2 Geometry of 1D and 2D Integers

The unit line, or line of length 1, is the fundamental geometric scalar that composes all integer elements in the 1D universe (i.e., $2^1, 3^1, 4^1, 5^1 \ldots N^1$, where N is an integer number). As illustrated in Figure 5a, all integer lines are formed by an aggregate chain of unit lines. Likewise, the unit square, or square of side 1, is the fundamental geometric scalar that composes all integer elements in the 2D universe (i.e., $2^2, 3^2, 4^2, 5^2 \ldots N^2$). As illustrated in Figure 5b, all integer squares are formed by an aggregate chain of unit squares. Generally, one can conclude from Figure 5 that in order for an integer element to exist, it needs to be completely split into multiples of the fundamental unit scalar particular to that dimension (i.e., unit line in 1D or unit square in 2D). The necessity for an integer (in order to be an integer) to be split into a multiple of a fundamental scalar of the respective dimension is from now on termed Condition 2.

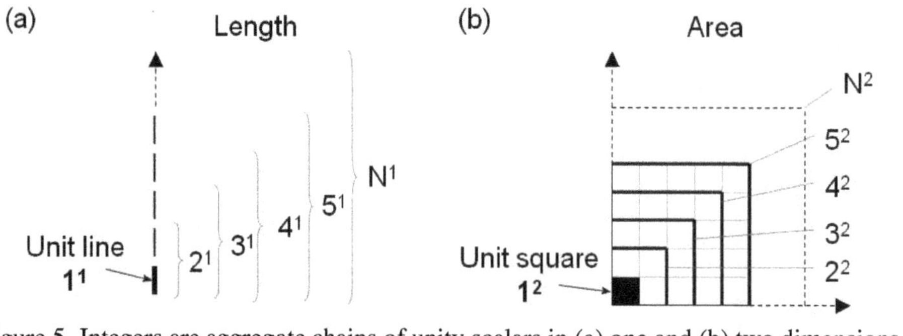

Figure 5. Integers are aggregate chains of unity scalars in (a) one and (b) two dimensions

The geometry of the scalar governs all integers that satisfy the Pythagoras' theorems, and consequently, all triples. The world of 1D triples is governed by the unit line (e.g., $1+2 = 3$), while for 2D triples, it is the unit square (e.g., $3^2+4^2 = 5^2$). Figure 6 shows the spiral of triples belonging to the Pythagoras' family, drawn using the central square approach (Teia, 2015), with a grid of unit squares superimposed.

It is seen that all squares that compose 2D triples (e.g., $3^2, 4^2, 5^2$) can be decomposed into a sum of unit squares. One can generalize this and say that all squares composing the integers of the 2D triple (x, y, z) can be expressed geometrically by aggregates of many unit squares, hence satisfying Condition 2.

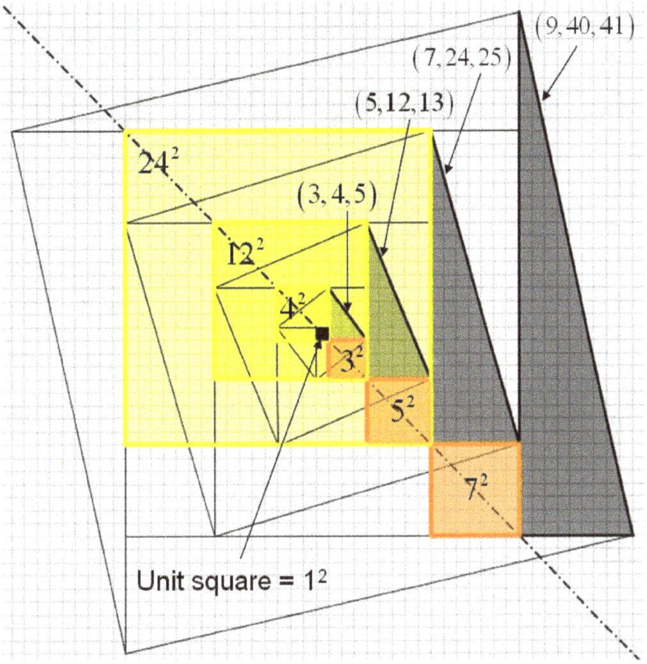

Figure 6. The Pythagoras' family of triples expressed via unit squares (Teia, 2015)

3. Hypothesis

A triple only exists, if all integer elements within that triple (e.g., $1^1, 2^1, 3^1$ for 1D, and $3^2, 4^2, 5^2$ for 2D) also exist. In turn, an integer element only exits if:

Condition 1. it satisfies the Pythagoras' theorem of the respective dimension,

Condition 2. it can be completely successfully split into multiple unit scalars.

One can therefore hypothesize that integer elements do not exist if either Condition 1 or 2 is not satisfied. By consequence, if the integers do not exist, then the associated triples also do not exist.

4. Development

4.1 Triples Transition from 1D, 2D to 3D

Integers in 1D and 2D Pythagoras' theorem have been shown to satisfy both conditions. Hence, all integers in 1D form triples (e.g., $1 + 1 = 2$, $1 + 2 = 3$, $1 + 4 = 5$) [Figure 7 left] and some integers in 2D form triples (e.g., $3^2 + 4^2 = 5^2$, $5^2 + 12^2 = 13^2$, $7^2 + 24^2 = 25^2$) [Figure 7 middle]. The aggregate of these triples forms the well known ternary tree of Pythagorean triples (Teia, 2016a). The question that rises then is – do integers in 3D satisfy Conditions 1 and 2 so that triples are possible? [Figure 7 right]. Or, is Fermat's conjecture right, and there are no triples in 3D?

4.2 The 3D Pythagoras' Theorem

This is sometimes confused with Fermat's Conjecture. Fermat's Last Theorem is a mathematical conjecture about integer numbers (Singh, 2002), while the 3D Pythagoras' theorem is a mathematical and geometrical proof about real numbers (Teia, 2015a). How does one build the 3D Pythagoras' theorem? This is now briefly explained. It is assumed that the process used to convert the 1D Pythagoras' theorem to 2D, shown previously in Figure 3, also applies from 2D to 3D.

The two steps for converting the 2D Pythagoras' theorem into 3D are:

Step 1. Rotation of the Z-square by 90 degrees (Figure 8a) forms a central octahedron [in blue] (Figure 8b) just as a central square [in blue] appeared naturally when passing from 1D to 2D (Figure 3). Both the central square (in 2D) and central octahedron (in 3D) are a direct and inherent consequence of the process. Hence, the geometric element that governs the 3D Pythagoras' theorem is an octahedron.

Step 2. Perpendicular extrusion along the axis of symmetry transforms squares into octahedrons. Finally, careful partition of the octahedrons gives the 3D Pythagoras' theorem (Figure 8c), which is equivalent to the 2D Pythagoras' theorem (Figure 8a). This partition is explained in detail in Teia (2015a).

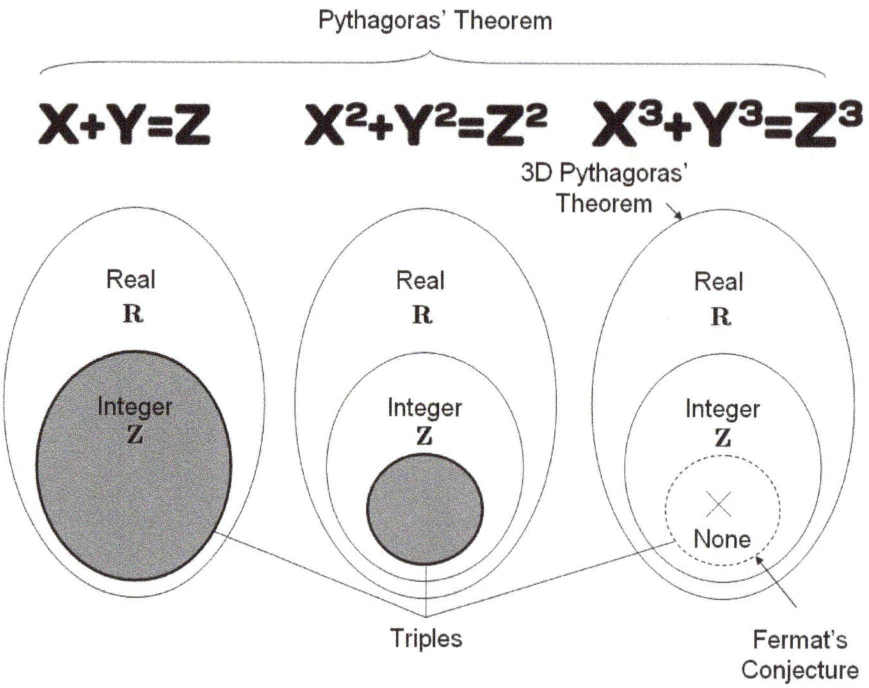

Figure 7. Triples in the world of one, two and three dimensional Pythagoras' theorem

An octahedron satisfies the 3D Pythagoras' theorem, and hence verifies Condition 1. Hence, in order for 3D integers to exist in the Pythagoras' world, octahedrons also need to satisfy Condition 2. The question then is – can an integer octahedron be split into multiple unit octahedrons?

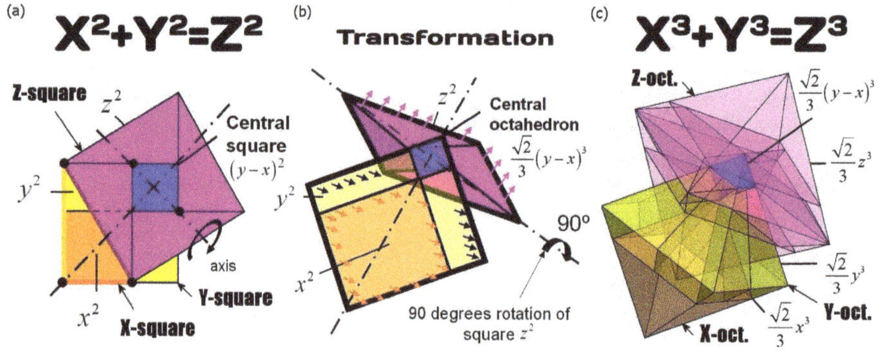

Figure 8. Transformation of the Pythagoras' theorem from 2D to 3D (Teia, 2015a)

4.3 Geometry of 3D Integers

The unit scalars in 1D (the unit line) and 2D (the unit square) have shown to correctly represent all integer elements in their respective dimension (Figure 5). For example, in 1D we have the integer 2^1 (the upper case 1 represents the dimension) defined by two unit lines, and in 2D we have 2^2 defined by four unit squares. Hence, the squares that compose the 2D Pythagoras' theorem can also be split into multiple unit squares (Figure 9a). The question now is – can the 3D Pythagoras' theorem be split into multiple unit octahedrons (Figure 9b)?

Let us try to build, using only unit octahedrons, the first 3D integer after 1^3 – the 2^3. In order to build an integer octahedron with side 2, it is necessary to link 6 unit octahedrons (Figure 10a). They do not completely satisfy the volume of 2^3, as there are tetrahedral gaps in the middle. Hence, eight additional unit tetrahedrons are required to complete the volume (Figure 10b).

This means that the overall integer octahedron 2^3 is defined by 6 unit octahedrons and 8 unit tetrahedrons (Figure 10c). This can be extrapolated to any integer octahedron N^3, where N is an integer. Take, as another example, the integer

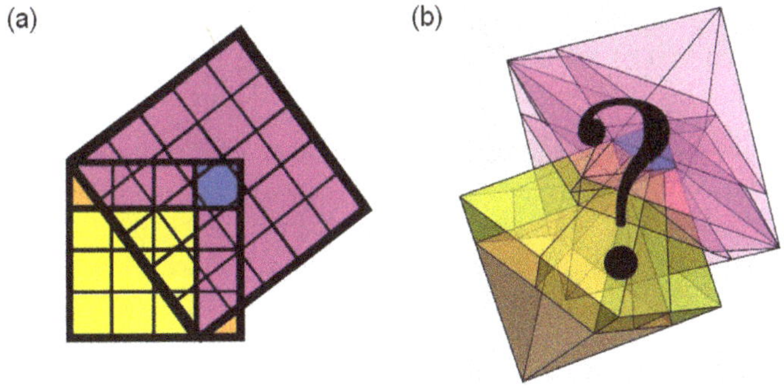

Figure 9. Partition of Pythagoras' theorem into multiple unit scalars (a) 2D and (b) 3D

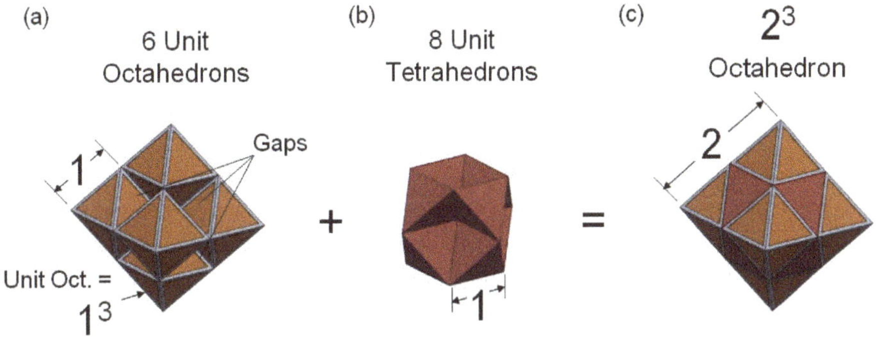

Figure 10. The octahedron 2^3: (a) unit octahedrons, (b) unit tetrahedrons and (c) all together

octahedron 3^3, which is an extension of the 2^3. The integer 2^3 is composed of 4 unit octahedrons at midplane, and 1 on either side up and down (as shown in Figure 11a). Adding four unit tetrahedrons on either side of the midplane completes the volume. The integer 3^3 (Figure 11b), in turn, is formed by adding to 2^3 a new layer of 9 unit octahedrons at midplane, and 12 unit tetrahedrons on either side. Therefore, the proportion of unit octahedrons to unit tetrahedrons changed to 19 and 32, respectively. This can be extrapolated to an octahedron side integer N. It can be shown that the number of unit octahedrons is $\sum_{i=1}^{N} i^2 + \sum_{i=1}^{N-1} i^2$ and unit tetrahedrons is $2\sum_{j=1}^{N-1}\sum_{i=1}^{j} 4i$ (Figure 11c).

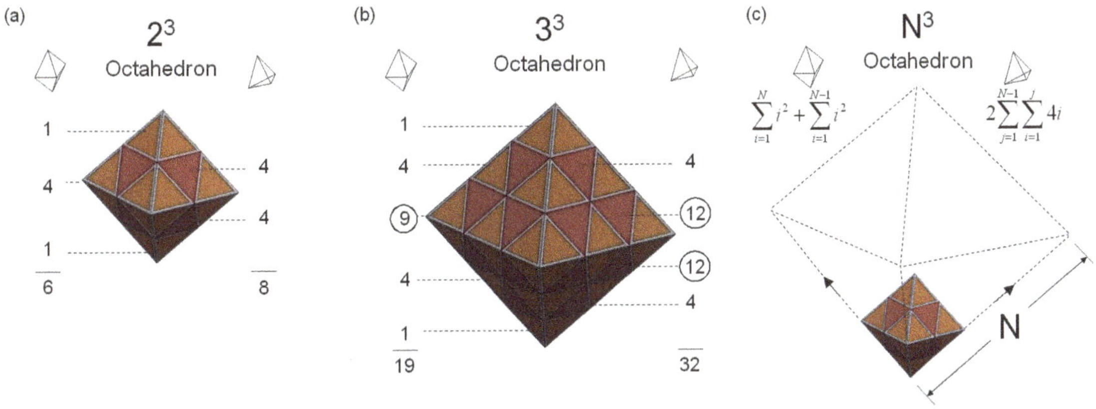

Figure 11. Composition of octahedrons (a) 2^3, (b) 3^3 and (c) N^3.

Integers are clear multiples of a unit. When a line is solely composed of multiple unit lines, the number it represents is an integer. When a square is solely composed of multiple unit squares, it also represents an integer. An octahedron is not solely composed of multiple unit octahedrons, hence it is not an integer. Therefore, even though octahedrons validate the 3D Pythagoras' theorem, and hence Condition 1, the integers ($2^3, 3^3, \ldots N^3$) do not exist in the context of Pythagoras' theorem as they don't satisfy Condition 2, i.e. not completely divisible into multiples of the fundamental geometrical unit.

This non-compliance extends to the entire universe of 3D integers, and validates the hypothesis that three dimensional integers do not exist within the context of the 3D Pythagoras' theorem. By consequence, if integers do not exist, then neither do their associated triples. This confirms Fermat's Conjecture for three dimensions.

Since integers from different dimensions are dependent on each other - for example, 2D integers are formed from 1D integers (i.e., squares of side 1 are formed from lines of length 1) - it would be expected integers in 3D to be formed from integers in 2D. However, it was shown that 3D integers do not exist within the 3D Pythagoras' theorem, it follows from this interdimensional dependency that if integers in 3D don't exist, then neither do in 4D (as they would be dependent on the 3D ones for geometrical construction) or any other higher dimension. This can be generaly stated as, integers within the Pythagoras' theorem for any dimension above 2 (i.e., for $n > 2$ where n is the number of dimensions) do not exist. If integers do no exist, then neither do their respective triples. Ultimately, this means that triples for dimensions $n > 2$ do not exist in the domain of the Pythagoras' equation $x^n + y^n = z^n$.

5. Conclusion

Fermat's Last Theorem is more than just about triples, it is about the fundamental nature of an integer, and it's mathematical and geometrical expression as a multiple of a unit. What are the conditions that define an integer in the context of the Pythagoras' theorem? One must first study how the Pythagoras' theorem is formed in all its dimensions. The 1D Pythagoras' theorem transforms to 2D in two steps: a 90 degree rotation, and a perpendicular extrusion. Condition 1 dictates that an integer within a triple must satisfy the geometrical part of the Pythagoras' theorem. All integers are formed by an aggregate chain of the fundamental unit specific to that dimension. That is, 1D integers like 2^1, 3^1 are defined in the Pythagoras' theorem by unit lines, and 2D integers like 2^2, 3^2 by unit squares. Condition 2 dictates that in order for an integer to be an integer, it must be successfully split into multiples of the fundamental geometric unit of the corresponding dimension. Condition 1 and 2 are valid in both 1D and 2D. The 2D Pythagoras' theorem transforms to 3D using the same steps as before. An octahedron emerges naturally during the process. This is the element that governs the geometry of the 3D Pythagoras' theorem, and verifies Condition 1. However, an integer octahedron cannot be built from aggregates of unit octahedrons, as gaps appear in the form of tetrahedrons, and hence Condition 2 is not verified. This means that the fundamental element that governs the 3D Pythagoras' theorem – the octahedron – cannot be used to construct 3D integers. Therefore, the very notion of integer elements like 2^3, 3^3 do not exist in the three dimensional Pythagoras' world, and consequentially neither do triples. This confirms Fermat's Last Theorem for three dimensions. The geometrical interdependency between integers in 1D and 2D suggests that all integers of higher dimensions are built, and hence are dependent, on the integers of lower dimensions. This interdependency coupled with the absence of integers in 3D suggests that there are no integers above $n > 2$, and therefore there are also no triples that satisfy $x^n + y^n = z^n$ for $n > 2$.

References

Clapham, C., & Nicholson, J. (2014). *The Concise Oxford Dictionary of Mathematics*, 5th ed., Oxford University Press.

Cooke, R. (2005), *The History of Mathematics*, New York Wiley-Interscience. https://doi.org/10.1002/9781118033098

Dickson, L. E. (1919). *History of the Theory of Numbers.* (Vol. 2). Diophantine Analysis. New York: Chelsea Publishing.

Euclid (300 BC). *Elements.* Book 1.

Singh, S. (2002). *Fermat's Last Theorem: The Story of a Riddle That Confounded The World's Greatest Minds for 358 Years*, published via Amazon.

Teia, L. (2015). Pythagoras' triples explained by central squares, *Australian Senior Mathematics Journal, 29*, 7–15.

Teia, L. (2015a). $X^3 + Y^3 = Z^3$: *The Proof*, published via Amazon.

Teia, L. (2016). Geometry of the 3D Pythagoras' theorem, *Journal of Mathematics Research, 8*, 78–84. https://doi.org/10.5539/jmr.v8n6p78

Teia, L. (2016a). Anatomy of the Pythagoras' tree, *Australian Senior Mathematics Journal, 30*, 38–47.

18

Types of Derivatives: Concepts and Applications (II)

Salma A. Khalil[1], Mohammed A. Basheer[2] & Tarig A. Abdelhaleem[3]

[1] Department of Mathematics, Faculty of sciences, Princess Nourah bint Abdulrahman University, Riyadh, KSA

[2] Department of Mathematics , Faculty of sciences, University of Alnillin, Khartoum, Sudan

[3] Department of Mathematics , Collage of Applied and Industrial sciences, University of Bahri , Khartoum, Sudan

Correspondence: Salma A. Khalil, Hepatobiliary Sciences &Liver transplant- 144National Guards Hospita Po box 22490 Riyadh 11426, KSA. E-mail: Salma-math@ windowslive.com

Abstract

The notion of differential geometry is known to have played a fundamental role in unifying aspects of the physics of particles and fields, and have completely transformed the study of classical mechanics.

In this paper we applied the definitions and concepts which we defined and derived in part (I) of our paper: Types of Derivatives: Concepts and Applications to problems arising in Geometry and Fluid Mechanics using exterior calculus. We analyzed this problem, using the geometrical formulation which is global and free of coordinates.

Keywords: differential geometry, exterior calculus, free of coordinates

1. Introduction

In our previous paper, we defined three types of major derivatives such as the Exterior Derivative, Lie Derivative and Covariant Derivative (Kolár, et al., 1999; Warner, 2013). We divided our work into two parts, where in the first section we started by defining a Differentiable Manifold structure, (Arkani-Hamed, et al., 2010) then The Tangent Bundle, where we built the Bundle from the tangent space defined on a Differentiable Manifold; we defined the Cotangent Bundle in a similar fashion to the Tangent Bundle, then considered Smooth Vector Fields and finally concluded our structure by defining Tensor fields and Riemannian Manifolds (Beig, R.).

In the second part, we defined The Covariant Derivative in which we defined Covariant derivatives of covectors and Tensors, Lie Derivatives:Lie Derivatives of tensor fields and Differential Forms. Finally, in the third part we defined The Exterior Derivative and shed light on the properties of the respective Derivative.

Exterior calculus is a concise formalism to express differential and integral equation on smooth and curved spaces in a consistent manner, while revealing the geometrical invariants at play. One of the main goals of developing a geometric theory of fluid is to put all the existing computational techniques in one abstract setting. This rationalization of computational mechanics will be theoretically interesting for its own sake.

Rewriting equations of fluid mechanics in terms of differential forms enables one to clearly see the geometric features of the fluid field theory.

2. Main Formulas

2.1 Definition (Kobayashi, S., & Nomizu, K., 1963)

Let M be a smooth differentiable manifold of dimension m. The Tangent Bundle TM is defined to be

$$TM = \cup_{p\in M} T_pM = \{(p,\upsilon) | p \in M, \upsilon \in T_pM\}. \quad (1)$$

2.2 Definition (Lang, S., 1999)

AVector field υ on M is a section of the tangent bundle TM, ie $\upsilon: M \to TM$ such that $\pi \circ \upsilon(p)$for every $p \in M$. In other words Let $x: U \to R^m$ be a local chart of M, and $p \in U$, then $\upsilon(p) = \sum_{i=1}^{m} \upsilon_p (x^i) \frac{\partial}{\partial x^i}|_p$. avector field on M is a map υ which assigns to each point $p \in M$ a tangent vector $\upsilon(p) = \upsilon_p \in T_p(M)$.

$$g_{UV}(p) = \left(\frac{\partial y^i}{\partial x^j}\right)_{1\le i,j\le m} |_{\phi_U(p)}. \quad (2)$$

2.3 Definition

A covariant derivative is an operator V on tensor fields which satisfies the following conditions:

1) If T is of rank (r, s), then ∇T is of rank $(r, s + 1)$; the covariant rank increases by 1.

2) For any function $f, \nabla(f) = df. [\nabla_a(f) = \partial_a(f)]$

3) For any function f and tensor $T, \nabla(fT) = df \otimes T + f\nabla T. [\nabla_a(fT^{...}_{...}) = \partial_a(f)T^{...}_{...} + f\nabla_a T^{...}_{...}]$.

4) More generally, for any tensors Sand $T, \nabla(S \otimes T) = \nabla S \otimes T + S \otimes \nabla T$.

2.4 Definition Lie Derivative (O'neill, B., 1983)

If φ is a local diffeomorphism $M \rightarrow N$, we may define a pull- back map

$\varphi^* : \Gamma T_{r,s} N \rightarrow \Gamma T_{r,s}$ Mon mixed field as follows. For $T \in \Gamma T_{r,s} N$ we define

$$\varphi^*(T)_m := (d_m \varphi)^* T_{\varphi(m)}.$$

2.5 Definition

Let $X \in \mathfrak{X}(M)$ be smooth vector field. Then for every $m \in M$we denote by $t \mapsto \varphi_X^t(m)$ the (maximal) integral curve for X with initial point m. the domain of this integral curve is an open interval $I_{X,m}$ containing 0. Let $T \in \Gamma T_{r,s}$ M and we define the Lie derivative of T with respect to X by

$$(\mathcal{L} \times T)_m := \frac{d}{dt}|_{t=0}[(\varphi_X^t)^* T]_m$$

note that φ_X^t is a diffeomorphism from a neighborhood of m onto a neighborhood of $\varphi_X^t(m)$. Accordingly, the expression

$$[(\varphi_X^t)^* T]_m$$

is a well-defined element of $(T_m M)_{r,s}$ which depends smoothly on t (in a neighborhood of 0). Accordingly, $(\mathcal{L} \times T)_m$ defines a tensor in $(T_m M)_{r,s}$. Moreover, by the smoothness of the flow of the vector field X it follows that the section $\mathcal{L} \times T$ of the tensor bundle $T_{r,s} M$ thus defined is smooth. In other words, we have defined a linear map.

$$\mathcal{L}_X : \Gamma T_{r,s} M \rightarrow \Gamma T_{r,s} M,$$

called The Lie derivative. In a similar way it is seen that the Lie derivative defines a linear map $\mathcal{L}_X : E_k(M) \rightarrow E_k(M)$

(Cartan's formula) let X be a smooth vector field on M. then on E(M),

$$\mathcal{L}_X = i(X) \circ d + d \circ i(X).$$

The proof was given in paper (I)Applications.

3. The Geometric Setup

We give a geometric model of the basic kinematics used in modeling the fluid flow.

3.1 The Fluid Space (Haller, G., 2001)

Assume our fluid flows in a smooth manifold M(M denotes a differentiable n-manifold).

A fluid particle is a point in the manifold. Points in a domain $D \subset M$ represent the geometric positions of material particles;these points are denoted by $x \in M$ and called particle labels.

3.2 Fluid Motion (Geometric Notion of the Fluid Motion)

The Fluid moves in a manifold whose points represent the fluid particles. Let $x \in M$ be a point in M (M is the space in which the fluid moves) and consider the particle of fluid moving through x at time t = 0. As t increases, we denote by$\phi_t(x)$ the curve followed by the fluid particle, which is initially at $x \in M$. For fixed t, each ϕ_t will be a diffeomorphism of M. Thus the fluid motion is a smooth one parameter family of diffeomorphisms ϕ_t: M→M; with ϕ_0 =Id.

$t \rightarrow \phi(t)$ is a one parameter family of diffeomorophisms of M.For each value oft,we define a vector field $X_t \in \mathfrak{x}(M)$ as follows:

For $x \in M$, $X_t(x)$ is the tangent vector to the curve $u \rightarrow \phi(u)\phi(t)^{-1}x$ at

u = t, X, is the velocity field corresponding to the fluid motion defined by $\phi(t)$: that is, $X_t(x)$ is the velocity vector of the particle that, at time t, is at the point x.

$$X(\phi_t(x), t) := \frac{\partial \phi_t}{\partial t}(x). \tag{3}$$

$$X_t(f)(X) = \frac{d}{du} f\left(\phi_u \phi_t^{-1}(x)\right)|_{u=t} = \frac{d}{du}(\phi_u \phi_t^{-1})(f)(x)|_{u=t}$$

$$= \frac{d}{du}((\phi_t^{-1})^* \phi_u^*)(f)(x)|_{u=t} = (\phi_t^{-1})^* (\frac{d}{du}\phi_u^*(f)(x)|_{u=t} \tag{4}$$

$$\text{Or } \phi_t^*(X_t(f))(x)|_{u=t} = \frac{d}{du}\phi_u^*(f)(x) = \frac{\partial}{\partial t}\phi_t^*(f)(x) \tag{5}$$

$$\text{Therefore } \quad X_t(f) = (\phi_t^{-1})^* \frac{\partial}{\partial t}\phi_t^*(f); \text{ For } f \in F(M) \tag{6}$$

Note: the inverse maps be computed by reversing time, $\phi_t^{-1} = \phi_{-t}$.
A flow is called steady (or stationary)if its vector field satisfies:

$$\frac{\partial X}{\partial t} = 0, \tag{7}$$

i.e. the "shape" of the fluid flow is not changing. Even if each particle is moving under the flow, the global configuration of the fluid does not change.

Let the one-form $\alpha \in \Omega^1(M)$ describe the velocity of a fluid. We define vorticity as

$$\varpi = d\alpha \tag{8}$$

The vorticity is a 2-form UJ is dual to the vorticity vector field (.The trajectories of vorticity field are called vortex lines.A flow is called irrotational if

$$\varpi = 0 \tag{9}$$

From the Poincare lemmaon some open subset $U \subset M$, there exist $\varphi \in \mathcal{F}_U$ such that

$$\alpha = d\varphi \tag{10}$$

$\varphi \in \mathcal{F}_U$is called velocity potential.

4. Continuity Equation (Erbar, M., 2010; Bajura, R., & Jones, E., 1976)

4.1 Continuity Equation on Manifolds

Reformulating the continuity equation from the point of view of vector fields and differential forms on manifolds.

Let $D \subset M$ be a sub region of MConsider a fluid moving in a domain $D \subset M$ and suppose ω is a fixed¬volume element differential form on M, which is point-wise nonvanishing. Then, the total mass of the fluid in the region D at time t is

$$m(D, t) = \int_D \rho_t \omega\,. \tag{11}$$

Where ρ_t describes the mass-density of the fluid at time t. from the principle of mass conservation in fluid dynamics, the total mass of the fluid, which at time t = 0 occupied a region D,remains unchanged after time t.

Thus, the total mass of the fluid at time t = 0 occupinga region D is maintained with time, i.e.

$$\int_{\phi_t(D)} \rho_t \omega = \int_D \rho_0 \omega\,. \tag{12}$$

Where $\phi_t$the one-parameter family of diffeomorophisms.Equation (12) isis the integral invariant for conservation of mass.

By the change of variable formula, the left hand side of this relation is equal to

$$\int_D \phi_t^*(\rho_t \omega) = \int_D \rho_0 \omega\,. \tag{13}$$

differentiating,with respect to t, we get

$$\frac{\partial}{\partial t}\int_D \phi_t^*(\rho_t \omega) = \int_D \frac{\partial}{\partial t}(\phi_t^*(\rho_t \omega)) = 0\,. \tag{14}$$

Since$\frac{\partial}{\partial t}(\phi_t^*(\rho_t \omega)) = \phi_t^*(L_{X_t}(\rho_t \omega)) + \phi_t^*(\frac{\partial}{\partial t}(\rho_t)\omega)$

Then (14) takes the form

$$\int_D [\phi_t^*(L_{X_t}(\rho_t\omega) + \frac{\partial}{\partial t}(\rho_t)\omega)] = 0. \quad (15)$$

by change of variable formula

$$\int_{\phi_t(D)} [L_{X_t}(\rho_t\omega) + \frac{\partial}{\partial t}(\rho_t)\omega] = 0; \ \forall D \quad (16)$$

Since D is an arbitrary open set, this can be true only if the integrand is zero; that is

$$L_{X_t}(\rho_t\omega) + \frac{\partial}{\partial t}(\rho_t)\omega = 0 \quad (17)$$

This is the equation of continuity in invariant form.Equation (17) can be written as

$$L_{X_t}\omega_t + \frac{\partial\omega_t}{\partial t} = 0 \quad (18)$$

where$\omega_t := \rho_t\omega$ is a 1-parameter family of n-forms on M.

(17) takes the following form

$$L_X(\rho\omega) = 0 \quad (19)$$

From properties of the Lie derivative this reduces to

$$(L_X\rho)\omega + \rho L_X\omega = 0 \quad (20)$$

Since $\rho$is a constant function,the first term on the left-hand side of equation (20) is vanish, then equation (20) take the following form

$$L_X\omega = 0 \quad (21)$$

This is the geometric form of equation of continuity for incompressible fluid.According to (Cartan's formula) the Lie-derivative applied to volume form can may be written as

$$L_X\omega = i_x d\omega + di_x\omega = di_x\omega \quad (22)$$

Since $d\omega = 0$. Using (22), we find the alternative invariant formulation of continuity equation for incompressible fluid

$$di_{X_t}\omega = 0 \quad (23)$$

These two formulations are equivalent.

Now, whenever $M = \mathbb{R}^3$. In fact

$$\left.\begin{aligned} L_{X_t}(\rho_t\omega) = (L_{X_t}\rho_t)\omega + \rho_t L_{X_t}\omega \quad &= X_t(\rho_t)\omega + \rho_t(\mathrm{div}X_t)\omega \\ X_t(\rho_t) &= \sum v^i(x,t)\frac{\partial\rho}{\partial x^i}(16) \\ \mathrm{div}X_t &= \frac{\partial v^i}{\partial x^i} \\ \frac{\partial}{\partial t}(\rho_t)\omega &= \frac{\partial\rho}{\partial t}\omega. \end{aligned}\right\} \quad (24)$$

substituting (24)into the continuity equation (17), we get the following

$$\frac{\partial\rho}{\partial t}\omega + \sum v^i\frac{\partial\rho}{\partial x^i}\omega + \rho\frac{\partial v^i}{\partial x^i}\omega = 0 \quad (25)$$

Since$\omega = dx^1dx^2dx^3$ (volume form), therefore

$$\frac{\partial\rho}{\partial t} + \mathrm{div}(\rho v) = 0 \quad (26)$$

Which is the usual equation of continuity.

4.2 Continuity Equation as Exterior Differential System (Verhulst, F., 2006)

In this sub section, we use the Cartan's theory to show that it is possible to rewrite the continuity equation as an exterior differential system.

A. Compressible (general) case:

To reformulate the continuity equation as an exterior differential system(alternate geometric approach), we set.

$$\left.\begin{aligned}\theta^1 &= dx^1 - v^1 dt\\ \theta^2 &= dx^2 - v^2 dt\\ \theta^3 &= dx^3 - v^3 dt\end{aligned}\right\} \tag{27}$$

We define the 3-form:

$$\begin{aligned}\omega &= \rho(dx^1 - v^1 dt) \wedge (dx^2 - v^2 dt) \wedge (dx^3 - v^3)\\ &= \rho dx^1 dx^2 dx^3 - \rho v^1 dx^1 dx^2 dx^3 dt + \rho v^2 dx^3 dt dx^1 - \rho v^3 dt dx^1 dx^2\end{aligned} \tag{28}$$

By applying the exterior derivative of both sides in (28)

$$\begin{aligned}d\omega &= d[\rho dx^1 dx^2 dx^3 - \rho v^1 dx^1 dx^2 dx^3 dt + \rho v^2 dx^3 dt dx^1 - \rho v^3 dt dx^1 dx^2]\\ &= d\rho \wedge dx^1 dx^2 dx^3 - d(\rho v^1) \wedge dx^2 dx^3 dt + d(\rho v^2) \wedge dx^3 dt dx^1 - d(\rho v^3) \wedge dt dx^1 dx^2]\\ &= \left(\frac{\partial \rho}{\partial t} dt + \frac{\partial \rho}{\partial x^1} dx^1 + \frac{\partial \rho}{\partial x^2} dx^2 + \frac{\partial \rho}{\partial x^3} dx^3\right) dx^1 dx^2 dx^3\\ &- \left(\frac{\partial \rho v^1}{\partial t} dt + \frac{\partial \rho v^1}{\partial x^1} dx^1 + \frac{\partial \rho v^1}{\partial x^2} dx^2 + \frac{\partial \rho v^1}{\partial x^3} dx^3\right) dx^2 dx^3 dt\\ &+ \left(\frac{\partial \rho v^2}{\partial t} dt + \frac{\partial \rho v^2}{\partial x^1} dx^1 + \frac{\partial \rho v^2}{\partial x^2} dx^2 + \frac{\partial \rho v^2}{\partial x^3} dx^3\right) dx^3 dt dx^1\\ &- \left(\frac{\partial \rho v^3}{\partial t} dt + \frac{\partial \rho v^3}{\partial x^1} dx^1 + \frac{\partial \rho v^3}{\partial x^2} dx^2 + \frac{\partial \rho v^3}{\partial x^3} dx^3\right) dt dx^1 dx^2\end{aligned} \tag{29}$$

Using the properties of wedge product, this becomes

$$\begin{aligned}d\omega &= \frac{\partial \rho}{\partial t} dt dx^1 dx^2 dx^3 - \frac{\partial \rho v^1}{\partial x^1} dx^1 dx^2 dx^3 dt + \frac{\partial \rho v^2}{\partial x^2} dx^2 dx^3 dtd - \frac{\partial \rho v^3}{\partial x^3} dx^3 dt dx^1 dx^2\\ &= \left[\frac{\partial \rho}{\partial t} + \frac{\partial \rho v^1}{\partial x^1} + \frac{\partial \rho v^2}{\partial x^2} + \frac{\partial \rho v^3}{\partial x^3}\right] dt dx^1 dx^2 dx^3\end{aligned} \tag{30}$$

Therefore divω= 0 corresponds to

$$\frac{\partial \rho}{\partial t} + \frac{\partial \rho v^1}{\partial x^1} + \frac{\partial \rho v^2}{\partial x^2} + \frac{\partial \rho v^3}{\partial x^3} = 0 \tag{31}$$

Thus the continuity equation correspond to the closed form

$$d\omega = 0 \tag{32}$$

Suppose we express the flow in terms of initial conditions (or other parameters) by

$$x = x(t, \alpha^1, \dots, \alpha^3), \tag{33}$$

so that the a' are the parameters and

$$\frac{\partial x}{\partial t} = v, \tag{34}$$

$$\text{Thus} dx^i - v^i dt = \left(\frac{\partial x^i}{\partial t} dt + \Sigma \frac{\partial x^i}{\partial \alpha^j} d\alpha^j\right) - v^i dt = \Sigma \frac{\partial x^i}{\partial \alpha^j} d\alpha^j \tag{35}$$

So that

$$\omega = \rho \frac{\partial(x^1, x^2, x^3)}{\partial(\alpha^1, \alpha^2, \alpha^3)} d\alpha^1 d\alpha^2 d\alpha^3 = K(t, \alpha) d\alpha^1 d\alpha^2 d\alpha^3 \tag{36}$$

Since $d\omega = 0$ we deduce that

$$\frac{\partial k}{\partial t} = 0 \tag{37}$$

Therefore

$$\omega = A(\alpha)d\alpha^1 d\alpha^2 d\alpha^3. \tag{38}$$

This means that the 3-form $\omega$is an integral-invariant for the flow ϕ_t, which represents property of conservation of mass Equation (32).We consequently have the following result. Given two 3-chains $C_3, C_3^{'} \in C_3(M)$ which are in 1 - 1 correspondence such that corresponding points lie on the same trajectory of the flow $\{\phi_t\}$, then:

$$\int_{C_3} \omega = \int_{C_3^{'}} \omega, \quad C_3^{'} = (\phi_t)_* C_3 \tag{39}$$

If now $C_3 = C_3^{(0)}$ $at\ t_0 = 0$, then, by (2.37),

$$\omega = \int_{C_3^{'}} \omega|_{t=t_0} = \int_{C_3} \rho dx^1 dx^2 dx^3, \tag{40}$$

And following up $C_3^{(0)} = C_3^{(1)}$ at time t_1, we have:

$$\int_{C_3^{(0)}} \rho dx^1 dx^2 dx^3 = \int_{C_3^{(1)}} \rho dx^1 dx^2 dx^3 \tag{41}$$

Which expresses that mass is preserved in the flow $\{\phi_t\}$, another form of the conservation of mass.

B. Inviscid, incompressible, and irrotational (potential flow) case:

A potential flow describes what the flow would be like if it were inviscid, incompressible, and irrotational. which in vector calculus is described by Laplace equation $\nabla^2\varphi =$ .Reformulating the problem in a differential geometry terms we consider the scalar function $\varphi\ which\ is\ a\ 0 - form\ or\ a\ 2 - form$, following contact variables

$$p = \frac{\partial \varphi}{\partial x}, q = \frac{\partial \varphi}{\partial y}, r = \tag{42}$$

Let M be a manifold with variables $(x, y, z, \varphi, p, q, r)$, on M we define the contact form

$$\alpha = d\varphi - pdx - qdy - rdz, \tag{43}$$

And the 2-form

$$\alpha \ = pdydz + qdzdx + rdxdy, \tag{44}$$

By taking the exterior derivative of (44) and using the anti-symmetric property of wedge product we obtain

$$d\sigma = d[pdydz + qdzdx + rdzdy] \ = dp \wedge dydz + dq \wedge dzdx + dr \wedge dxdy$$

$$= \left(\frac{\partial p}{\partial x}dx + \frac{\partial p}{\partial y}dy + \frac{\partial p}{\partial z}dz\right) dydz + \left(\frac{\partial q}{\partial x}dx + \frac{\partial q}{\partial y}dy + \frac{\partial q}{\partial z}dz\right) dzdx$$

$$+ \left(\frac{\partial r}{\partial x}dx + \frac{\partial r}{\partial y}dy + \frac{\partial r}{\partial z}dz\right) dxdy \tag{45}$$

By using the anti-symmetric property of wedge product, this becomes

$$d\sigma = \frac{\partial p}{\partial x}dxdydz + \frac{\partial q}{\partial y}dydzdy + \frac{\partial r}{\partial z}dzdxdy \ = \left[\frac{\partial^2 \varphi}{\partial x^2} + \frac{\partial^2 \varphi}{\partial y^2} + \frac{\partial^2 \varphi}{\partial z^2}\right] dxdydz \tag{46}$$

Note that $d\sigma = 0$. Then Laplace equation is equivalent to the relation

$$d\sigma = 0 \tag{47}$$

This is the coordinate-free version.

5. Equations of Motion

5.1 Momentum Equation on Manifolds (Datta, A., & Majumdar, A., 1980)

Navier-Stokes equation is the most general equation for description of fluid phenomena , which as special case comprises Euler's equation of motion.Let M be a differentiable n-manifold, θ be a differential 1-form on M, co be a volume-element differential form on Mand D be a domain in M.

Consider a fluid moving in a domain $D \subset M$. For any continuum there are two types of forces acting on a piece of material.

1) First there are external or body forces:

whereF is a vector field representing the volume forces on a domain D, the θ-component of the volume forces acting on the domain D are

$$Force_{vol} = \int_D \rho^{\theta}(F)\omega, \tag{48}$$

2) Second there are stress forces (discipline of continuum mechanics can also encounter forces that come from the region surrounding a bit of fluid, expressed by the stress tensor):

Phrasing this stress force a differential geometric language.

Let T be a tensor field on M representing this stress. At $x \in M$, T is a skew-symmetric tensor field on M define in the following form:

$$\theta^{n-1}(X_1, \dots, X_{n-1}) = \theta\big(T(X_1, \dots, X_{n-1})\big); X_i \in T_x \tag{49}$$

Then we define the stress tensor of the fluid inside $D \subset M$ in terms of the multilinear map:

$$T = T(X_1, X_2, \dots, X_{n-1}): T_x(M) \times \dots \times T_x(M) \longrightarrow T_x(M), \tag{50}$$

For $x \in M,\ X_1, X_2, \dots, X_{n-1} \in T_x(M)$

$\big(T(X_1, \dots, X_{n-1})\big)$is an $(n-1)$ covector on M. Then $\theta(T)$ defines for each θ an $(n-1)$-differential form on M. If D is a domain in M with boundary ∂D, then the total θ-component of the stress force is

$$Force_{str} = \int_{\partial D} \theta(T), \tag{51}$$

Applying Stokes' theorem, to $\int_{\partial D} \theta(T)$ we get

$$Force_{str} = \int_{\partial D} \theta(T) = \int_D d\theta(T). \tag{52}$$

From (11) and (15) the 8-component of the force acting on the domain D at fixed time is

$$Force_{tot} = \int_{\partial D} [\rho\theta(F)\omega + d\theta(T) = \int_{\phi_t(D)} [\rho_t\theta(F_t)\omega + d(\theta(T(z_t)(i_z\omega)))] \tag{53}$$

Where$\phi_t(D)$ is a volume preserving fluid flow with evolution operator ϕ_t.

Suppose a group of particles making up the fluid start out at t = 0 to occupy the domain D. At time t, they will be in domain $\phi_t(D)$, their θ- component of total momentum will be

$$Momentum_{tot} = \int_{\phi_t(D)} \theta(X_t)\rho_t\omega, \tag{54}$$

by the change of variables theorem, this equals

$$Momentum_{tot} = \int_D \phi_t^*(\theta(X_t)\rho_t\omega). \tag{55}$$

The rate of change of momentum

$$\frac{\partial}{\partial t}\int_D \phi_t^*(\theta(X_t)\rho_t\omega) = \int_D \frac{\partial}{\partial t}[\phi_t^*(\theta(X_t)\rho_t\omega)]$$

$$= \int_D \phi_t^*\left(L_{X_t}(\theta(X_t)\rho_t\omega)\right) + \int_D \phi_t^*\left(\theta\left(\frac{\partial X_t}{\partial t}\right)\rho_t\omega + \theta(X_t)\frac{\partial \rho_t}{\partial t}\omega\right) \tag{56}$$

By the change of variables theorem, we get

$$\frac{\partial}{\partial t}\int_D \phi_t^*(\theta(X_t)\rho_t\omega) = \int_{\phi_t(D)} \left[L_{X_t}(\theta(X_t)\rho_t\omega) + \theta\left(\frac{\partial X_t}{\partial t}\right)\rho_t\omega + \theta(X_t)\frac{\partial \rho_t}{\partial t}\omega\right] \tag{57}$$

By applying Newton's law of motion (principle of balance of momentum): the rate of momentum of a portion of the fluid equals the total force applied to it. By equating (16) to the expression (20) for the force acting in this region, we get

$$\int_{\phi_t(D)} \left[\rho_t\theta(F_t)\omega + d\left(\theta\big(T(z_t)(i_z\omega)\big)\right)\right]$$

$$= \int_{\phi_t(D)} \left[L_{X_t}(\theta(X_t)\rho_t\omega) + \theta\left(\frac{\partial X_t}{\partial t}\right)\rho_t\omega + \theta(X_t)\frac{\partial \rho_t}{\partial t}\omega\right] \tag{58}$$

Since this relation is to hold for all domainsD, we have

$$\rho_t\theta(F_t)\omega + d\theta(T_t) = L_{X_t}(\theta(X_t)\rho_t\omega) + \theta\left(\frac{\partial X_t}{\partial t}\right)\rho_t\omega + \theta(X_t)\frac{\partial\rho_t}{\partial t}\omega. \quad (59)$$

This is the geometric version of momentum equation of fluid motion, which is coordinate-free.

If M= $\mathbb{R}^3$, with coordinates $x_i, i = 1,2,3$.

$$X_i = v_i(x,t)\frac{\partial}{\partial x_i}, \qquad \rho_t = \rho(x,t), \quad (60a)$$

$$\omega = dx_1 \wedge dx_2 \wedge dx_3, \qquad \theta = dx_k, \quad (60b)$$

$$F = F^k\frac{\partial}{\partial x_i}, \qquad \theta(\mathrm{F}) = dx_k(F) = dx_k\left(F^i\frac{\partial}{\partial x_i}\right) = F^k \quad (60c)$$

We consider the bilinear map:

$$T_x(M) \times T_x(M) \to T_x(M)$$

$$(X,Y) \to T(X,Y)(x) = \sum X^iY^j\,T\left(\frac{\partial}{\partial x^i}, \frac{\partial}{\partial x^j}\right)$$

Defined by

$$T_x(X(x),Y(x)) = T(X,Y)(x), \quad (61)$$

Where

$$X = \sum X^i\frac{\partial}{\partial x^i}, \quad Y = \sum Y^j\frac{\partial}{\partial x^j}, \qquad X,Y \in \mathfrak{x}(M)$$

$$T\left(\frac{\partial}{\partial x^i}, \frac{\partial}{\partial x^j}\right) = T_{ijk}\frac{\partial}{\partial x^k}, (2.3.15)$$

Then

$$T(X,Y)(x) = X^iY^jT\left(\frac{\partial}{\partial x^i}, \frac{\partial}{\partial x^j}\right)(x) = X^i(x)Y^j(x)T_{ijk}(x)\frac{\partial}{\partial x^k}, \quad (62)$$

$$\text{And } T_{ijk} = \frac{1}{2}\in_{ij\imath} T_{\imath k}, \quad (63)$$

Where$\in_{ij\imath}$is the 3-index, skew-symmetric tensor (with $\in_{123} = 1$). Equations (63) and (64) correspond to the conventional continuous approach where:

$$T_k(dS) = T_{ij\imath}dS^{ij}, \quad (64)$$

Now

$$d(\theta(T)) + \frac{1}{2}\frac{\partial T_{ijk}}{\partial x_h}dx_h \wedge dx_j = \frac{1}{2}\left(\in_{hij}\frac{\partial T_{ijk}}{\partial x_h}\right)\omega$$

$$= \frac{1}{2}\in_{hij}\in_{ij\imath}\frac{\partial T_{ik}}{\partial x_h}\omega = \delta_{h\imath}\frac{\partial T_{ik}}{\partial x_h}\omega = \frac{\partial T_{ik}}{\partial x_h}, \quad (65)$$

$$\text{And } X_t(\omega) = \frac{\partial v_i}{\partial x_i}\omega \quad (66)$$

Hence (66) becomes

$$\rho F_k + 3(\omega) = \frac{\partial T_{ik}}{\partial x_i} = v_i\frac{\partial}{\partial x_i}(v_k\rho) + v_k\rho\frac{\partial v_i}{\partial x_i} + \frac{\partial v_k}{\partial t}\rho + v_k\frac{\partial\rho}{\partial t}, \quad (67)$$

Which is the usual momentum equation of motion.

5.2 Euler Equation on Riemannian Manifold (Gilkey, P. B., 1975; Yamabe, H., 1960)

In this section we express the Euler's equation in the language of differential one-forms on Riemannian manifold. The formulation is obtained in the absence of body forces.

Consider an n-dimensional manifold M with a Riemannian metric g. By identifying the differential forms with their dual vector fields, we rewrite Euler equation in terms of differential 1-forms. Let $\alpha \in \Omega^1(M)$ be the one form associated to X (α is the one-form dual to X), describe the velocity of an ideal fluid. We seek an invariant meaning for the sum of the last three terms on the left-hand side of equation (26). For fixed i this expression equals

$$v_j \frac{\partial v_i}{\partial x^j} = v_j \frac{\partial v_i}{\partial x^j} + v_j \frac{\partial v_j}{\partial x^i} - v_j \frac{\partial v_j}{\partial x^i} = L_x \alpha - \frac{1}{2} d(\alpha(x)) (27) \text{Also} \frac{\partial p}{\partial x_i} = dp \quad (68)$$

So substituting (67) and (68) into the Euler equations (65), we get the following exterior differential system

$$\frac{\partial \alpha}{\partial t} + L_x \alpha - \frac{1}{2} d(\alpha(x)) + \frac{1}{\rho} dP = 0 \quad (69)$$

This is the Euler equation in terms of differential 1-forms. According to (Cartan's magic formula) the Lie-derivative applied to one-form a may be written as

$$\begin{aligned} L_X \alpha &= i_X d\alpha + d i_X \alpha \\ &= i_X d\alpha + d(\alpha(x)) \\ &= i_X \varpi + d(\alpha(x)) \end{aligned} \quad (70)$$

Since $d\alpha = \varpi$. So substituting (70) into the Euler equation (69), we get the following alternative form

$$\frac{\partial \alpha}{\partial t} + i_X \varpi + \frac{1}{2} d(\alpha(x)) + \frac{1}{\rho} dP = 0 \quad (71)$$

On the basis of the Euler equation (69)and exterior calculus, we derive in the following a set of invariant fluid equations.

A. Bernoulli equation.

For a steady (time-independent) flow of a perfect (an inviscid and incompressible) fluid, Euler equation (69) becomes

$$L_X \alpha = -d\left(\frac{1}{\rho} P - \frac{1}{2} \alpha(x)\right) \quad (72)$$

By using Cartan Magic formula

$$\begin{aligned} L_X \alpha &= i_X d\alpha + d i_X \alpha \\ &= i_X d\alpha + d(\alpha(x)) \end{aligned} \quad (73)$$

Substituting (73) into (72) we get

$$i_X d\alpha = -d\left(\frac{1}{\rho} P - \frac{1}{2} \alpha(X)\right) \quad (74)$$

$$\text{Where} \quad f = \frac{1}{\rho} P - \frac{1}{2} \alpha(X) \quad (75)$$

From (74) we have

$$i_X d\left(\frac{1}{\rho} P + \frac{1}{2} \alpha(X)\right) = 0 \quad (76)$$

Since $i_X i_X = 0$, so that the quantity $\frac{1}{\rho} P + \frac{1}{2} \alpha(X)$ is constant along each streamline (integral curves of X) of the fluid flow. This is the Bernoulli's principle. The function f is known as the Bernoulli function of X.

Also

$$d(i_X d\alpha) = L_X d\alpha = d(-df) = 0 \quad (77)$$

meaning that $d\alpha$ is preserved by X.

$$\text{Since} i_X d\alpha - df, \text{ we get} L_X f = -i_X i_X d\alpha = 0 \quad (78)$$

Thus f is invariant under the flow of X. We will call Euler vector fields to the solutions to the equations of an ideal steady incompressible fluid on a manifold.

If the fluid is irrotational ($d\alpha$= 0) then Poincare lemma tells us that

$$\alpha = d\varphi \quad (79)$$

Where$\varphi \in \mathcal{F}_U$is a velocity potential, $U \subset M$. Substituting (79) into (71) we obtain

$$\frac{\partial d\varphi}{\partial t}+\frac{1}{2}d(\alpha(X))=-\frac{1}{\rho}dP \tag{80}$$

Therefore,

$$d\left[\frac{\partial d\varphi}{\partial t}+\frac{1}{2}(\alpha(X))+\frac{1}{\rho}P\right]=0 \tag{81}$$

Thus we have the Unsteady Bernoulli Equation form from the geometric point of view

$$\frac{\partial\varphi}{\partial t}+\frac{1}{2}(\alpha(X))+\frac{1}{\rho}P+c, \tag{82}$$

$$\text{on } U\subset M,\text{ where } c\in\mathcal{F}_U \text{ satisfied } \quad dc=0 \tag{83}$$

B. Vorticity Equation and Conservation Properties

On the basis of the Euler equation (69) we show conservation of vorticity

$$\varpi=d\alpha\in\Omega^2(M)$$

By taking the exterior derivative of (69),

$$d\left[\frac{\partial\alpha}{\partial t}+L_x\alpha-\frac{1}{2}d(\alpha(X))+\frac{1}{\rho}dP\right]=0 \tag{84}$$

As the Lie-derivative and the exterior derivative commute, we get the following

$$\frac{\partial d\alpha}{\partial t}+L_x d\alpha-\frac{1}{2}dd(\alpha(X))+d\left(\frac{1}{\rho}\right)ddP=0 \tag{85}$$

For the case of incompressible flows

$$\frac{\partial\varpi}{\partial t}+\mathrm{L}_x\varpi=0. \tag{86}$$

This equation is called vorticity equation From (85) it follows that

$$:\frac{\partial}{\partial t}(\phi_t^*\varpi_t)=0,$$

So

$$\phi_t^*\varpi_t=\varpi_0, \tag{88}$$

Showing that vorticity moves with the fluid.This is, via stokes theorem another way of phrasing Kelvin's theorem.

5.2.1 Kelvin Circulation Theorem. (*Marsden, J., & Weinstein, A., 1983*)

Let M be a manifold and $l\subset M$a smooth closed loop that is, a compact one-manifold. Let X_t solve the Euler equations on M for ideal isotropic compressible or homogeneous incompressible flow and $l(t)$ be the image of l at time t when each particlemoves under the flow $\phi_t$of X_t; that is, $l(t)=\phi_t(l)$. Then the circulation is constant in time; that is,

$$\frac{d}{dt}\int_{l(t)}\alpha=0. \tag{89}$$

Proof. Let ϕ_t be the flow of X_t. Then $(t)=\phi_t(l)$, and by change of variables theorem, (89) becomes

$$\frac{d}{dt}\int_{\phi_t(l)}\alpha=\frac{d}{dt}\int_l\phi_t^*\alpha=\int_l[\phi_t^*(L_X\alpha)+\phi_t^*\left(\frac{\partial\alpha}{\partial t}\right)]. \tag{90}$$

However, $L_X\alpha+{}^{\partial\alpha}/_{\partial t}$ is exact from the equations of motion and the integral of an exact form over a closed loop is zero. We now use Stokes theorem, which will bring in the vorticity. If Σ is a surface (a two-dimensional submanifold of M) whose boundary is a closed contour C, then Stokes theorem yields

$$\Gamma_C=\int_C\alpha=\int_\Sigma d\alpha=\int_\Sigma\varpi \tag{91}$$

Thus, as a corollary of the circulation theorem, we can conclude:

5.2.2 Helmholtz Theorem.

The flux of vorticity across a surface moving with the fluid is constant in time

6. Classification of Fluids According to Diffeomorophisms (Kiehn, R., 2002)

We will now sketch a conceptual frame work for some of the geometric structure of fluids.

6.1 Compressible Fluid

Let M and N be differentiable manifolds of the same dimension. Let L denote an interval of positive real numbers. Let $\Gamma_{fl}(M \times L, N)$denote the set of all maps:

$$\phi: M \times L \to N \quad : (x,t) \longmapsto \phi(x,t), \tag{92}$$

Such that¢ ϕ is smooth. For each t EL, the map

$$\phi_t: M \to N \quad : x \longmapsto \phi(x,t), \tag{93}$$

is a diffeomorphism.

6.1.1 Definition

$\Gamma_{fl}(M \times L, N)$is called the space of smooth fluid flows in N parameterized by M. geometrically, $t \longmapsto \phi_t$ can be considered as a one parameter deformation in the space of diffeomorphisms, we can write:

$$\phi_t = \phi^\circ c_t, \tag{94}$$

Where $t \longmapsto c_t$ ∈Diff(N) is a curve in the group of all diffeomorphisms of N. recall the ideas of the jet bundle calculus. We do this in a form that accords with intuition surrounding what is called the Eulerian picture in fluid mechanics.

Let M and N be finite dimensional, infinitely-differentiable manifold. Let $\phi_{fl}(M \times L, N)$be as above, introduce the following equivalence relation on $M \times L \times \Gamma_{fl}(M \times L, N)$:

$$\left.\begin{array}{c}(x,t,\phi) = (x',t',\phi')iff \\ x = x' \\ t = t' \\ \phi(x,t) = \phi'(x,t)\end{array}\right\} \tag{95}$$

The curves $s \to \phi(x,s)$ and$s \to \phi'(x,s)$ have the same tangent vector at $s = t$.

6.1.2 Definition

Let $E(M,N)$ be the quotient set of $M \times L \times \Gamma_{fl}(M \times L, N)$ with respect to this equivalence relation. $E(M,N)$is called the Eulerian velocity bundle for motion of fluids in N parameterized by M. It is readily seen to be a manifold.

6.1.3 Definition

Given $(x,t,\phi) \in M \times L \times \Gamma_{fl}(M \times L, N)$, the equivalence class to which the pair (x,ϕ) belongs is denoted as $v_{(\phi)(x,t)}$.as(x,t) varies, we obtain a map.

Called the Eulerian velocity field associated with the fluid flow ϕ. Consider the target map

$$v_\phi: M \times L \to E(M,N), \tag{7}$$

Called the Eulerian velocity field associated with the fluid flow ϕ. Consider the target map:

$$\begin{array}{c} M \times L \times \Gamma_{fl}(M \times L, N) \to N \times L \\ : (x,t,\phi) \longmapsto \phi(x,t) \end{array} \tag{8}$$

This map is constant on the equivalence relation (6), hence defines a quotient map that we will label as follows:

$$\pi_{tar} E(M,N) \to N \times L$$

The fibers of π_{tar} are tangent vectors to N, hence π_{tar} defines E(M, N) as a vector bundle over $N \times L$ where TN is the tangent bundle to N.

6.2 Incompressible Fluids

Let us define the geometry of an incompressible fluid. Let η and ω denote fixed volume-element differential forms on N and M, respectively.

Let $\Gamma_{infl}(M \times L, N)$denote the set of all maps

$$\begin{array}{c}\phi: M \times L \to N \\ : (x,t) \longmapsto \phi(x,t),\end{array} \tag{10}$$

Such that $\phi \in \Gamma_{fl}(M \times L, N)$, for each $t \in L$, the map:

$$\phi_t: M \to N \quad : x \longmapsto \phi(x,t), \tag{96}$$

Is a diffeomorphism, and satisfies: $\phi_t^*(\eta) = \omega$.

References

Arkani-Hamed, N., Cachazo, F., & Kaplan, J. (2010). What is the simplest quantum field theory? *Journal of High Energy Physics, 9*, 1-92.

Bajura, R., & Jones, E. (1976) Flow distribution manifolds. *Journal of Fluids engineering, 98*(4), 654-665.

Beig, R. *Differential Geometry And Lie Groups Application.* Initiativkolleg Der Universitat Wien.

Datta, A., & Majumdar, A. (1980). Flow distribution in parallel and reverse flow manifolds. *International Journal of Heat and Fluid Flow, 2*(4), 253-262.

Erbar, M. (2010). The heat equation on manifolds as a gradient flow in the Wasserstein space. *In Annales de l'institut Henri Poincaré (B).*

Gilkey, P. B. (1975). The boundary integrand in the formula for the signature and Euler characteristic of a Riemannian manifold with boundary. *Advances in Mathematics,15*(3), 334-360.

Haller, G. (2001). Distinguished material surfaces and coherent structures in three-dimensional fluid flows. *Physica D: Nonlinear Phenomena, 149*(4), 248-277.

Kiehn, R. (2002). Holonomic and Anholonomic Constraints and Coordinates. *Frobenius Integrability and Torsion of Various Types. Technical Report, University of Houston.*

Kobayashi, S., & Nomizu, K. (1963). *Foundations of differential geometry.*

Kolár, I., Slovák, J., & P. W. (1999). Michor, Natural operations in differential geometry.

Lang, S. (1999). Fundamentals of Differential Geometry. *Graduate Texts in Mathematics., Springer-Verlag, 191.*

Marsden, J., & Weinstein, A. (1983). Coadjoint orbits, vortices, and Clebsch variables for incompressible fluids. *Physica D: Nonlinear Phenomena, 7*(1), 305-323.

O'neill, B. (1983). *Semi-Riemannian Geometry With Applications to Relativity, 103.* Academic press.

Verhulst, F. (2006). Nonlinear differential equations and dynamical systems. *Springer Science & Business Media.*

Warner, F. W. (2013). Foundations of differentiable manifolds and Lie groups. *Springer Science and Business Media, 94.*

Yamabe, H. (1960). On a deformation of Riemannian structures on compact manifolds.

Permissions

All chapters in this book were first published in JMR, by Canadian Center of Science and Education; hereby published with permission under the Creative Commons Attribution License or equivalent. Every chapter published in this book has been scrutinized by our experts. Their significance has been extensively debated. The topics covered herein carry significant findings which will fuel the growth of the discipline. They may even be implemented as practical applications or may be referred to as a beginning point for another development.

The contributors of this book come from diverse backgrounds, making this book a truly international effort. This book will bring forth new frontiers with its revolutionizing research information and detailed analysis of the nascent developments around the world.

We would like to thank all the contributing authors for lending their expertise to make the book truly unique. They have played a crucial role in the development of this book. Without their invaluable contributions this book wouldn't have been possible. They have made vital efforts to compile up to date information on the varied aspects of this subject to make this book a valuable addition to the collection of many professionals and students.

This book was conceptualized with the vision of imparting up-to-date information and advanced data in this field. To ensure the same, a matchless editorial board was set up. Every individual on the board went through rigorous rounds of assessment to prove their worth. After which they invested a large part of their time researching and compiling the most relevant data for our readers.

The editorial board has been involved in producing this book since its inception. They have spent rigorous hours researching and exploring the diverse topics which have resulted in the successful publishing of this book. They have passed on their knowledge of decades through this book. To expedite this challenging task, the publisher supported the team at every step. A small team of assistant editors was also appointed to further simplify the editing procedure and attain best results for the readers.

Apart from the editorial board, the designing team has also invested a significant amount of their time in understanding the subject and creating the most relevant covers. They scrutinized every image to scout for the most suitable representation of the subject and create an appropriate cover for the book.

The publishing team has been an ardent support to the editorial, designing and production team. Their endless efforts to recruit the best for this project, has resulted in the accomplishment of this book. They are a veteran in the field of academics and their pool of knowledge is as vast as their experience in printing. Their expertise and guidance has proved useful at every step. Their uncompromising quality standards have made this book an exceptional effort. Their encouragement from time to time has been an inspiration for everyone.

The publisher and the editorial board hope that this book will prove to be a valuable piece of knowledge for researchers, students, practitioners and scholars across the globe.

List of Contributors

Samuel K. Amponsah
Department of Mathematics, Kwame Nkrumah University of Science & Technology, Kumasi, Ghana

Elvis K. Donkoh
Department of Mathematics & Statistics, University of Energy & Natural Resources, Sunyani, Ghana

James A. Ansere and Kusi A. Bonsu
Electrical/Electronics Department, Sunyani Polytechnic, Sunyani, Ghana

V. Sadhasivam and A. Santhi
PG and Research Department of Mathematics, Thiruvalluvar Government Arts College, Rasipuram, Namakkal - 637 401, Tamil Nadu, India

Pon. Sundar
Om Muruga College of Arts and Science, Salem - 636 303, Tamil Nadu, India

Yirang Yuan, Jiuping Li and Tongjun Sun
Institute of Mathematics, Shandong University, Jinan, P. R. China, 250100

Changfeng Li
Institute of Mathematics, Shandong University, Jinan, P. R. China, 250100
School of Economics, Shandong University, Jinan, P. R. China, 250100

Youngsoo Kim
Department of Mathematics, Tuskegee University, United States

Byunghoon Lee
Department of Mathematics, Tuskegee University, United States

Supaporn Saduakdee and Varanoot Khemmani
Department of Mathematics, Srinakharinwirot University, Bangkok, Thailand

Mohammad Ali Bashir
Department of Mathematics, Faculty of Sciences, University of Alnillin, Khartoum, Sudan.
Academy of Engineering Sciences, Khartoum, Sudan.

Tarig Abdelazeem Abdelhaleem
Department of Mathematics, collage of Applied and Industrial Science, University of Bahri, Khartoum, Sudan

Alassane Diouf
Dèpartement de Mathèmatiques et Informatiques, Facultèdes Sciences et Techniques, Universitè Cheikh Anta Diop, Dakar, Sènègal

Moumouni Diallo
FSEG, Universitè des SSG. BP: 2575 Bamako, Rèpublique du Mali

Diakarya Barro
Universitè Ouaga II. BP: 417 Ouagadougou 12, Burkina Faso

Kemal Toker
Faculty Of Arts And Sciences Department Of Mathematics, Harran University, S¸ anlıurfa-Turkey

Mohammad Zannon and Hussam Alrabaiah
Department of Mathematics, Tafila Technical University, Tafila, Jordan

Ali Moghani
Department of Computer Science, William Paterson University, Wayne, NJ

Michel Fortin
Dèpartement de Mathèmatiques et Statistique , Universitè Laval, Quèbec, G1K 7P4, Canada

Abdellatif Serghini Mounim
Department of Mathematics and Computer Science, Laurentian University, Sudbury, Ontario, P3E 2C6, Canada

Mbakiso Fix Mothebe, Professor Kaelo and Orebonye Ramatebele
Department of Mathematics, University of Botswana, Pvt Bag 00704, Gaborone, Botswana

Iwundu M. P.
Department of Mathematics and Statistics, Faculty of Science, University of Port Harcourt, Port Harcourt, Nigeria

Qilong Cheng
Department of Mechanical Engineering, Tsinghua University, 100084, China

Tiancheng Yu and Jingkai Yan
Department of Electronic Engineering, Tsinghua University, 100084, China

Ru Wang
School of Foreign Languages, Jiangxi Normal University, 330022, China

Ehmet Kasim
College of Mathematics and Systems Science, Xinjiang University, Urumqi 830046, P.R.China

Luis Teia
Department of Energy Sciences, Lund University Ole R¨omers V¨ag 1, M-Building SE-22100 Lund, Sweden

Salma A. Khalil
Department of Mathematics, Faculty of sciences, Princess Nourah bint Abdulrahman University, Riyadh, KSA

Mohammed A. Basheer
Department of Mathematics , Faculty of sciences, University of Alnillin, Khartoum, Sudan

Tarig A. Abdelhaleem
Department of Mathematics , Collage of Applied and Industrial sciences, University of Bahri , Khartoum, Sudan

Index

A

A-optimality Criterion, 150-151
Absolute Valued Algebra, 25-28
Archimedean Copulas, 65, 72
Asymptotic Behavior, 187, 191, 195, 200-201

C

Cauchy Problem, 50, 57-59, 64
Cell Planning, 1-2, 4, 12-13
Cell Range, 1-2, 4, 12
Central Composite Design, 150-151, 162-163, 165-166, 168, 170, 173
Chromatic Index, 102, 105
Circular Cells, 1-2
Classical Thin Plate, 54
Clique Number, 102, 105
Coefficient, 14-15, 17, 19, 21, 23, 54, 65-66, 72, 76, 81, 89, 91-92, 115, 118, 125, 133, 135
Conical Shell, 51-52, 54-56
Conjugacy Class, 107
Convection-diffusion, 92, 134-135, 137, 141, 143, 147-149
Copulas Analysis, 66
Cyclic Subgroups, 107-108

D

Domination Number, 102-103
Duplication Process, 25-28

E

Electrical Impedance Tomography, 75, 77, 79, 81, 83, 89, 91-93
Electrical Resistivity, 75-76, 83, 90-91
Electrodes, 75-76, 81-83, 89-90
Element Mass Matrix, 134
Exponential Variogram, 65, 70, 72-74

F

F -derivation, 14-16, 18-19
Factorial Design, 150-151, 158-160
Finite Dimensional, 25-26, 219
Finite Element, 75-77, 79, 81, 83, 89, 91-93, 134-135, 137, 141, 143, 147-149
Finite Volume, 76, 83, 86, 89-90, 92, 134, 147-149
Frequency, 1-2, 4, 54, 117
Fundamental Equation, 51

G

Gathering Model, 114, 116, 132
Geometrical, 57, 202-203, 205, 207-209
Geophysical Exploration, 75, 91
Geostatistical, 65
Green's Function, 57, 64
Gsm Antenna, 1, 4, 13
Gsm Masts, 1, 4, 7, 9, 12
Gsm Network, 1-2, 4, 6-7, 9, 11-13

H

Hexagonal, 1-2, 4, 6-7, 9, 11-13
Hyper Surface, 57-59
Hyperbolic Spacetime, 58, 62
Hypothesis, 99, 132, 205, 208

I

Inverse Problem, 75-77, 83, 86, 88-89
Isomorphism, 16, 27-28, 181

J

Jacobi Matrix, 75-76, 78, 81-83, 89-90

L

Lie Derivation, 14-15, 18
Lie Triple Derivation, 14, 19
Linear Operators, 29

M

Mathematical Theory, 51, 92
Multidimensional Extension, 134

N

Numerical Integrals, 76, 81-82
Numerical Method, 75-76, 83, 93, 201
Numerical Simulation, 75-76, 78, 83, 88-90, 93

O

Open Mapping Theorem, 57, 64
Oscillations, 51
Overlap Difference, 1-2, 4, 6-7, 12-13

P

Perfect Graph, 102, 105-106
Plackett-burman Design, 150, 152, 154, 156
Polygon, 1-2, 4, 13
Polynomial, 14, 178-179, 181, 183
Polynomial Algebra, 178-179, 181, 183

Q

Q-conjugacy Character, 107-108, 110, 113

Quasi-element, 75-77, 81, 91
Quasilinear, 148, 187-188, 191, 195, 201

R
Refugee Immigration, 114-115, 117, 119, 121, 125, 127, 129, 131, 133

S
Semi-riemannian Metric, 57-58
Shear Deformation, 51, 54-56
Spatial Statistics, 65-66
Steenrod Squares, 178
Stochastic Process, 65
Strongly Left Unit, 25-28

T
Tessellation, 1-2, 4, 6-7, 9, 11-13
Theorem, 17-19, 23, 25-28, 42, 44, 57-59, 62, 64, 86, 95-96, 99, 103-106, 108, 110, 137-138, 178-181, 188, 190, 195, 198, 200, 202-208, 218
Trapezoidal Quadrature, 134
Triangular Algebra, 14-15, 18-19

U
Upwinding, 134

V
Vacation Interruption, 29, 37, 50
Variogram, 65-66, 70-74

W
Working Vacation, 29, 37, 50

Z
Zero-divisor Graph, 102, 106

www.ingramcontent.com/pod-product-compliance
Lightning Source LLC
Chambersburg PA
CBHW080254230326
41458CB00097B/4452
* 9 7 8 1 6 3 2 3 8 5 8 1 9 *